M. Zimberlin

MARINE SCIENCE

Marine Biology and Oceanography

SECOND EDITION

THOMAS F. GREENE

Adjunct Professor, Oceanography
Kingsborough Community College
Brooklyn, New York

Past President, NYS Marine Educators Association

Former Assistant Principal, Science
Fort Hamilton High School
Brooklyn, New York

AMSCO SCHOOL PUBLICATIONS, INC.
315 Hudson Street, New York, N.Y. 10013

The author wishes to acknowledge the helpful contributions of the
following consultants in the preparation of this book:

Edna M. Black
Science Teacher
Manalapan High School, Englishtown, NJ

Ed Braddy
Science Teacher
J. W. Mitchell High School, Pasco, FL

Gertrude Karabas
Assistant Principal, Curriculum & Instruction
Teaneck High School, Teaneck, NJ

Lynne Popper
Biology and Marine Science Teacher
Newtown High School, Queens, NY

Amelia Quillen
Biology and Oceanography Teacher
Smyrna High School, Smyrna, DE

Carl M. Raab, Project Coordinator
Former Director of Academic Initiatives
New York City Board of Education, NY

Jeffrey Schley
Marine Science Teacher
Elsinore High School, Wildomar, CA

Suzanne Tansey
Science Teacher
John Ehret High School, Marrero, LA

Tim Tindol
Former Marine Science Teacher
Mission High School, San Francisco, CA

Robert Zottoli, Ph.D.
Professor of Biology
Fitchburg State College, Fitchburg, MA

Edited by   Carol Davidson Hagarty

Cover and Text Design by   Merrill Haber

Art by   Eric Hieber/Tech Graphics

Composition by   Brad Walrod/High Text Graphics, Inc.

Photo Research by   Tobi Zausner

Please visit our Web site at: www.amscopub.com

When ordering this book, please specify:
R 770 P *or* MARINE SCIENCE Softbound
or
R 770 H *or* MARINE SCIENCE Hardbound

ISBN 978-0-87720-938-6 Softbound
ISBN 978-0-87720-939-3 Hardbound

Printed in the United States of America

4 5 6 7 8 9 10   08
6 7 8 9 10   08

To the Student

You live on land, but most of Earth is covered with water. Water is everywhere. Look at any photograph of Earth taken from space. A color photograph will show that our planet is mostly blue—the color of the ocean. How much of the world is actually covered by ocean? Close your eyes in front of a globe of the Earth. Spin the globe. When it stops spinning, touch the globe's surface. Odds are you will put your finger on a part of Earth's ocean. In fact, about 70 percent of Earth's surface is ocean.

How deep is the sea? Why is it salty? Do strange creatures really live in the deepest parts of the sea? These questions, and many others, may come to mind when you think about the ocean. Today, we can answer many of these questions thanks to the work of oceanographers, scientists who study the sea. However, we cannot answer every question. Scientists continue to study the ocean, so that we can understand more of its mysteries.

In the first unit of *Marine Science, Second Edition*, you will be introduced to the history and methodology of the study of the ocean and to the great variety of marine environments. The next three units concentrate on life in the sea—the branch of science called *marine biology*. The fifth and sixth units deal primarily with the science of *oceanography*, the study of the physical geology characteristics of the sea. In fact, oceanography is a mix of many areas of scientific study, including biology, chemistry, geology, and physics. Together, marine biology and oceanography make up the body of knowledge known as *marine science*. In the last unit of this textbook, you will explore how the physical and biological characteristics of the ocean interact to affect marine ecology.

At the end of each chapter, you will find a Chapter Review section that includes question sets and a suggestion for a research project or activity. The question sets help you review the material you have learned. The research projects or activities help you expand your knowledge and understanding of marine biology and oceanography. You will find that doing a research project can be a challenging and creative endeavor, often one that will help you to think like a scientist. You will ask questions, make observations, develop explanations, and draw conclusions. We hope you enjoy doing these projects. Now, let us begin our exploration of the marine world.

Contents

UNIT 3
MARINE INVERTEBRATES

UNIT 4
MARINE VERTEBRATES

UNIT 5
THE WATER PLANET

UNIT 6
ENERGY IN THE OCEAN

UNIT 7
MARINE ECOLOGY

UNIT 1

INTRODUCTION TO MARINE SCIENCE

The ocean has been called one of the last frontiers on Earth. People have sailed the seas for thousands of years. Yet we actually know more about the surface of the moon than we know about the bottom of the sea.

In Chapter 1, you will begin your study of marine science by examining the history of ocean exploration.

In Chapter 2, you will learn about the process scientists use in their work. Discoveries in marine science are made by scientists who work in the field and in laboratories.

Most life on Earth is in the ocean, which contains a variety of environments. In Chapter 3, you will learn about the characteristics and inhabitants of these diverse marine environments.

1 Exploring the Oceans

When you have completed this chapter, you should be able to:

EXPLAIN what happened to the *Titanic* when it hit the iceberg.

DISCUSS some of the important people and discoveries in the field of oceanography.

DESCRIBE some of the important events and developments in the history of ocean exploration.

On April 14, 1912, at twenty minutes to midnight, a magnificent passenger ship, the R.M.S. *Titanic,* struck an iceberg in the North Atlantic, 564 kilometers (km) southeast of Newfoundland, Canada. This huge ship—at 269 meters, almost as long as three football fields—was making its maiden voyage, a trip from Southampton, England, to New York. After it struck the iceberg, water poured into the ship. As it took on more and more water, the ship's bow, or front, gradually moved below the frigid surface of the ocean. By twenty minutes past two on that dark night, the great ship assumed an almost vertical position in the water, with the stern, or back, of the ship pointing skyward. Within a few more minutes, the ship sank beneath the Atlantic's surface.

For more than seven decades, the *Titanic* lay on the bottom of the ocean, its grave unmarked. In 1985, the *Titanic's* resting place was found. The technology developed in the years since the ship sank played a crucial role in its discovery. In this chapter, you will discover how science and technology have been used to increase our understanding and knowledge of the ocean.

1.1 THE "UNSINKABLE" SHIP

The sinking of the *Titanic* was among the worst human disasters that ever occurred at sea. Of the 2224 passengers on board the ship, 1513 drowned. Most of the victims were swept into the freezing water when the ship upended. The 711 people who survived in lifeboats were later picked up by rescue ships that responded to the *Titanic's* distress signals.

Shipbuilders described the *Titanic* as "unsinkable" because it had 16 watertight compartments designed to keep the ship afloat. The builders were so convinced that their "giant floating palace" could not sink, that they provided only 20 lifeboats with a total capacity of 1178 seats—not enough for everyone on the huge vessel. Ironically, Captain E. J. Smith, who went down with his ship, said, "I cannot conceive of any vital disaster happening to this vessel."

Buoyancy

Why were the builders and the captain of this ship so wrong? All large ships, like the *Titanic,* have hulls made of steel. You may already know that a steel anchor sinks to the bottom of the ocean. Why do some objects sink, while others float? The Greek scientist Archimedes, who lived from 287 to 212 B.C., discovered that floating objects are supported by an upward force called **buoyancy**. According to legend, Archimedes suddenly discovered the nature of this force while taking a bath. He jumped out of the tub and ran through the street yelling, "Eureka, eureka!," meaning "I found it."

Archimedes discovered that (1) the buoyant force on any object is equal to the weight of the liquid that the object displaces, or pushes aside; and (2) a body immersed in a liquid seems to lose weight, and the apparent loss in weight is equal to the weight of the liquid displaced. These observations are known as **Archimedes' principle**, and are shown in Figure 1-1 on page 6. The word equation for Archimedes' principle can be summarized as follows: buoyant force = weight of liquid displaced = loss of weight in liquid. For example, as shown in Figure 1-1, 20 grams of buoyant force of liquid = 20 grams of weight of liquid displaced (25g mass – 5g container weight) = 20 grams of apparent loss in weight (35g weight in

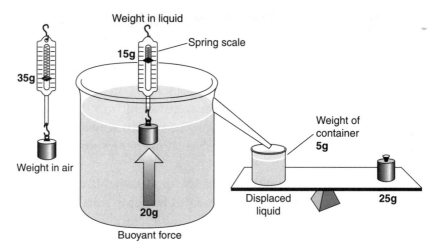

Figure 1-1 An illustration of Archimedes' principle: buoyant force = weight of liquid displaced = loss of weight in liquid.

Weight in liquid

Spring scale

15g

35g

Weight in air

Weight of container
5g

Displaced liquid

25g

20g

Buoyant force

air – 15g weight in water). Another way of expressing this principle is: buoyant force = weight of object in air – weight of object in liquid. Thus, buoyant force = 35 grams – 15 grams = 20 grams (the apparent loss in weight).

A ship floats because it is constructed with many air-filled compartments. Air, as you know, weighs less than water. Thus, the weight of air and steel in the ship is less than the weight of the water displaced by the ship, so the ship floats. An anchor sinks because it weighs more than the water it displaces. The *Titanic* sank almost three hours after it collided with the iceberg. At that time, experts believed that the impact of the collision caused a huge gash in the ship's hull that punctured a number of watertight compartments. The ship filled with water. In time, the combined weight of the water and weight of the ship became greater than the buoyant force supporting the ship, thereby causing the ship to sink.

The *Titanic* sank to the ocean floor in water that was about 3600 meters deep. For years, oceanographers searched in vain for the *Titanic*. In 1985, it was finally located. Finding the *Titanic* was a triumph in the use of modern technology. How was this lost ship located? What is left of its remains? Before we answer these questions, we need to learn more about the history of ocean exploration that led to this discovery.

1.1 SECTION REVIEW

1. Why do some objects sink while others float?

2. What damage did people first think caused the *Titanic's* sinking?

3. What is the relationship between a buoyant object and the water it displaces?

1.2 VOYAGES OF DISCOVERY

Since early times, oceans were used for transport and travel. Merchants found that it was often cheaper to transport cargo by sea than by land. Human settlements sprang up along the sea or along waterways that led to the sea. The earliest boats were log rafts and canoes made from hollowed tree trunks. These small crafts were powered by people using paddles or oars.

The first sailing vessels were probably made about 3000 B.C. Powered by the wind, a sailboat could move faster and travel farther offshore than other types of small boats. By using sailing vessels, people became seafarers, and the ocean became a highway for trade.

The earliest recorded trade routes were established by the Phoenicians, about 2000 B.C. The Phoenicians were skilled seafarers who lived in the coastal area of the Mediterranean where present-day Syria and Lebanon are located. By 700 B.C., the Phoenicians had sailed completely around Africa. By A.D. 150, they had sailed north to present-day Great Britain and the North Sea. In that year, based on knowledge gained from the Phoenicians' voyages, the Greek geographer Ptolemy prepared a fairly accurate map of the world.

After the fall of the Roman Empire and the decline of civilization in the Mediterranean, seafaring and sea exploration almost came to a halt—with one major exception. The Vikings, a Norse people living in the region now known as Norway, Sweden, and Denmark, invaded and plundered much of Europe. Raiding voyages were even made as far away as North Africa and North America.

Why were the Vikings such successful seafarers? Part of the answer is that the ships they built were superior to anything afloat at the time. The remains of several Viking vessels have been excavated and reconstructed by marine archeologists. Viking ships were about 15 to 30 meters long, made almost entirely of oak, and powered by sails and oars. The main sail powered the boat when the winds were strong. In times when no winds blew, 32 people rowing together could move the ship quickly through the water. Its flattened bottom allowed the ship to move safely through even shallow water, where other ships would have run aground.

The era of Viking exploration lasted from about A.D. 800 to 1100. During this period, the Vikings discovered new lands. They landed in Greenland by A.D. 1000. Shortly thereafter, they landed in Newfoundland, on the eastern coast of Canada. The Viking era came to an end during the middle of the so-called Dark Ages.

The end of the Dark Ages signaled a rebirth in the exploration of the seas. Expeditions were mounted to discover new trading routes. In 1488, the Portuguese navigator Bartholomeu Dias (1450–1500) sailed around the southern tip of Africa, which was later named the Cape of Good Hope. In 1492, Christopher Columbus set sail from Spain looking for a new route to India by traveling west. He crossed the Atlantic and landed in what is now the Bahamas in the Caribbean. On three subsequent voyages, he reached the coasts of what are now called Central America and South America.

Other navigators and explorers followed Columbus. In 1497, John Cabot sailed from England to explore the coast of North America. He mapped the coast from Labrador to as far south as the present-day state of Delaware. The Italian navigator Amerigo Vespucci, upon landing on the continent of North America, was the first person to realize that this land was not part of Asia. He named the land "Mundus Novus," Latin for "New World." On later voyages, Vespucci explored the coasts of present-day Argentina and Uruguay. Vespucci was responsible for exploring more than 9600 km of coastline in the New World. Mapmakers in Europe named the continents of the New World the "Americas" in his honor.

Word of the Americas spread, beginning a new wave of exploration. In 1513, the Spanish explorer Vasco Núñez de Balboa traveled across Panama and saw "el Mar del Sud," the South Sea—later to be called the Pacific Ocean.

Ferdinand Magellan (1480–1521), a Portuguese navigator, was the first person to attempt to **circumnavigate**, or sail completely around, the Earth. (See Figure 1-2.) In 1519, Magellan sailed west from Spain with five ships and 290 men. The journey was to take three years to complete. In 1522, only one ship with 18 surviving sailors made it home to Spain. (Magellan was not one of the sur-

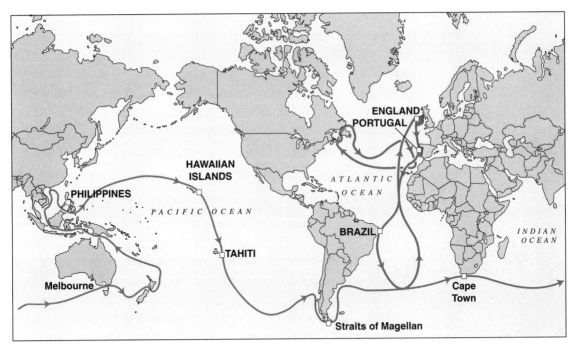

Figure 1-2 Magellan's path in his voyage around the world.

vivors; he was killed in the Philippine Islands. Most of the initial crew were killed in shipwrecks, in mutinous fights, and by disease.)

During the 1500s and 1600s, explorers and navigators continued to map the coastline of the Americas. In 1524, the Italian navigator Giovanni da Verrazano explored the coast of North America from Georgia to Massachusetts. Between 1534 and 1541, Jacques Cartier, a French explorer, traveled along the St. Lawrence River and the eastern coast of Canada. From 1607 to 1611, Henry Hudson, an English navigator commanding a Dutch ship, explored and mapped the river that was later to be named in his honor. During the next 100 years, further coastal explorations resulted in more detailed and accurate maps of the oceans and the coastlines that bordered them.

The spirit of adventure, the quest for discoveries, and the hope of untold riches were the motivating forces behind these navigators and explorers. By the middle of the eighteenth century, much was known about the geography of the ocean. However, scientific knowledge of the ocean was lacking. How deep is the sea? What causes ocean currents? What natural resources does the ocean contain? There were many scientific questions about the ocean that the early explorers did not answer. Another kind of explorer, the scientist, was beginning to uncover the answer to these kinds of questions. Scientific exploration of the oceans followed the work of the explorers and navigators.

1.2 SECTION REVIEW

1. Why are many early settlements situated along coasts?
2. What special characteristics did Viking ships have?
3. How did Magellan's voyage prove that Earth is round?

1.3 SCIENTIFIC EXPLORATION

Scientific exploration of the oceans began in the mid-1700s with the voyages of Captain James Cook, a British navigator and explorer. On his first voyage, Cook explored the South Pacific in search of a southern continent. Although Cook was not trained as a physician, he was very concerned with the health of his crew. On his long voyages, Cook observed that when the crew ate citrus fruits—oranges, lemons, and limes—they did not develop scurvy. Scurvy is a very serious disease that killed a number of sailors on earlier voyages. Cook made sure that the crew ate a varied diet that included citrus fruits. As a result, on Cook's first voyage, only one sailor out of a total crew of 118 died. Today we know that scurvy is caused by a lack of vitamin C in the diet. This vitamin is found in citrus fruits. So Captain Cook's observations protected the health of his crew, even though he did not understand the reasons why eating citrus fruits worked. (His work proved that, through careful observation, scientific discoveries can be made by nonscientists.)

Being a skilled astronomer, Cook made important astronomical observations. Cook used his knowledge of latitude and longitude to determine his location and he was able to map many islands in the South Pacific. Unfortunately, he did not complete his dream of finding the southern continent of Australia. His life came to an untimely end when he was killed while exploring in Hawaii.

The American statesman Benjamin Franklin (1706–1790) also made several important scientific contributions. For example, he experimented with electricity and invented a heating device. According to popular history, while serving as U.S. postmaster general in 1770, Franklin wanted to find out why mail delivery from Europe to the American colonies took longer than mail delivery from the colonies to Europe. Since all mail traveled by ship, Franklin asked whalers about their experience. In this way, he learned about a water current that moved up along the east coast from the Gulf of Mexico. Franklin had a map drawn of this ocean current, which he called the Gulf Stream; the current flows northeast, then turns and heads east across the Atlantic Ocean. Franklin reasoned that if ships heading west from Europe to the colonies got caught in the Gulf Stream, they would be moving against the current, and therefore the trip would take longer than for ships that were traveling east with the current. Modern oceanographers marvel at the accuracy of Franklin's map of the Gulf Stream, considering the fact that he was not trained as an oceanographer and hardly spent any time at sea.

Scientific exploration of the ocean accelerated greatly in the nineteenth century. In 1831, a British sailing vessel, the H.M.S. *Beagle,* sailed for the western coast of South America on a mission to map the coastline and collect biological specimens. On board was Charles Darwin (1809–1882), the ship's naturalist. During a 5-year period, Darwin collected and observed animals and plants that lived in South America. He also visited the Galápagos Islands, located 965 km off the coast of Ecuador. On these islands, Darwin discovered unique animal species, including the giant tortoise and the marine iguana. In 1859, Darwin published his famous book, *On the Origin of Species by Natural Selection.*

The Beginning of Oceanography

By the middle of the 1800s, a wealth of scientific information on ocean depth, ocean currents, and winds had been collected by ships at sea. An American naval officer, Matthew Fontaine Maury (1806–1873), analyzed the data collected from ships' logbooks. In 1855, Maury published one of the first books on physical oceanography, *The Physical Geography of the Sea.* This book proved to be a valuable guide in navigation for shipmasters and a good reference text because of the technical information it contained.

DISCOVERY
A Highway for Cultural Dispersion

On April 28, 1947, the Norwegian sailor and adventurer Thor Heyerdahl (1914–2002) set sail from the coast of Peru on board the *Kon-Tiki*, a raft made of balsa wood (see photo). Exactly 101 days later, he ran aground on a reef near the island of Tahiti—6935 km away. Heyerdahl wrote about his epic voyage in an adventure book called *Kon-Tiki*, which became an international bestseller. The purpose of Heyerdahl's expedition was to confirm his theory that inhabitants of South America, such as the Peruvian Indians, were capable of crossing the Pacific Ocean to populate the Polynesian islands—centuries before Christopher Columbus dared to cross the much smaller Atlantic Ocean.

Heyerdahl's theory that some ancient human cultures were dispersed across oceans to distant shores—just as driftwood floats on ocean currents to faraway beaches—was put to the test again in 1970. He built a papyrus reed boat, called *Ra* after the Egyptian sun god. Heyerdahl successfully sailed his papyrus boat across the Atlantic Ocean, from the west coast of North Africa to the Caribbean islands, to show that ancient Egyptians could have done the same. Heyerdahl wanted to prove that the Egyptians might have introduced pyramid-building techniques to the Mayan Indians of Central America.

Today, most scholars who study human cultures and societies do not endorse Heyerdahl's theory that all cultural similarities between peoples on distant continents are the result of oceanic dispersion. Instead, they rely on linguistic, archaeological, and genetic evidence to understand cultural similarities between distant peoples. Still, Heyerdahl's efforts were valuable for confirming some ideas about ancient seafaring peoples. Thousands of years ago, similar long-distance voyages across the open ocean did enable the dispersal of people, and their cultures, throughout the widely scattered islands of the Pacific.

QUESTIONS

1. Compare and contrast the main purposes and destinations of the *Kon-Tiki* and *Ra* voyages.

2. Why did Heyerdahl cross the oceans on primitive sailing vessels rather than on modern boats?

3. In what way were Heyerdahl's voyages valuable for an understanding of early ocean travel and cultural dispersion?

However, there was so much more to learn about the oceans. Countries began to mount ocean expeditions devoted exclusively to gathering information in a scientific way. One of the most successful was the voyage of H.M.S. *Challenger*. This British sailing vessel was redesigned into a laboratory ship. From 1873 to 1876, the *Challenger* crossed the major oceans while carrying out a host of scientific tests. For example, water samples were taken and analyzed chemically. Sediments were dredged up from the ocean bottom and studied. Recordings of temperature and pressure were made at many depths in the water column. More than 4700 new species of marine organisms were discovered, described, and cataloged. Data on tides, currents, and wave action were also collected. Enough scientific information was compiled to fill a 50-volume *Challenger Report*. The *Challenger Report* was the most comprehensive study ever completed in the field of oceanography. The organizer of the *Challenger* expedition and the director of research was British zoologist Sir Charles Wyville Thompson. Because of his work, Sir Charles Thompson is credited by many as being the "founder of oceanography."

Many oceanographic expeditions followed. From 1893 to 1896, a Norwegian vessel, the *Fram,* explored the relatively unknown Arctic Ocean. On board was the Norwegian explorer and scientist Fridtjof Nansen (1861–1930), who invented a water-sampling bottle. This device, named the **Nansen bottle** in his honor, is still in use today. (See Figure 1-3.) It is used to collect water samples from different depths in the water column.

By the early 1900s, much of the ocean floor remained unmapped. In 1925, a German research ship, the *Meteor,* cruised the South Atlantic for 25 months. Using a new device called **sonar** (**so**und **na**vigation **r**anging), which emits and receives sounds, the *Meteor* took continuous readings of the seafloor. The map that was prepared from these sonar readings revealed a seafloor that had many varying depths and features.

In the late 1940s, the study of the world's oceans expanded rapidly. Many countries launched their own oceanographic vessels, and some countries cooperated in joint ventures of exploration. Expensive oceanographic vessels, staffed by marine biologists and oceanographers, were built to learn more about the oceans. Research facilities were founded that conducted both field and laboratory experiments in marine biology and oceanography.

Figure 1-3 The Nansen bottle is used to collect water samples.

1.3 SECTION REVIEW

1. Describe Benjamin Franklin's contributions to the field of oceanography.

2. Why is Sir Charles Thompson considered by many to be the "founder of oceanography"?

3. What do navigators and explorers have in common with scientists?

1.4 EXPLORING INNER SPACE

The early explorers and navigators, on their voyages of discovery, mapped the oceans of the world, increasing our knowledge of the geography of Earth's surface. The world that is found beneath the ocean's surface, called **inner space**, also caught the human imagination. Underwater exploration paralleled the voyages of discovery being made on the surface.

The earliest records of underwater exploration show that the ancient Greeks dove for ornamental shells thousands of years ago. Unfortunately, a diver's vision is blurred underwater; objects cannot be sharply seen. By 2500 B.C., when glass was developed, divers made crude face masks that contained two small pieces of flat glass. The ability to see underwater improved dramatically with the use of such a face mask.

In the Mediterranean Sea, the ancient Greeks dove for pearls, sponges, and black coral. The Greek historian Herodotus (484–425 B.C.) wrote about Persian divers who rescued treasures from their own ships that had been sunk by the Greeks. Early divers also took part in military operations. The Greek poet Homer said that divers were used during the Trojan Wars, which took place between 1194 and 1184 B.C. And it was said that the Macedonian king Alexander the Great (356–323 B.C.) descended into the sea inside a special container to watch the destruction of enemy fortifications.

Figure 1-4 Early diving chambers held a supply of air in a barrel.

Air

Diving Devices

How were early divers able to remain underwater for extended periods of time? A device called a **diving chamber** was used. The diving chamber contained a supply of air. (See Figure 1-4.) Halley's chamber, developed by Edmond Halley (1656–1742), was able to hold a larger supply of air than did the earlier diving chambers. This device had a reserve air supply in a barrel, connected by a hose to the chamber.

Underwater movement in diving chambers was awkward and limited. In time, a more suitable apparatus called a **diving suit** was developed. The diving suit was made of watertight canvas and had a heavy metal helmet, or hard-hat. Weighted boots enabled the diver to walk on the ocean bottom. Air was pumped from the surface through a tube and into the helmet. (See Figure 1-5.) The hard-hat diving suit was a great improvement over the diving chamber. During the era of the diving suit, the sponge industry flourished because the divers could work alone for long periods of time underwater. The diving suit even protected the diver from cold temperatures. However, there was one major disadvantage. The diver had limited movement underwater because of the necessary air hose to the surface.

Divers knew that they could increase their time and mobility underwater if they carried their air supply along with them. In 1808, Friedrich von Drieberg invented a device that had a supply of air strapped to the diver's back. But it could be used for only short periods of time. In 1865, French engineer Benoit Rouquayrol and French naval officer Auguste Denayrouze invented a breathing device that contained compressed air. Their device was worn around the waist.

It remained for French ocean explorer Captain Jacques-Yves Cousteau and French engineer Émile Gagnan to improve this underwater breathing apparatus to make it more efficient. In 1943, Cousteau and Gagnan invented the modern **aqua-lung**, which is a tank of compressed air that is strapped to the diver's back. (See Figure 1-6.) The diver breathes air from the tank through a mouthpiece device called a *regulator*. The regulator (which is connected by a hose to the tank) adjusts the air in the tank to the correct pressure that a diver can safely breathe at any given depth. The aqua-lung is also called a **scuba tank**. Scuba is an acronym that stands for *s*elf-contained *u*nderwater *b*reathing *a*pparatus.

Diving Vessels

The scientists who explore outer space are called *astronauts*. Scientists who explore inner space are called **aquanauts**. Using scuba

Figure 1-5 The diving suit had a helmet and heavy boots; air was pumped from the surface through a tube.

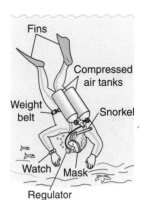

Figure 1-6 The modern aqua-lung, or scuba tank, allows a diver greater mobility and more time underwater than did earlier diving devices.

gear, scientists can explore the ocean. However, using this type of equipment has its limitations. The average depth of the ocean is about 3600 meters. The deepest dive made by a scuba diver is only about 135 meters. Because of the great water pressure, the depth to which a scuba diver can descend is severely limited.

Only a specially constructed steel chamber can protect human explorers at great depth. In 1934, the American oceanographer Dr. William Beebe reached a depth of about 1000 meters in a round steel chamber called a **bathysphere** (derived from the Greek word *bathos,* meaning "deep"), which had a thick glass porthole for viewing and room for only two people. It was attached to a ship on the surface by a length of very strong cable.

In the 1950s, the Swiss father-and-son team of Auguste and Jacques Piccard developed a much deeper diving vessel called the **bathyscaphe.** In January 1960, Jacques Piccard (with co-pilot Don Walsh) made the deepest dive ever recorded in his bathyscaphe. Named *Trieste,* the vessel descended to a depth of 10,852 meters into the Mariana Trench in the Pacific Ocean. This undersea trench, the deepest known on Earth, is deeper than Mount Everest is tall. It took more than four hours to make the trip to the bottom.

Beebe and Piccard were pioneers in the use of underwater vessels for scientific research. Since the 1960s, many underwater vehicles have been developed for the scientific exploration of the ocean. These small, research submarines, such as the "souscoup" invented by Cousteau, are called **submersibles**.

Modern Submersibles

A submersible that has logged more than 1000 dives is the *Alvin,* shown in Figure 1-7. This self-propelled submersible can carry a crew of three and has been used on a variety of missions. For example, in 1966, the *Alvin* recovered a hydrogen bomb in 1830 meters of water. The bomb fell from a bomber that had crashed in the ocean off the coast of Spain. Floodlights are used to illuminate the darkness at great depths, and the *Alvin* uses mechanical arms to pick up objects from the seafloor. In addition, the *Alvin* has been used to investigate hydrothermal vent communities along the mid-Atlantic Ridge (see Chapter 16), and has photographed living organisms at depths of more than 3900 meters.

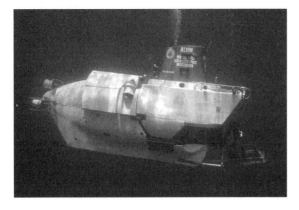

Figure 1-7 The *Alvin* submersible has been used for research on more than 1000 deep dives.

With each passing year, humans are exploring more of the ocean depths. In 1979, American scientist Dr. Sylvia Earle set a record for the deepest dive. She made a solo dive to a depth of 380 meters off the Hawaiian coast. In this magnificent effort, Dr. Earle wore a special "space-age" dive suit, illustrated in Figure 1-8. This high-tech suit was named the *Jim suit* in honor of Earle's colleague, diver Jim Jarratt. It is in many ways like a diver's own personal submersible. The suit is made from the durable metal titanium and is strong enough to protect the diver from the crushing effects of great pressure. Although it contains its own air supply, the suit must remain connected by cable to a ship that is on the surface.

Oceanographers are also exploring the ocean using **robots**. Robot vehicles do not carry any people on board. One important undersea robot is the *Jason*. The *Jason* is about two meters long. It is controlled by computers and is packed with lights, cameras, and scientific sensors. The *Jason* is tethered, or connected, to a sled that is suspended by cables from a surface ship, or to a submersible (such as the *Alvin*). Among other tasks, the *Jason* was used to explore the Gulf of California, 1.5 km below the surface of the sea. Photographs from the *Jason* revealed an alien world of unusual creatures and geological formations.

The technology of underwater explorations is changing rapidly. Untethered robots also have been developed and are now being used. One such robot, called the *Deep Drone*, is a remotely operated vehicle (ROV) that is designed to locate and recover objects on the deep ocean floor. By using robots to explore the world beneath the sea, researchers have increased our knowledge of the ocean floor and its inhabitants. With every dive made by a robot, a new region

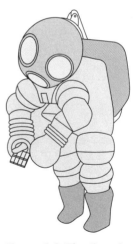

Figure 1-8 The *Jim suit* can withstand the crushing pressure of the ocean depths.

yields its secrets to the probing camera eye. Like the ancient explorers and navigators, the modern-day aquanauts are also making voyages of discovery.

1.4 SECTION REVIEW

1. How did ancient people explore the undersea world?
2. What advantages do submersibles have over scuba gear?
3. What advantages do robots have over submersibles?

1.5 FINDING THE *TITANIC*

Scientific discoveries are often made by teams of people working together. On September 1, 1985, a joint United States–French expedition located the *Titanic* about 560 km southeast of Newfoundland, Canada. The ship was lying on the seafloor, covered by more than 3600 meters of water. Using advanced technology, marine scientists Dr. Robert D. Ballard and Jean-Louis Michel of the United States–French team searched the area where the *Titanic* was believed to have sunk. As their ships crisscrossed the target area, sonar was beamed from the sea surface to the bottom to help pinpoint the wreck. A robot vehicle called the *Argo,* which contained side-scan sonar and video cameras, searched along the ocean floor. The *Argo* was tethered to the support ship at the surface.

The first video pictures of the *Titanic* lasted only a few minutes. They showed a rusting steel hull of the ship that was remarkably well preserved and covered by a thin film of sediment. On subsequent dives, the team used the search vehicle *Argo* to pass over the *Titanic* to photograph the whole wreck. The video pictures showed that the ship had broken into two parts, with the halves lying about 600 meters apart. A great deal of debris was scattered on the seafloor between the two sections. The bow of the great ship was buried about 15 meters into the mud and sediment. Part of the *Titanic's* stern, lying exposed on the ocean floor, is shown in Figure 1-9.

To get a closer look at the wreck, the submersible *Alvin* was used. Tethered to the *Alvin* was the *Jason.* Signals sent from the scientists aboard the *Alvin* directed the *Jason* into and around the *Titanic.* No

Figure 1-9 Part of the *Titanic's* rusting hull and railing are shown here in a photograph taken by a robot vehicle more than three kilometers deep in the North Atlantic.

remains of passengers could be found. All dead organic matter, including the wooden deck and fixtures, had long since been eaten by marine organisms. Out of respect for the victims of this disaster, the discoverers of the *Titanic* made no attempt to raise the hull or any of the objects found with it.

However, the video pictures they took provided important information. They did not show the huge gash in the ship that was once believed to have been the cause of the ship's sinking. Instead, they showed that several steel plates were dented. Another United States–French research team, in 1996, used sonar to examine the part of the bow that was buried in mud. They discovered a series of six narrow openings along the right side of the hull. Based on the new evidence, marine scientists discovered that the collision with the iceberg tore the hull in several places, breaking the steel rivets that held together the steel plates covering the ship. The compartments inside the ship's hull had watertight **bulkheads**, or walls. But water poured in over the bulkheads, flooding 6 out of 16 compartments in succession, in much the same way that water runs over the sections of an ice cube tray when you tilt it as you fill it. Eventually, the ship held enough water to cause it to sink.

Often, something good comes from tragedy. In this case,

reforms were instituted after the *Titanic* sank. All ocean liners are now required to provide sufficient lifeboats to accommodate everyone on board in case of emergency. The sea-lanes used in the winter by ships sailing in the North Atlantic were moved farther south. An International Ice Patrol was also established to locate and monitor drifting icebergs. But these reforms were paid for at a very high price indeed.

1.5 SECTION REVIEW

1. How was the *Titanic* finally located?
2. Can a ship truly be made unsinkable? Why or why not?
3. What actually caused the sinking of the *Titanic*?

Laboratory Investigation 1

Making Ocean Water

PROBLEM: How can we make seawater for an aquarium?

SKILLS: Using a graduated cylinder; using a triple-beam balance.

MATERIALS: 1000-mL graduated cylinder, large beaker, triple-beam balance, sea salts, spatula, labels, stirrer.

PROCEDURE

1. Fill the graduated cylinder to the 1000-mL mark with tap water.

2. Pour the tap water from the graduated cylinder into a 1000-mL beaker or large container.

3. Set up your triple-beam balance to measure out the sea salts. Use the diagram in Figure 1-10 as a guide. Begin by moving all three riders on the three scales to the notches at the far left.

4. Check that the scale is balanced. The pointer should be at the zero mark. If the pointer is off-center, turn the counterweight screw located under the pan until it comes into balance.

5. Place a piece of paper on the pan to hold the salts. Since the paper has weight (mass), you need to bring the scale back into balance by moving the smallest rider until it balances.

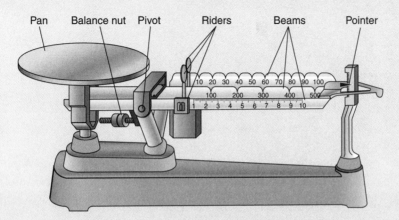

Figure 1-10 A triple-beam balance.

6. Weigh out 35 grams of salt. Move the middle rider to the 30-gram mark and then move the lowest rider up to 5 grams.

7. Use the spatula to put sea salts on the paper. Keep adding the sea salts until the scale comes into balance at the zero mark.

8. Carefully transfer the sea salts from the pan into the jar containing the 1000 mL of tap water. Stir. Label the jar Artificial Seawater 3.5%, and put your name and the date on it.

9. Tap water contains chlorine, which may be harmful to living things in your aquarium. Let the container of seawater stand uncovered overnight. This will de-chlorinate the water (the chlorine gas leaves the water) before it is poured into your tank.

OBSERVATIONS AND ANALYSES

1. A student measured out 350 grams of salt. How much tap water would have to be added to make ocean water containing 3.5% salt?

2. Why should artificial seawater sit overnight before being poured into an aquarium tank?

3. Why is it important to balance the scale before using it?

Chapter 1 Review

Answer the following questions on a separate sheet of paper.

Vocabulary

The following list contains all the boldface terms in this chapter.

aqua-lung, aquanauts, Archimedes' principle, bathyscaphe, bathysphere, bulkheads, buoyancy, circumnavigate, diving chamber, diving suit, inner space, Nansen bottle, robots, scuba tank, sonar, submersibles

Fill In

Use one of the vocabulary terms listed above to complete each sentence.

1. Floating objects are supported by a force called _____.

2. Magellan attempted to _____ Earth by ship.

3. A _____ device uses sound to make readings of the seafloor.

4. Modern divers breathe compressed air from an _____.

5. Diving vessels such as the *Alvin* are types of _____.

Think and Write

Use the information in this chapter to respond to these items.

6. Describe the advantages of the diving suit over the diving chamber. How is the aqua-lung superior to both of these?

7. What were some of the factors that motivated the early ocean explorers? How were they similar to those of ocean explorers today?

8. The *Titanic* was the largest passenger ship built at its time. Explain how its particular construction enabled it to float— and to sink—according to Archimedes' principle.

Inquiry

Base your answers to questions 9 through 12 on the diagram below, which shows an experiment that a student performed in order to measure the buoyant force on a submerged object.

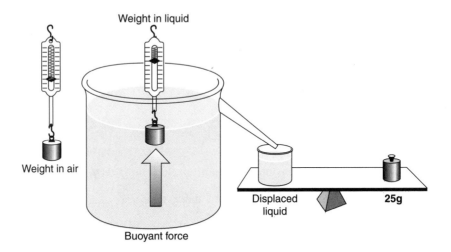

Weight in liquid

Weight in air

Displaced liquid

25g

Buoyant force

Results of the student's experiment are recorded in the following data table.

Weight of collecting can or beaker	5 grams
Weight of metal object in air	35 grams
Weight of the collecting can plus water	25 grams

9. What is the weight of the displaced liquid?

10. What is the apparent weight of the submerged object?

11. What is the buoyant force on the submerged object, as measured in grams?

12. Describe the relationship between a submerged object and the liquid it displaces.

Multiple Choice

Choose the response that best completes the sentence or answers the question.

13. The *Titanic* sank after it hit an iceberg because   *a.* the buoyant force on the ship was greater than the weight of the ship   *b.* the buoyant force on the ship was less than the weight of the ship   *c.* the weight of the ice pushed the ship below the surface, filling it with water   *d.* the ship had too many passengers and not enough lifeboats.

14. The ancient seafarers used the ocean primarily for *a.* shipping and trading   *b.* scientific exploration *c.* underwater research   *d.* recreational travel.

15. A coral stone placed in a saltwater tank caused the water to overflow. The weight of the coral stone underwater was 20 grams. The weight of the displaced liquid was 25 grams. What was the weight of the coral stone in the air? *a.* 20 grams   *b.* 25 grams   *c.* 35 grams   *d.* 45 grams

16. The best explanation for the failure to find all sunken ships is that   *a.* sunken ships cannot be found in very deep water *b.* the technology to find ships is too primitive   *c.* the ocean is too vast and largely unexplored *d.* sunken ships completely decay underwater.

17. The founder of the science of oceanography is considered to be   *a.* Sir Charles Thompson   *b.* Captain James Cook *c.* Christopher Columbus   *d.* Benjamin Franklin.

18. An important advantage that a submersible has over scuba gear is that the submersible can   *a.* be used to observe the environment   *b.* have greater maneuverability   *c.* collect items from coral reefs   *d.* be used at greater depths.

Base your answers to questions 19 and 20 on the different diving devices shown below and on your knowledge of marine science.

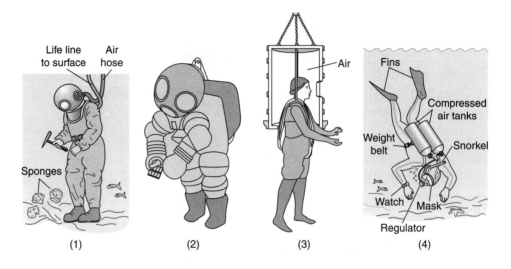

Life line to surface Air hose

Sponges

Air

Fins

Compressed air tanks

Weight belt

Snorkel

Watch Mask

Regulator

(1) (2) (3) (4)

19. What is the proper sequence for the development and use of the diving devices shown (going from past to present)?
    a. 1, 2, 3, 4 *b.* 4, 3, 2, 1 *c.* 3, 1, 4, 2 *d.* 2, 1, 3, 4

20. A diver using which device can descend to the greatest depths? *a.* 1 *b.* 2 *c.* 3 *d.* 4

21. An example of a vessel that is a submersible is the
    a. Challenger *b. Alvin* *c. Beagle* *d. Meteor.*

22. The scientist responsible for locating the *Titanic* was
    a. Sylvia Earle *b.* Charles Darwin *c.* Robert Ballard
    d. Jacques Piccard.

23. The co-inventor of the modern aqua-lung was
    a. Fridtjof Nansen *b.* Sylvia Earle *c.* Edmond Halley
    d. Jacques-Yves Cousteau.

Research/Activity

Write a report about a known shipwreck. Include the causes for the shipwreck and your impressions of what the experience must have been like.

2 Marine Scientists at Work

When you have completed this chapter, you should be able to:

IDENTIFY the steps of the scientific method.

DEVELOP an experimental procedure.

APPLY the basic units and tools of scientific measurement.

COMPARE and CONTRAST the different types of microscopes.

For students of science, these are exciting times. Every day, new discoveries are being made in marine biology and oceanography. You know about the spectacular discovery of the *Titanic* in the deep waters of the North Atlantic. Around the world, the research efforts of scientists are producing a wealth of information about the ocean and its inhabitants.

Important discoveries continue to be made by scientists working both in the field and in the laboratory. To carry out their work, scientists use a problem-solving approach called the *scientific method*. In this procedure, experiments are performed, observations are made, data are gathered, and conclusions are drawn. You also use the scientific method, perhaps without realizing it, in everyday situations. However, scientists use a much more formal, organized approach. In this chapter, you will learn about some of the procedures and tools scientists use to conduct their research, and how the scientific method can be used to study marine science.

2.1 MARINE SCIENCE TODAY

Throughout the past century, improvements in technology have greatly expanded our knowledge of marine science. Sophisticated instruments have been developed that observe and monitor the ocean world, from far above Earth in outer space to the depths of the ocean's inner space.

Satellites and Remote Sensing

The oceans are monitored by satellites (in orbit around Earth) by means of their **remote sensors**—instruments that gather information on the features of Earth without being in physical contact with it. These satellites send back information on ice cover, ocean depth, water temperature, and weather conditions.

The *TOPEX/Poseidon* satellite, which was launched in 1992, monitors global ocean circulation, sea surface temperatures, and sea surface height. Sea surface temperatures are shown in computer-generated colors; red and yellow represent warm water, while blue and green represent cold water. Computer-generated colors also reveal the presence of plankton, an indicator of ocean productivity. (See Figure 2-1.) In 1997, NASA launched the *SEASTAR*, which carries a

Figure 2-1 Satellites such as the *TOPEX/Poseidon* are used to study Earth's oceans.

color scanner that measures the ocean's entire oxygen-producing plankton population on a weekly basis. NASA's latest Earth-observing satellite, *AQUA*, launched in 2002, gathers information on precipitation and evaporation, in order to determine if Earth's water cycle is being affected by climate change.

Remote-sensing satellites are also used in search-and-rescue operations. The National Oceanic and Atmospheric Administration (NOAA), a federal agency that monitors the ocean, operates SARSAT (*S*earch-*a*nd-*R*escue *S*atellite-*A*ided *T*racking), which locates people in distress on land and on sea. As of March 2002, nearly 13,000 people had been rescued by SARSAT worldwide.

Probing the Depths

Geologists are scientists who study the physical structure of Earth. **Marine geologists** study the characteristics of, as well as any changes that occur in, the seafloor. They drill into the seafloor and remove rock samples. The first drilling ship, the *Glomar Challenger,* began taking core rock samples in 1968. This drilling ship has been replaced by the modern vessel *Resolution,* which can drill into the seafloor at water depths of 6500 meters. Remote sensors also gather data on the deep ocean floor. These sensors replace submersibles operated by people, which are much more expensive to use. The Remote Underwater Manipulator (RUM III) can pick up objects from the ocean floor, take video pictures, and use sonar to determine water depths. The RUM III stays tethered to a research vessel on the surface. Smaller remote-controlled robots have been used to examine underwater shipwrecks such as the *Titanic* (described in Chapter 1).

Centers of Learning

As they investigate the marine world, scientists are always learning. New discoveries are reported in the scientific literature. The *Journal of Marine Biology* and the *Journal of Marine Science* are two of many scientific journals that publish original research and make it available to the scientific community as well as to the public. Much research is done in college and university laboratories around the country. Undergraduate and graduate programs in marine science

Figure 2-2 The FLIP vessel is used for oceanographic research.

are offered at the University of Miami, the University of Texas, the University of Hawaii, Oregon State University, and the University of Rhode Island, to name a few.

Research institutes are important in the field of marine science. Three of the most famous centers of learning devoted to marine research are the Scripps Institution of Oceanography in La Jolla, California, the Woods Hole Oceanographic Institution in Woods Hole (Cape Cod), Massachusetts, and the Lamont-Doherty Geological Observatory (of Columbia University) in New York. The submersible *Alvin*, which has been used to collect volumes of information about the undersea world, is maintained and operated by Woods Hole.

Since 1963, Scripps, Woods Hole, and Columbia University have cooperated in a joint venture in underwater research by using the Floating Instrument Platform (FLIP), shown in Figure 2-2. FLIP is a 100-meter hollow tube with a research station at one end. After FLIP is towed out to a research site, its ballast tanks are filled with water. The floating instrument platform then "flips" over on itself. When FLIP is in a stable vertical position, scientists can conduct a variety of underwater tests.

Monitoring Our Marine Resources

Environmental concerns are a very important part of marine science. How can we protect the marine environment and conserve its resources? The National Oceanic and Atmospheric Administration (part of the United States Department of Commerce) conducts research, monitors the atmosphere and the oceans, and helps conserve, manage, and protect our marine resources.

Popular Marine Science

The marine world has become an important part of popular culture. You can snorkel or scuba dive around shipwrecks and coral reefs, swim with captive dolphins in Florida, or view a fantastic assortment of marine creatures in public aquariums. Whale-watching trips in the Atlantic and the Pacific have become increasingly popular as a means of getting close to marine life in its natural setting. (See Figure 2-3.) Recently, small passenger submersibles have been developed. These vessels can take you below the surface to view sea life, while you remain in dry comfort.

Aquariums have been built in many cities to satisfy people's interest in the marine world. Public aquariums not only display a great variety of marine organisms, they also offer educational programs to students and interested adults. Some public aquariums also conduct scientific research. The fields of marine biology and oceanography have been made popular in recent years due to the achievements of modern pioneers such as Captain Jacques-Yves Cousteau, Dr. Robert D. Ballard, and Dr. Sylvia Earle. (See Figure 2-4 on page 32.) Like the astronauts who explore outer space, these aquanauts who explore the inner space of the ocean have helped to uncover and explain many of its mysteries. Who are these ocean

Figure 2-3 Passengers on a whale-watch boat view a humpback whale as it dives below the ocean's surface.

pioneers and what have they achieved? See the chart below for a list of some of their important accomplishments.

PIONEERS IN OCEAN EXPLORATION

Captain Jacques-Yves Cousteau
(1910–1997)

Accomplishments: French Naval Officer for 27 years; co-inventor of the aqua-lung (scuba-diving gear); invented one-person submersible; renowned ocean explorer; award-winning undersea filmmaker; author of many books on marine topics; founder of environmental organization, The Cousteau Society, Inc.

Quote: "The oceans of the world are our inheritance, to preserve and protect for all time."

Dr. Robert D. Ballard
(1942–)

Accomplishments: U.S. Naval Officer; oceanographer; discovered tube worms in hydrothermal vents; founder of the Institute for Exploration; author of more than 50 scientific articles; *Jason* Project Director; located resting places of the *Titanic,* the *Bismarck,* and the *Lusitania.*

Quote: "I grew up wanting to be Captain Nemo from *Twenty Thousand Leagues Under the Sea.*"

Dr. Sylvia Earle
(1935–)

Accomplishments: World record holder for the deepest dive (in *Jim* suit); former Chief Scientist and Project Director at NOAA; author of more than 125 scientific publications; public speaker for conservation of marine resources.

Quote: "I was swept off my feet when I was three and I have been in love with the sea ever since."

Figure 2-4 Three important pioneers in ocean exploration.

2.1 SECTION REVIEW

1. Describe how remote-sensing devices are used to investigate the oceans.

2. What contributions have institutions of higher learning made to marine science?

3. Describe three examples of marine science in the popular culture.

2.2 THE SCIENTIFIC METHOD

In many cases, we learn best by doing. Marine biologists learn about life in the sea by performing experiments. An experiment is an investigation that is conducted according to an organized step-by-step problem-solving approach called the **scientific method**. The steps in the scientific method are: state the problem, collect relevant information, form your hypothesis, test your hypothesis using selected materials and methods, record observations, tabulate results, and draw a conclusion.

The first step in the scientific method is actually to observe a selected part of nature and then state the problem you wish to solve in the form of a testable hypothesis. You select a problem that requires you to make observations, which are descriptions of events in the environment that are based on the use of your senses. The experiment that tests the hypothesis generates data that can be analyzed.

Stating the Problem

Let's use an example to help you understand how the scientific method works. Students in one class observed the behavior of marine snails in an aquarium. The tank contained several snails at one end and food (an opened mussel) at the other end. The students recorded their observations in a **logbook**, a notebook in which such data are kept. Read the following entries made by one student.

Nov. 1—Observed five snails climbing up the side of the aquarium tank.

Nov. 2—Opened a mussel and put it into the aquarium. In a few minutes, snails swarmed all over the food.

Based on these observations, can you suggest a problem to solve? Remember, it must be testable and in the form of a question. One student wondered whether snails would respond in some measurable way to the presence of food. She came up with the following problem to study: Does the presence of food affect the rate of movement in snails?

Forming and Testing the Hypothesis

Identifying a problem to study is the first important step in conducting an experiment. When a problem has been identified, the scientist should do library research to learn what, if anything, has already been discovered about that topic. The next step in the scientific method is to offer a possible solution to the problem. A possible solution to a problem is called a **hypothesis**. The student suggested the following hypothesis:

Hypothesis: If food is present in the aquarium, then snails will move with greater speed (toward the food).

A hypothesis is often called an "educated guess" because previous knowledge is used to formulate it. (Scientific advances always build on a foundation of previously learned knowledge.) The hypothesis is often written with the words *if* and *then*. When the hypothesis is stated in this form, it is easier to devise an experiment to test it.

Testing the hypothesis is the third step in the scientific method. To find out if a hypothesis is correct, an experiment is designed and carried out. However, before you can do the experiment, you must first select the appropriate materials.

Selecting the Materials

Materials are the items or pieces of equipment needed to carry out an experiment. In our example, one student chose the following materials: aquarium, seawater, food, graduated cylinders, snails, and a metric ruler. Then the equipment must be assembled in the proper manner so that the hypothesis can be tested. The part of the scientific method in which materials are set up is called the method, or procedure. In this part of the project, you can use your creativity and originality in developing a good experimental design.

Method (Procedure)

Look at Figure 2-5 to see how the student scientist designed her experiment. She wanted to see if food affected snail locomotion. In this investigation, food is called the variable. A **variable** is any factor that could affect the outcome of an experiment. In this investigation, the seawater could also be a variable. The salt content and temperature of the water might affect the movement of a snail. The size and shape of the container could also be a variable. There are many possible variables in an experiment. However, you can test only one variable at a time, and all other variables must be kept the same (constant). Otherwise you could not be certain which variable, or combination of variables, produced the effects you observed.

To determine if a particular variable affects the outcome of an investigation, a scientist carries out a **controlled experiment** in which two groups are tested: one that is exposed to the variable and one that is not. The group that is exposed to the variable is called the **experimental group**. The experimental group in Figure 2-5 is the group of snails in the container with the food. The scientist must also set up a **control group**. This would be the group of snails in the container without the food. The control group contains exactly the same conditions as the experimental group, except for

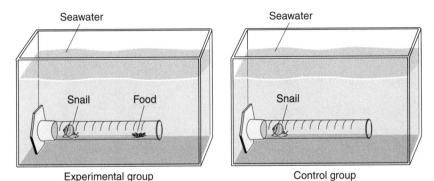

Seawater

Snail   Food

Experimental group

Seawater

Snail

Control group

Figure 2-5 An experimental set-up in which food is the variable.

the variable being tested—in this case, the food. Since all other conditions are kept the same in both the experimental and the control groups, any difference in snail locomotion would have to be caused by the presence of food.

To understand how the materials were set up to measure snail locomotion, read the instructions listed in the procedure below, which were developed by the student who carried out this experiment.

1. Using a metric ruler, mark off the length of two graduated cylinders in centimeters.
2. Fill the cylinders with seawater. Label one cylinder "experimental" and the other "control."
3. Place a snail at the bottom of each cylinder. Submerge the graduated cylinders in the seawater in each aquarium. Be sure to place the cylinders flat on their sides on the bottom of each aquarium, and be sure the centimeter measurements are visible.
4. Put a piece of food at the entrance of the cylinder labeled "experimental." Remember that in this experiment the control cylinder receives no food.
5. Start timing the snails as soon as they begin to move. In each case, record the time it takes for the snail to move the length of the cylinder. Enter the time and distance traveled in Table 2-1. Perform as many trials as you can in the allotted time. Try to use different snails each time, so that you have a control group and an experimental group.

TABLE 2-1 SNAIL LOCOMOTION IN THE PRESENCE AND ABSENCE OF FOOD

Trial	Experimental Group (With Food)			Control Group (Without Food)		
	Time (min)	Distance (cm)	Speed (cm/min)	Time (min)	Distance (cm)	Speed (cm/min)
1	3.5	20	5.7	7.0	20	2.9
2	4.0	20	5.0	9.5	20	2.2
3	6.2	20	3.2	8.0	20	2.5
4	2.5	20	8.0	11.0	20	1.8
5	5.0	20	4.0	8.5	20	2.4
6	3.0	20	6.7	7.5	20	2.7
Total	24.2	120	32.6	51.5	120	14.5
Av. Speed			5.4			2.4

Observations and Results

Accuracy is most important in making observations on snail loco-motion. Saying that one snail moves faster or slower than another snail is not very precise. How fast is fast? (In the case of a snail, not very fast at all!) When you use exact measurements, you are being more precise. Measurements that are recorded during an experiment are called *data*. The data collected make up the *results* of your experiment, which can be analyzed. The results represent the next step in the scientific method.

Numerical data from an experiment are best recorded in a table, which is organized and easy to read. Look at the data collected in the student's experiment, shown in Table 2-1.

For the sake of accuracy, experiments are usually repeated. Each time that an experiment is carried out, it is referred to as a **trial**. (The average of several trials will give the most accurate results.) In this case, there were six trials conducted.

The results of an experiment also include calculations. In this experiment, the speed of the snail is computed by using the following formula:

Speed = Distance/Time

Scientists use mathematics in solving problems. Compute the speed of the snail in Trial 1 of the experimental group as shown in Table 2-1. Substituting data from the table into the formula, you get speed in centimeters per minute (cm/min):

Speed = Distance/Time

Speed = 20 cm/3.5 min

Speed = 5.7 cm/min

The average speed for the six trials was computed for the experimental and control groups. To compute the average speed, find the sum of the speeds in all the trials and then divide by the number of trials. In this case, the average speed for the experimental group was 5.4 cm/min. The control group had an average speed of 2.4 cm/min.

The data from an experiment can also be displayed in the form of a graph. A **graph** is a pictorial representation of data that shows relationships at a glance. Suppose you want to compare the average

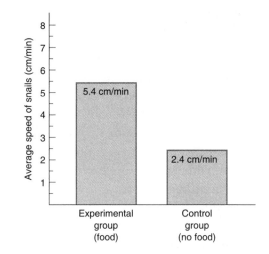

Figure 2-6 A bar graph is used to compare data and show relationships at a glance.

speed of snails in the presence and absence of food. Using the data from Table 2-1, you can construct a bar graph. Look at the bar graph shown in Figure 2-6. The scale on the vertical axis is numbered from 1 to 8. These numbers represent the range of snail speeds, in cm/min, taken from Table 2-1 on page 36.

According to the bar graph, which group of snails has the higher average speed? The bar graph clearly shows that the experimental group with food moves faster than the control group without food.

In another experiment, a student wanted to know if water temperature could affect the speed of snail movement. The student formulated the following hypothesis:

Hypothesis: If the temperature is increased, then the snails will move more slowly.

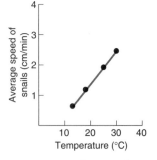

Figure 2-7 A line graph shows the relationship between variables.

The average speed of snails at varying temperatures was collected, as shown in Table 2-2. The data from the table can be plotted on a line graph. Look at the line graph shown in Figure 2-7. The line graph contains two lines, or axes. Temperature data are placed on the horizontal axis, and snail speed data are placed on the vertical axis. The temperature is the "independent variable," while the snail's speed, which is affected by it, is the "dependent variable." Notice that the units on each axis are marked off in equal intervals from the lowest to the highest value. To plot the line graph, look at Table 2-2; the average speed for each temperature is given at the bottom. Plot the first point by locating 13 on the horizontal axis. Then

TABLE 2-2 AVERAGE SPEED OF SNAILS AT DIFFERENT TEMPERATURES (CM/MIN)

Trial	13°C	18°C	24°C	30°C
1	.75	1.0	2.0	2.5
2	.90	1.1	1.8	2.8
3	.50	1.2	1.9	2.4
4	.25	1.3	2.0	2.3
5	.80	1.1	2.1	2.2
6	.65	1.3	2.2	2.0
Total	3.85	7.0	12.0	14.2
Av. Speed	.64	1.2	2.0	2.4

find (or approximate) 0.64 on the vertical axis. Make a dot on the graph where the two points intersect. Do the same for the other readings. Connect the dots by drawing a straight line between them with a ruler.

Drawing Conclusions

Now that you have recorded data in tables and displayed data on graphs, you are ready to analyze and interpret this information. The part of the scientific method in which data are analyzed and interpreted is called the conclusion. What do you conclude from your results? First, you must determine if the results in the experiment support your hypothesis. In the first experiment, it was hypothesized that if food is present, the snails will move faster. Now look at the results in Table 2-1 to see if the hypothesis was supported by the data. According to the results, the average speed of the experimental group (with food) was 5.4 cm/min. The average speed of the control group was 2.4 cm/min. What do you conclude from these results?

The results show that snails move faster in the presence of food. The student would therefore conclude that her hypothesis was supported by the data. A hypothesis that is supported by the data is called a *valid hypothesis*. To check the validity of a hypothesis, scientists will repeat the same controlled experiments several times to see if they obtain similar results. An important tool used in the process of drawing conclusions is statistical analysis. To determine

A marine aquarium is an aquatic environment that is set up to house marine organisms for observation and study. You will not be able to duplicate the ocean environment exactly, but the more your aquarium resembles the natural qualities of the ocean, the greater will be the survival of your specimens. Here is a brief description of the kinds of materials, methods, and equipment that are required for setting up a marine aquarium.

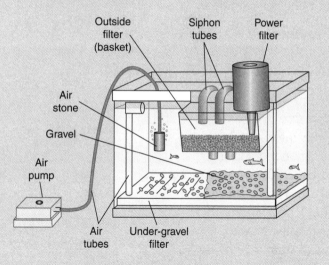

Outside filter (basket) — Siphon tubes — Power filter — Air stone — Gravel — Air pump — Air tubes — Under-gravel filter

The Aquarium Tank Marine organisms can be kept in a variety of containers, or aquarium tanks, of different sizes and shapes. The containers are made of either plastic (if small) or glass. Some tanks are unfiltered, while others contain filter systems; those with filters can house organisms for a longer time.

Ocean Water Two types of ocean water may be used, natural ocean water or commercially prepared saltwater. If you live near the ocean and are studying local specimens, it is best to use the ocean water. You can also pur-chase sea salts from a pet-supply or aquarium shop and mix the salts with tap water to produce "instant ocean." Use a hydrometer to check the salinity.

Filtration In the close confines of an aquarium tank, waste products from the animals' bodies build up quickly. These waste products, which include ammonia, can become toxic to the organisms. Removal of these waste products is called *filtration*. In the filtered tank, there are two kinds of filtration: biological filtration and mechanical filtration.

Biological Filtration The breaking down of waste products by bacteria present in the gravel is called *biological filtration*. One to two inches of gravel at the bottom of the tank will provide a surface area on which bacteria will grow. When an under-gravel filter and air-lift tubes are used, the water is drawn through the gravel and acted upon by bacteria, which decompose the nitrogenous wastes into nitrates. Aquatic plants in the tank can use these nitrates to make proteins. Gravel should be in the form of dolomite or crushed limestone, which contains calcium-magnesium carbonate, a natural substance in ocean water. Dolomite also helps to maintain the proper pH of the aquarium water, which should be slightly alkaline (at a pH of 8).

Mechanical Filtration The separation and removal of suspended particles from the water is called *mechanical filtration*. The suspended particles, which include wastes, excess food, and debris stirred up from the bottom, will make the tank water look cloudy. Mechanical filtration will change a cloudy tank into a clear one through the use of an outside filter. An outside filter system pumps water through a cartridge of filtering material or filter floss that is located in the outside filter basket. The cartridge should be cleaned of debris every week by removing it from the basket and running tap water through it.

Aeration Nearly 20 percent of the air in our atmosphere is oxygen, which is vital to marine animals and plants. The process of getting sufficient air into aquarium water is called *aeration*. An air pump is used to bubble air into an aquarium tank through a plastic tube. An air stone attached to the end of the delivery tube breaks up the air bubbles into an even finer stream of bubbles, allowing more oxygen to dissolve in the water. In tanks with under-gravel filters, the rising air bubbles in the corner air-lift tubes also deliver oxygen to the water.

Temperature Marine organisms should be kept at a temperature similar to that of their natural habitat. For tropical marine organisms, the temperature should be kept between 22°C and 26°C. The floating thermometer must be checked daily and the thermostatic heater must be kept in good working order. Organisms from temperate waters should be kept at or below room temperature.

Salinity A floating hydrometer is used to measure *salinity*, the amount of salt dissolved in water. The average salinity of ocean water is 3.5 percent and has a hydrometer reading of 1.025. Salinity should be checked daily. A cover on the aquarium tank prevents evaporation, which increases salinity.

pH The pH of ocean water is approximately 8.0. You can check the pH level of your tank by using a special pH paper, a pH colorometric test, or a pH meter. If your aquarium's pH level drops below 8, it may mean that acids are building up in the tank. Check for dead organisms or unconsumed food. Clean the tank by siphoning out the wastes and check the pH again. There are aquarium kits that can be purchased and used to adjust the water's pH level.

Illumination Overhead room lights are usually sufficient. Some tanks come with plastic covers that include a light, which can be turned on as needed. Keep tanks away from windows where there is direct sunlight. Too much light, from either the sun or a bulb, can cause excessive growth of algae, which turns the water green.

Results The goal is to maintain clean, clear tank water that has a stable pH, temperature, and salinity in order to ensure the survival of the marine organisms you want to observe and study.

Conclusions The following is a summary of important techniques for managing a saltwater aquarium: be sure that you stock your tank with compatible species of organisms; do not overstock (too many marine animals will overwhelm the filtering system with wastes, and overcrowding can be stressful for animals in captivity); do not overfeed your specimens (excess food will foul the tank); do a partial water change at least once a month by replacing one-quarter of the tank's water with clean saltwater; use commercial sea salt, not regular table salt, when making up ocean water; keep tanks away from windows, since direct sunlight causes growth of algae; place aquarium tanks on a sturdy table or special aquarium stand.

Caution Be careful when handling electrical units near aquarium tanks.

whether certain results are valid, statistics may be used to calculate the significance of the various numbers obtained. Figures that appear insignificant at first may prove to show important differences when analyzed with statistics.

In the second investigation, it was hypothesized that if the temperature of the water is increased, then the snails will move more slowly. Look at the results shown in Table 2-2 and Figure 2-7. Do the data support this hypothesis? You can see in the table and in the graph that as temperature increases, the average speed of the snails also increases. Therefore, the hypothesis would have to be rejected because it is not supported by the data.

Discussion

The rejection of a hypothesis does not mean that the experiment is a failure. Experimental results, whether they confirm or reject a hypothesis, are important information because they add to our knowledge. Additional hypotheses can be proposed to test if other factors can affect snail locomotion. One student suggested that light might be a factor in snail locomotion. Another student thought that gravity could affect snail locomotion. (That hypothesis would be very difficult to test in a classroom lab!) An important part of the conclusion section in a research paper that describes an experiment is offering new ideas for further research. It is important for scientists to communicate their experimental results in journals and books.

Doing a Science Project and Research Paper

A science project is a do-it-yourself experiment that is designed using the scientific method. You can develop your own science project. There are many projects involving marine biology and oceanography that can be done. Your project can be written up as a research paper or displayed as an exhibit at a science fair. If you decide to do a science project, your teacher is a good source of information and useful suggestions. Talk over your ideas before you begin.

A written account of your experiment can be prepared in the

form of a research paper. It is challenging to write a research paper that is interesting, informative, and easy to read. The research paper follows the format of the scientific method and has the following sections: Title Page, Abstract, Introduction, Materials, Procedure, Results, Conclusion, and Sources of Information (Bibliography). Each section should begin on a new page. Some sections will be longer than others; the procedure, results, and conclusion would contain the most information. The paper should be typed and double-spaced. The format of a research paper is shown below.

Title Page: Title, student's name and address, the date.

Abstract: Concise summary of the paper, including important items from procedure, results, and conclusion sections.

Introduction: Explain purpose of paper, state hypothesis, include brief background information, define terms.

Materials: List equipment used, include specifications for all materials, use the scientific names for organisms involved.

Procedure: Describe experimental set-up, illustrate with labeled diagrams, identify control and experimental groups.

Results: Organize numerical data into tables and graphs, use labeled diagrams (or photos) where necessary.

Conclusion: Explain and interpret results, state if hypothesis is supported, suggest new hypothesis or further research.

Bibliography: Cite (list alphabetically) numerous references used from science journals, magazines, texts.

No project can be done without knowing something about the topic in advance. You need to go to the library to gather information about your project. You can also do research in a computer database or on-line service. Information can be obtained from textbooks, encyclopedias, science magazines, and scientific journals. A listing of your sources of information is called a *bibliography,* which should list the literature cited in the experiment. The bibliography makes up the last section of a research paper. The format of an entry in a bibliography is usually as follows: author's last name, first name, and middle initial; year of publication; title of book or titles of article and journal or magazine in which the article was printed; name of publisher (for book); city and state of publication (for book); volume number (for journal/magazine); and the page numbers on which the information cited was found.

The goal of a researcher is to aim for quality, not quantity. When your research paper is completed, you may submit it to a variety of local and national science fairs. Many of them give out awards. Some, like the Intel Talent Search Competition, provide college scholarships. Ask your science teacher for details on local and national science fair competitions.

Exhibiting a Science Project

Some science fairs accept research papers only. Other science fairs require you to make a visual display of your project. A typical science fair exhibit would be a display board that contains (similar to a research paper): the project's title; an introduction; an abstract; a description of the procedure; the results (in tables and graphs); a conclusion; and a bibliography. A good science fair exhibit is visually appealing—the board should contain a nice balance of written text, photos and/or diagrams, and tables and graphs. (See Figure 2-8.)

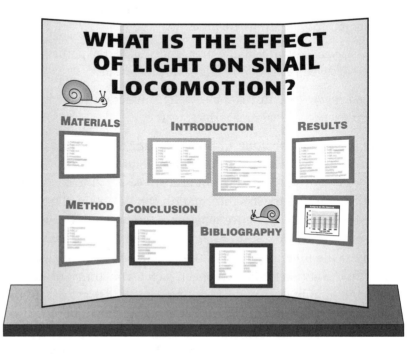

Figure 2-8 A typical science fair presentation board.

Doing a science project is the most creative way of learning science. Displaying your project also gives you the special recognition you deserve.

2.2 SECTION REVIEW

1. Why is the scientific method a useful way to solve problems?

2. What is the value of a control group in an experiment?

3. Why is library research an important part of a science project?

2.3 SCIENTIFIC MEASUREMENTS

Scientific research is based, in large part, on accurate measurements. One of the workplaces of the marine scientist is the laboratory. The laboratory is where detailed observations are made and experiments are carried out. However, accuracy is also important in fieldwork. Good observations depend on accurate descriptions of what is seen. For example, two marine biology students walking along the beach made the following observations:

Student A: "I saw some dead fish among the seaweed on the beach. They were big."

Student B: "I saw four dead fish in the seaweed at the strandline. I measured the fish with a ruler. Each fish measured one meter in length."

Which student provided more precise information? Student A used words like *big* and *some.* But how big is "big" and how many is "some"? These adjectives do not convey to the reader the exact number and size of the fish. Student B, however, gave a more accurate picture because exact quantities and measurements were used. Measurements are precise observations based on numerical descriptions. One of the most important tasks of the scientist is taking measurements. The **metric system** is the system of measurement used in science. It is important to learn the metric system because it is used around the world, for everyday measurements and in all scientific work. The metric system is convenient to use because its units are based on multiples of 10. Metric units are used in the laboratory to determine length, volume, mass, and temperature, and to calculate density.

Length

The measurement of distance from one point to another point is called *length*. The meter (m) is the basic unit of length in the metric system. Rulers are instruments used to measure length. One kind of ruler is the meterstick. It can be used to measure the length of a medium-sized object, such as a fish. The meterstick is divided into units called *centimeters* (cm). *Centi* means "hundredth," so the meterstick is 100 cm long.

The metric ruler is used to measure the length of smaller objects, such as a snail. The metric ruler is also divided into centimeters. Each centimeter is further divided into smaller units called millimeters (mm). (See Figure 2-9.) *Milli* means "thousandth," so there are 1000 millimeters in each meter and 10 mm in 1 cm. The units of length in the metric system are summarized in Table 2-3.

Figure 2-9 A metric ruler is marked off in centimeters and millimeters.

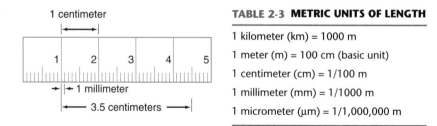

TABLE 2-3 **METRIC UNITS OF LENGTH**

1 kilometer (km) = 1000 m
1 meter (m) = 100 cm (basic unit)
1 centimeter (cm) = 1/100 m
1 millimeter (mm) = 1/1000 m
1 micrometer (μm) = 1/1,000,000 m

Volume

All objects occupy space. The amount of space occupied by an object is called its *volume*. Suppose you wanted to measure the volume of an aquarium tank. You can find the volume by using the formula V = l × w × h. (See Figure 2-10.) The tank's measurements are: length = 60 cm; width = 30 cm; and height = 45 cm. Substitute these numbers in the formula: Volume = 60 cm × 30 cm × 45 cm. Multiply the three measurements and you get the volume in cubic centimeters, written as cc, or cm³. The volume of water the aquarium tank holds is 81,000 cm³.

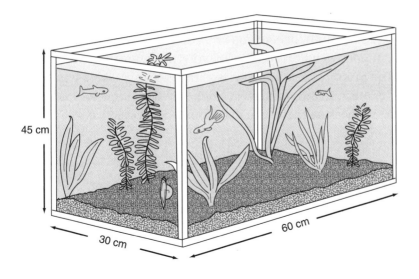

Figure 2-10 You can calculate the volume of an aquarium tank by using the formula V = l × w × h.

In the laboratory, the volume of a liquid is measured using a graduated cylinder. Graduated cylinders are made of plastic or glass and are available in many different sizes, such as 100 mL, 250 mL, and 1000 mL. Look at the 100-mL graduated cylinder shown in Figure 2-11. It is marked off in lines, or gradations. Each line represents 1 milliliter (mL). The basic metric unit of measurement for the volume of a liquid is the liter (L). There are 1000 milliliters in 1 liter. (*Note*: 1 mL is equal to 1 cm^3 in volume.) Notice that the liquid in the graduated cylinder is curved upward where it meets the wall of the cylinder. This curved surface is called the **meniscus**. To measure accurately the amount of liquid in the graduated cylinder, you have to take your reading at the level of the bottom of the meniscus, not at its curved edges. What is the volume of liquid shown in Figure 2-11? There are 70 mL of liquid in the cylinder.

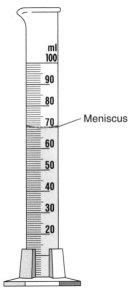

Figure 2-11 A graduated cylinder is used to measure the volume of a liquid.

Mass and Weight

All objects are made up of matter. The amount of matter in an object is called its **mass.** Your body has mass. A boat has mass. Air, as light as it seems, has mass.

The weight of an object depends not only on its mass, but also on the pulling force of gravity. Gravity is stronger on Earth than on the moon. So if you could weigh yourself on the moon, your weight would be much less than it is on Earth. However, your mass remains the same no matter where you are. On Earth or on the moon, even though your weight has changed, your mass remains the same.

The units for measuring mass in the metric system are kilograms (kg), grams (g), and milligrams (mg). The basic unit for mass is the gram. To measure mass, you use a triple-beam balance.

Density

Floating in water is fun. Is it easier to float in a lake or in the ocean? You may know from experience that it is easier to float in the ocean. Floating in the ocean is easier because the ocean contains salt. Salt increases the density of water. **Density** is defined as mass per unit volume. The denser ocean water can support your body's weight better than freshwater can.

How much denser is ocean water than freshwater? You can determine the density of a liquid by using an instrument called a **hydrometer**. The hydrometer is an empty glass tube, with a weight at its bottom to keep the instrument upright in liquid. You can observe, in Figure 2-12, that the hydrometer floats higher in cylinder B than in cylinder A. Cylinder B contains ocean water, whereas cylinder A contains freshwater. The salt in the ocean water increases its density, making it easier for the hydrometer to float.

Notice the scale inside the hydrometers in Figure 2-12. The numbers on the scale represent the range of densities of water. The hydrometer reading for distilled water (pure freshwater) is 1.000. The value of 1.000 represents the density and the **specific gravity** of distilled water. Specific gravity (SG) is defined as the ratio of the density of a substance to the density of distilled water.

What is the density of the salt water in cylinder B? The hydrometer floats at the 1.025 level. The higher the hydrometer reading is, the greater the density of a liquid. Ocean water has an average density of 1.025. Therefore, ocean water is denser than freshwater.

Salinity

The amount of salt dissolved in water is called **salinity**. The average salinity of the ocean is about 3.5 percent. Salinity is also

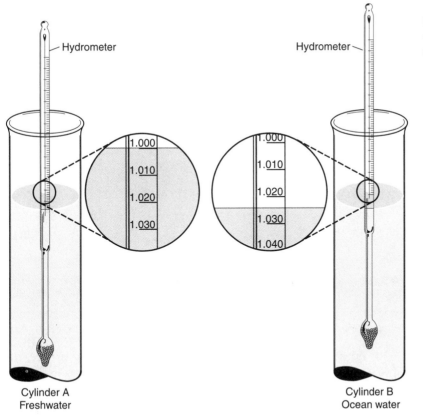

Figure 2-12 A hydrometer is used to determine the density of freshwater and salt water.

Hydrometer

1.000
1.010
1.020
1.030

1.000
1.010
1.020
1.030
1.040

Hydrometer

Cylinder A
Freshwater

Cylinder B
Ocean water

expressed in parts per thousand (ppt). A salinity of 3.5 percent is equivalent to 35 ppt, also written as 35‰. It is advisable to keep the salinity of a saltwater aquarium at about this level.

You can check the salinity of an aquarium using the hydrometer. Recall that the hydrometer is used to measure density. Note that salinity and density are often used interchangeably, since the density of ocean water is due almost entirely to the presence of salt. Temperature and pressure also have an effect on water density. You will learn more about this later.

Since salinity is given in percent, and the hydrometer uses specific gravity units, a conversion table is needed to convert the units. Look at Table 2-4, on page 50, which converts readings on the hydrometer into parts per thousand (ppt) of salt. What is the salinity of a sample of ocean water that has a hydrometer reading of 1.025? As you can see from the table, the salinity is 35 ppt, or 3.5 percent—the average salinity of ocean water, as stated above.

Suppose the hydrometer in your aquarium has a reading of

TABLE 2-4 DENSITY-SALINITY CONVERSION CHART
(PARTS PER THOUSAND AT 20°C)

Density	Salinity	Density	Salinity	Density	Salinity
.998	0	1.009	14	1.020	29
.999	1	1.010	15	1.021	30
1.000	2	1.011	17	1.022	31
1.001	4	1.012	18	1.023	33
1.002	5	1.013	19	1.024	34
1.003	6	1.014	21	1.025	35
1.004	8	1.015	22	1.026	37
1.005	9	1.016	23	1.027	38
1.006	10	1.017	24	1.028	39
1.007	11	1.018	26	1.029	41
1.008	13	1.019	27	1.030	42

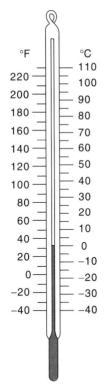

°F   °C
— 110
220 — — 100
200 — — 90
180 — — 80
160 — — 70
140 — — 60
120 — — 50
100 — — 40
— 30
80 — — 20
60 — — 10
40 — — 0
20 — — −10
0 — — −20
−20 — — −30
−40 — — −40

Figure 2-13 A thermometer measures temperature in °F and °C.

1.030. What would be the salinity of the water? As you can see, the salinity would be 42 ppt, or 4.2 percent. A salinity of 4.2 percent is high for a saltwater aquarium. How can you lower the salinity to the recommended 3.5 percent? You can add freshwater to the aquarium, while checking the density with the hydrometer, until the desired level of 3.5 percent is reached.

Temperature

It is important for the marine biologist—and for the student who has an aquarium—to accurately measure air and water temperature. *Temperature* is a measure of the heat energy in a substance. The instrument used to measure temperature is the thermometer. Look at the thermometer shown in Figure 2-13. The unit of measurement used for temperature is the degree (shown with the symbol °). The thermometer is marked off in lines representing degrees that form a scale. The thermometer has two scales, the Celsius (C) and the Fahrenheit (F). Scientists use the Celsius scale, which, like other metric measurements, is based on units of 10. Liquid mercury or alcohol inside a thermometer either expands or contracts, respectively, in response to a rise or fall in the surrounding air or water temperature.

On the thermometer, locate the freezing point and boiling point of water. Water normally freezes at 0°C and boils at 100°C. Average room temperature is about 24°C, and normal human body temperature is about 37°C.

2.3 SECTION REVIEW

1. State two reasons why scientists use the metric system.

2. How is the salinity of ocean water measured?

3. Why is ocean water more dense than freshwater?

2.4 VIEWING THE MICROSCOPIC WORLD

Many of the organisms you will study in marine biology cannot be seen with the unaided eye. To observe these tiny animals and plants, you need to enlarge their images through a process called magnification. One instrument used to magnify objects is the microscope. It is an important tool of the marine biologist.

The Simple Microscope

The hand lens is a simple kind of microscope that has a single lens. A lens is a curved piece of glass that bends light. For example, a hand lens can be used to magnify the image of a piece of seaweed. How much bigger than the actual seaweed is image of the seaweed seen through a hand lens? It is easy to find out. On the hand lens, you might see the number 5×. This is the magnification number. The magnification number means that you see the seaweed five times larger than it actually is by using this specific hand lens. However, the image may be blurred. To get a sharp image, move the hand lens back and forth until the image is clear. This process is called *focusing*.

The Compound Light Microscope

Suppose you wanted to see the cells in a piece of seaweed. Unfortunately, the hand lens does not provide enough magnification to see objects as small as cells. What is required is a **compound light microscope**, such as the one shown in Figure 2-14. When you use a compound microscope, you are looking through at least two lenses.

How much bigger are cells when they are seen through a compound light microscope? To find out, you need to know the magnifications of the lenses that are used. One of the lenses is called the ocular, or eyepiece. Find the ocular in Figure 2-14. The ocular has a magnification of 10×, which means that, by itself, this lens magnifies an object to appear 10 times bigger than it actually is.

Next, locate the objective lenses in the figure. The objective lenses are attached to a rotating nosepiece. One objective lens is called the low-power objective and has a magnification of 10×.

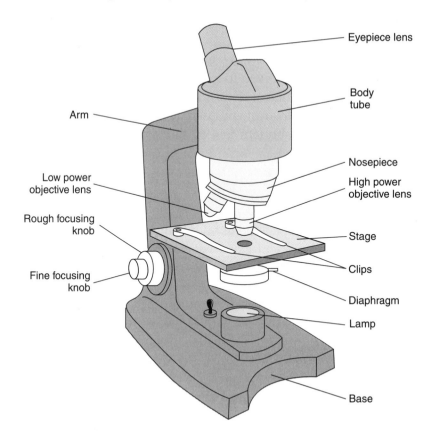

Figure 2-14 The compound light microscope is used to observe both living and dead cells.

The other is called the high-power objective; it has a magnification of 40×.

When you view an object through the microscope, you are looking through the ocular and the objective. To calculate the total magnification, multiply the magnification of the ocular (10×) by the magnification of the objective lens (10× or 40×). What is the total magnification of cells viewed under low power? The total magnification is 100×. What is the total magnification if a student switched to high power? The total magnification is 400×.

Some microscopes have a third objective lens called an oil immersion lens. The magnification of the oil immersion lens is 97×. To observe cells under an oil immersion lens, a drop of clear oil is placed on the cover slip. The oil immersion lens is then carefully lowered into the oil. A 10× ocular used with this oil immersion lens will provide a total magnification of 970×.

The Dissecting Microscope

Objects that are too big for the compound microscope but that require some magnification can be observed through a **dissecting microscope**. When you look through a dissecting microscope, you get a three-dimensional, or stereoscopic, view of an object. The dissecting microscope comes equipped with built-in illumination and a large stage on which the specimens can be examined or dissected.

The Research Microscope

Laboratories in many schools are equipped with other types of microscopes. Research microscopes have special features that enable students to do more advanced work. For example, some research microscopes have two eyepieces for viewing. This type of microscope is called a **stereoscopic** (or binocular) **microscope**. Such microscopes are equipped with built-in illumination. A video camera can be mounted on one ocular, and microscopic images of living cells can displayed on a monitor for viewing by the class. Sharp images of the cells can be stored in a computer and printed out for research reports.

The Electron Microscope

The highest magnification that can be obtained with the light microscope is about 1000×. When scientists want to examine an organelle such as a chloroplast or a mitochondrion (very small parts of a cell), they need to use much greater magnification. The **electron microscope** can provide such high magnification.

Instead of light, the modern electron microscope uses a different energy source—a beam of electrons. With an electron beam, a much higher level of magnification can be obtained. In fact, magnification in the hundreds of thousands can be obtained with the electron microscope. However, there is one drawback to using an electron microscope. The electron beam operates in a vacuum, so living things cannot be observed. However, dead cells and cell structures can be viewed. The specimen is usually coated with a reflective substance. An image of the specimen is formed by means of measuring the pattern of reflected electrons. The microscope can be used to scan the surface of an object to produce a detailed picture. Electron microscopes are used in universities and research laboratories.

2.4 SECTION REVIEW

1. A marine biology student placed a frozen fish on the stage of a dissecting microscope. He wanted to observe cells in the fish's eye, but he saw nothing. Explain why.

2. How is the total magnification of a cell determined when seen under the compound light microscope?

3. What is one advantage and one disadvantage of using the electron microscope instead of the light microscope?

Laboratory Investigation 2

Measuring Snail Speed

PROBLEM: How fast (or slow!) is a marine snail?

SKILLS: Measuring distance; calculating rate of movement.

MATERIALS: Graduated cylinder (or jar), seawater, metric ruler, marine snail, clock or watch.

PROCEDURE

1. Fill a graduated cylinder (or jar) with seawater. Use a metric ruler to measure the height of the water column in millimeters, so you have an idea of the distance the snail may travel in each trial.

2. Place a small snail at the bottom of the cylinder or jar of water. Start timing the snail when it begins to move from the bottom. Continue timing the snail as it moves up the side of the container. If the snail moves too slowly to record for the whole length of the cylinder, record it for a shorter distance.

3. Try to run six trials in the allotted time. Enter the distance and time for each trial in a copy of Table 2-5 in your notebook.

4. Compute the speed for each trial "run" of the snail by using the following formula: Speed = Distance (mm)/Time (min).

Enter your results for each trial in the Speed column of the table. Total these figures and divide by the number of trials to get an average rate of speed for the marine snail.

TABLE 2-5 MEASURING THE SPEED OF A MARINE SNAIL

Trial	Distance (mm)	Time (min)	Speed (mm/min)
1			
2			
3			
4			
5			
6			
Total			
Av. Speed			

OBSERVATIONS AND ANALYSES

1. Why do you think different groups doing this experiment may show different average speeds for their snails?

2. Why is it preferable to run six trials rather than just one trial in this type of experiment?

3. As you have observed, the snail is a slow-moving animal. Name one adaptive feature that helps the snail compensate for its slowness.

Chapter 2 | Review

Answer the following questions on a separate sheet of paper.

Vocabulary

The following list contains all the boldface terms in this chapter.

compound light microscope, control group, controlled experiment, density, dissecting microscope, electron microscope, experimental group, graph, hydrometer, hypothesis, marine geologists, mass, meniscus, metric system, remote sensors, salinity, scientific method, specific gravity, stereoscopic microscope, trial, variable

Fill In

Use one of the vocabulary terms listed above to complete each sentence.

1. The factor that affects the outcome of an experiment is called the _____.

2. A possible solution to a scientific problem is known as the _____.

3. Data on the sea surface and the deep ocean floor can be obtained by use of _____.

4. The _____ is the group that is exposed to the variable in a controlled experiment.

5. The ratio of the density of a substance to the density of distilled water is defined as its _____.

Think and Write

Use the information in this chapter to respond to these items.

6. Scientists report the precise materials and procedures they use in an experiment. Why is this important to the scientific process?

7. Explain why seawater has a higher hydrometer reading than that of freshwater.

8. Which tool would be best to examine growth bands in the scale of a fish—a hand lens, compound light microscope, dissecting microscope, or electron microscope? Explain.

Inquiry

Base your answers to questions 9 through 12 on the diagram below, which shows an experiment performed by a marine biology student who wanted to measure the speed of a snail, and on your knowledge of biology and the scientific method.

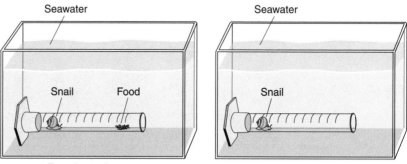

Experimental group   Control group

9. What measuring device would be needed in this experiment?
10. State a hypothesis that can be tested in this experiment.
11. Why wasn't food placed in the control group's water?
12. Why are graduated cylinders used in this experiment?

Multiple Choice

Choose the response that best completes the sentence or answers the question.

13. Which of the following represents the proper sequence of steps in the scientific method?   *a.* hypothesis, procedure, results, conclusion   *b.* procedure, results, hypothesis, conclusion   *c.* results, conclusion, hypothesis, procedure   *d.* conclusion, procedure, hypothesis, results

14. A hypothesis can be described as   *a.* an untested conclusion   *b.* a verified result   *c.* an educated guess   *d.* a rejected idea.

15. Which statement taken from a student's logbook would be considered an experimental result? *a.* The speed of a snail is affected by temperature. *b.* The average speed of the control group of snails is 2.4 cm/min. *c.* A mercury thermometer was used to measure temperature. *d.* If temperature increases, then the speed of the snail decreases.

16. The part of the scientific method in which the data are analyzed and interpreted is called the *a.* hypothesis *b.* results *c.* procedure *d.* conclusion.

17. The hydrometer readings for two classroom aquariums are as follows: Tank A = 1.000, Tank B = 1.030. Which tank(s) contain(s) ocean water? *a.* A *b.* B *c.* A and B *d.* neither A nor B

18. What would happen if water from Tank A were added to Tank B? *a.* The salinity in Tank A would be less. *b.* The salinity in Tank A would be higher. *c.* The salinity in Tank B would be less. *d.* The salinity in Tank B would be higher.

19. It took 10 minutes for a snail to travel a distance of 20 centimeters. How fast did the snail travel, in cm/min? *a.* 1 cm/min *b.* 2 cm/min *c.* 3 cm/min *d.* 4 cm/min

20. A marine science student placed a slide of living microorganisms on the stage of a microscope. He could best observe these organisms by using *a.* a stain on the cells *b.* an electron microscope *c.* a stereoscopic microscope *d.* a dissecting microscope.

Research/Activity

Using Internet sources such as NASA, research information about a remote-sensing satellite that has been launched. Report on some recent data about the oceans that have been gathered by the satellite.

3 Marine Environments

When you have completed this chapter, you should be able to:

DISTINGUISH between the different life zones along a shore.

DISCUSS the characteristics of a variety of marine environments.

DESCRIBE the typical inhabitants of these marine environments.

We all need a place to live. The place, or "home," in which an organism is typically found is called its *habitat*. The total surroundings of a living thing are called the *environment*. The ocean contains various kinds of environments; some are located along the coast, while others are found far out at sea.

Environments have living (biological) and nonliving (physical and chemical) components. The living (biotic) things in an environment are called the *biota*. A coral reef's biota, for example, includes algae, fish, crustaceans, cnidarians, sponges, bacteria, and any other forms of life that inhabit the area. The nonliving (abiotic) parts of a coral reef environment include the water chemistry, light, temperature, salinity, and water pressure, to name a few. Interactions among the biotic and abiotic factors characterize all environments. In this chapter, you will learn about important marine environments in order to understand how each provides a home for living things.

3.1 MARINE LIFE ZONES

Where in the vastness of the ocean can marine organisms be found? Fortunately, you don't have to travel very far. Many marine plants and animals live along the coast. Look at the profile of a coast shown in Figure 3-1 on page 62, which illustrates some life zones on a sandy beach. A **life zone** is a region that contains characteristic organisms that interact with one another and with their environment. One important coastal life zone is the intertidal zone.

The Intertidal Zone

The **intertidal zone** is the area located between high tide and low tide. At high tide the ocean reaches its highest point along a beach, and at low tide the ocean is at its lowest level. High tide is marked by the **strandline**, a long line of seaweed and debris deposited on the beach during each high tide. (See Figure 3-2 on page 63.) If you were to turn the seaweed over, tiny crustaceans called beach hoppers, or beach fleas, would dart and jump about. When the tide is low, you can walk in the intertidal zone and find a variety of marine invertebrates, including other crustaceans, worms, and mollusks. Organisms that live in the intertidal zone are well adapted to meeting the challenges of living in an area that has alternating periods of wet and dry, as the tides come in and go out each day.

The Supratidal Zone

When you go to the beach, you put your blanket or towel down in a life zone called the **supratidal zone**, which is the area above the intertidal zone, up to the sand dunes. Even though you may be a good distance from the ocean, you can smell it because the supratidal zone gets a fine mist of salt spray from the crashing waves. However, the salt spray limits the growth of plants in the lower supratidal zone. In the upper supratidal zone, where there is less salt spray, many species of grasses, shrubs, and trees grow. (Refer to Figure 3-1 on page 62.)

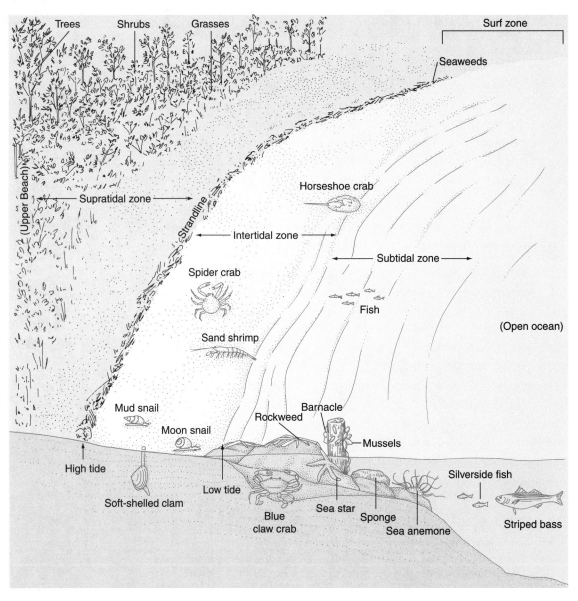

Trees   Shrubs   Grasses

Surf zone

Seaweeds

(Upper Beach)

Supratidal zone

Strandline

Horseshoe crab

Intertidal zone

Subtidal zone

Spider crab

Fish

(Open ocean)

Sand shrimp

Mud snail

Moon snail

Barnacle

Rockweed

Mussels

High tide

Low tide

Silverside fish

Soft-shelled clam

Blue
claw crab

Sea star

Sponge

Sea anemone

Striped bass

Figure 3-1 Characteristic organisms of the life zones on a sandy beach.

The Subtidal Zone

Below the low tide line is the subtidal zone, the coastal life zone that remains underwater. The **subtidal zone** includes an area of heavy wave impact, and the sandy area beyond that, which is affected by underwater turbulence. Some organisms in this zone have structures that help them cling to hard substrates; this pre-

Figure 3-2 The high tide deposits a strandline of seaweed and debris.

vents their being swept away by waves and currents. (Refer to Figure 3-1.) For example, encrusting sponges secrete an acid that enables them to bore into rocks and shells. These shells are often found on the beach, pockmarked with holes from the sponges. Another clinging animal is the sea star. Sea stars cling to hard surfaces by means of suction from their tube feet. Pounding waves cannot dislodge marine snails or sea anemones, which cling to hard surfaces using their muscular feet. Mussels and barnacles receive the full force of wave impact along sandy and rocky shores, but they manage to stay put. Mussels cling to the rocks by secreting tough, fibrous byssal threads that stick to hard substrates. The barnacle has the strongest attachment, because it literally cements itself with glue to rocks and other hard substrates. Clinging organisms also include marine algae, such as kelp and rockweed, which are anchored to rocky surfaces by a fibrous pad of tissue called a holdfast. (Organisms in the intertidal zone also experience the stress of wave impact; read about the surf zone below.)

Many organisms that live in the subtidal zone possess flattened bodies. A flat body minimizes exposure to wave impact. Flat fish such as the flounder avoid turbulence, as well as their enemies, by burying themselves in the sand. Only their gill cover and eyes poke through the sand. While buried, the flounder might happen to see its prey, the sand dollar, another flat inhabitant of the subtidal zone. The sand dollar uses its tube feet to move slowly along the sandy seafloor, where it feeds on algae and dead organic matter. Other inhabitants of the subtidal zone are crabs, shrimp, clams, snails, and worms, which are eaten by fish swimming above the sandy bottom.

The Pelagic Zone

The largest life zone in the ocean is the pelagic zone. (See Figure 3-3.) The **pelagic zone** covers the entire ocean of water above the sea bottom—that vast region where large schools of fish and pods of marine mammals swim freely. The pelagic zone includes the neritic zone (fewer than 200 meters in depth) and the oceanic zone (more than 200 meters in depth).

The Neritic Zone

Beyond the subtidal zone is a life zone called the neritic zone. The **neritic zone** is the region of water that lies above the **continental shelf**, the relatively shallow part of the seafloor that adjoins the continents. When people go deep-sea fishing, they are actually in the neritic zone. In fact, most of the world's commercial fishing takes

Figure 3-3 Cross section of the ocean's life zones; the pelagic includes the neritic and oceanic zones.

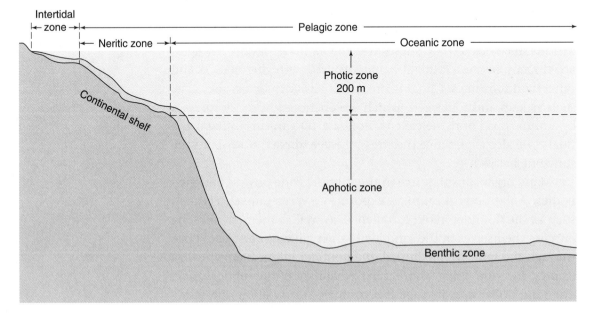

place in the neritic zone. Why is the neritic zone so productive? Rivers that contain runoff from the land flow into the neritic zone, thus providing nutrients for plankton. In addition, much of the neritic zone's depths are sunlit, so algae, phytoplankton, and marine plants can carry on **photosynthesis**, the food-making process on which most living things depend.

The Oceanic Zone

The **oceanic zone** is the life zone that extends beyond the neritic zone and includes most of the open ocean. The upper part of the oceanic zone receives light, whereas the lower part (most of the ocean) is in darkness. The part of the ocean that light penetrates is called the **photic** (meaning "light") **zone**, which is the area most suitable for supporting life. (This is also called the euphotic zone.) Most light penetrates to an average depth of about 100 meters, and to a maximum depth of about 200 meters, within the photic zone. Ninety-nine percent of all sunlight that enters the water is absorbed within the photic zone. The vast area below, which is dark, is called the **aphotic** (meaning "no light") **zone**. There is more life in the photic zone, because light promotes the growth of algae and plants, which provide a source of food for marine animals. As a result, more communities of organisms develop in the photic zone than in the aphotic zone. Communities are made up of populations of different species of organisms that interact with one another and with their environment.

Although there are fewer communities of organisms in the aphotic zone, numerous life-forms, such as fish, worms, squid, and crustaceans, have been observed and photographed in the great depths of the ocean. Deep-sea organisms are specially adapted to live in the depths of the aphotic zone. An **adaptation** is any characteristic of an organism that enables it to live successfully in its environment. The anglerfish, for example, is adapted to live in the deep ocean because it possesses a huge mouth and long sharp teeth to help it catch prey in the darkness. It even has a lure over its mouth that glows in the dark, making it possible for the fish to see and attract prey. (See Chapter 12.) Other deep-sea fishes have mouths that are pointed upward, which helps them catch the scraps of food that fall from shallower waters.

DISCOVERY

Biodiversity Among the Benthos

Marine biologists who specialize in deep-sea organisms are having a field day—they have recently discovered an oasis of life in the deep ocean, an area that had previously been thought to be a biological wasteland. The scientists have probed deep down in the Atlantic and Pacific oceans and found a surprising diversity of life among the benthos, or bottom-dwelling organisms. This biodiversity was unexpected because biologists did not think the seafloor had the features and conditions necessary to promote the evolution of so many different, yet closely related, organisms.

A great variety of benthos were collected by use of new sampling equipment controlled by ships at the surface. One piece of equipment consisted of a box corer (a device that looks like a cookie cutter), which sampled bottom sediments. A more typical tool, a fine mesh net, scooped up specimens from along the seafloor. Marine biologists can collect specimens in this manner from as far down as 1.3 km. One sample contained nearly 1600 different species, including various slugs, snails, crabs, worms, sea stars, and anemones. Many samples were collected that contained species never seen before. Scientists now estimate that the deep ocean may contain ten million or more species of invertebrates—much more than the number of invertebrate species known to live on land.

Many questions remain about the evolution and adaptations of deep-sea benthos. How do so many different types of organisms survive in this pitch-black region of near-freezing temperatures and crushing pressure? How does the environment provide nourishment to support the high numbers? These questions have even given rise to some speculation that much larger animals may exist in the deep that are still unknown to science. We may never know for certain. But it is likely that, in light of the recent findings, more comprehensive explorations of the deep seafloor will continue to be undertaken.

QUESTIONS

1. Why were marine scientists surprised by the biodiversity among the benthos?

2. What kinds of organisms have been found in the samples that were collected?

3. How do scientists collect invertebrates that live more than 1 km deep?

The deepest part of the ocean floor is the **ocean basin**, or abyssal plain, which is also home to a variety of organisms. Many fish and invertebrates inhabit the bottom and rarely swim near the surface. Bottom-dwelling organisms that live on the seafloor inhabit an area called the benthic zone. The **benthic zone** actually includes the entire ocean floor, from the shallow intertidal zone to the deep ocean basin. Organisms that inhabit the benthic zone are called **benthos**. The benthos that live in the ocean basin are adapted to regions of very low temperatures and very high pressure.

3.1 SECTION REVIEW

1. Which life zone is more productive, the neritic or the oceanic? Explain.

2. How are some benthic organisms adapted to live in turbulent waters?

3. Why might it be hard for deep-sea fish to find food? What adaptations do they have for this?

3.2 THE SANDY BEACH ENVIRONMENT

One of the most familiar environments along the shore is the **sandy beach**. Sandy beaches are composed of sand, a loose sediment that is easily shifted and moved about by wind and water. This battering by wind and water makes the sandy beach a rather harsh environment, yet it provides a variety of habitats for living things. If you were to walk from the upper beach area down to the shore, you would pass through some of the life zones described in Section 3.1. Each zone contains a particular group of organisms that share the habitat. This pattern of marine life, which forms distinct bands, or life zones, along the shore is called **zonation**. The upper beach contains a zone of beach plants that includes trees, shrubs, and grasses. The roots of these plants hold onto the sand and prevent its erosion by wind and water. The sand collects in small hills, or dunes. The trees occupy the highest elevations in the dunes, followed by shrubs growing along the slopes of the dunes and beach grasses at the lower levels. (See Figure 3-4 on page 68.)

Figure 3-4 Beach grasses and shrubs grow on sandy dunes.

If you continued your walk down the beach to the water, you would pass through the area of wet sand called the intertidal zone. When covered with water, the intertidal zone harbors a variety of marine animals. When the tide goes out, some of these animals, such as fish and crabs, retreat to deeper waters, while others, such as marine worms and the mud snail and other mollusks, stay behind and burrow in the moist sand.

The Surf Zone

The typical sandy beach has a region of crashing waves called the **surf zone**, which is white with foam as a result of air mixing with water as waves pound on the shore. The surf zone is not a fixed zone. Instead it moves with the tide as it advances and retreats on the slope of the beach, from the subtidal to the intertidal zones. Since the water in this zone is in constant motion, the sand is pushed and moved about by wave action. For creatures that live in the surf zone, life is like always being in a storm. (See Figure 3-5.)

Some small marine animals, such as the mole crab (*Emertia*) and the surf clam (*Spisula*), have managed to adapt to the turbulence of the surf zone. The mole crab avoids the waves by using its paddlelike appendages to dig into the sand. From there, it feeds on microscopic organisms by sticking its feathery appendages up into the water. Its smooth jelly-bean shape helps the mole crab swim with minimal resistance through swirling mixtures of sand and water. (See Figure

3-6.) The thick shell of the surf clam protects it against wave impact and erosion from moving sand; and the surf clam's large muscular foot enables it to dig quickly into the sand to avoid predators.

When the tide is out and you are walking in shallow water, you are in the subtidal zone. You would have to be careful where you stepped—you could be pinched by the sharp claws of a sand crab, such as the lady crab (*Ovalipes ocellatus*) or the blue claw crab (*Callinectes sapidus*). Like the mole crab, they swim and dig in the sand by means of their paddlelike appendages; and they hide from predators by burrowing in the sand, with only their eyestalks sticking up. These crabs usually feed on scraps of food, but occasionally they catch small fish such as silversides (*Menidia*), which swim in schools in the oxygenated waters of the surf zone. The silversides feed on invertebrates and crab eggs, and are, in turn, preyed on by larger fish such as the striped bass (*Morone saxatilis*), which swim into the shallows from the deeper sea.

3.2 SECTION REVIEW

1. What factors influence the zonation along a sandy beach?

2. Compare the habitat of the upper beach with that of the intertidal zone. What kinds of plants or animals are typical of each zone?

3. How are conditions for life in a sandy beach's intertidal zone different from those in its subtidal zone?

Mole crab

Figure 3-6 The mole crab survives the turbulence of the surf zone by using its paddlelike appendages to dig into the sand.

3.3 THE ROCKY COAST ENVIRONMENT

Many coastal states, such as Maine, California, Oregon, Washington, Alaska, and Hawaii, have shores composed of solid rock. Shores made up of solid rock are called **rocky coasts.** What kind of environment for living things is provided by a rocky coast? Look at the typical rocky coast shown in Figure 3-7. Rocks provide a surface on which marine organisms can attach themselves. When the tide is out, you can see seaweeds clinging to the rocks. Similar to that of a sandy beach, the rocky coast shows zones of habitats, with each made up of different communities of living things. Typically, four major bands, or zones, of life can be observed: the upper intertidal, mid-intertidal, lower intertidal, and subtidal zones. (See Figure 3-8.)

The Upper Intertidal Zone

The upper intertidal zone, or wave splash zone, is the area above the high-tide mark that gets moisture from the ocean spray of crashing waves. The moist rocks provide an environment for the growth of blue-green bacteria (formerly called blue-green algae), which form a thin film on the rocks. These photosynthetic organisms absorb water from the splashing waves and take in CO_2 from

Figure 3-7 The rocky coast shows distinct bands of different seaweeds.

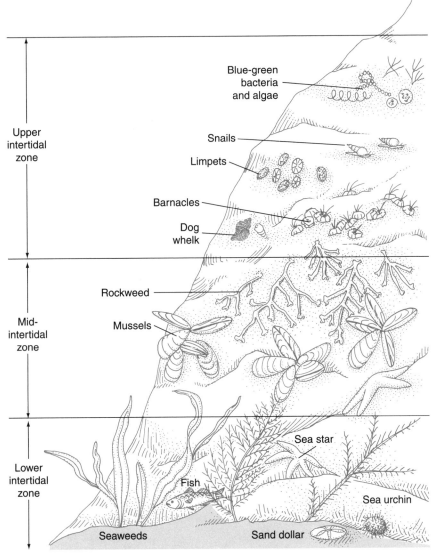

Figure 3-8 Characteristic organisms of the life zones on a rocky coast.

Upper intertidal zone

Blue-green bacteria and algae

Snails

Limpets

Barnacles

Dog whelk

Mid-intertidal zone

Rockweed

Mussels

Lower intertidal zone

Sea star

Fish

Sea urchin

Seaweeds

Sand dollar

the air to produce food and oxygen. When the bacteria die, they stain the rocks black, causing a discoloration that is often mistaken for an oil spill.

The periwinkle snail (*Littorina littorea*) feeds on algae in the upper intertidal zone. The snails scrape the algae off the rocks with their rasping mouthpart, called a radula. The limpet, which is another mollusk, also grazes on algae in this zone.

The Mid-Intertidal Zone

Below the upper intertidal zone lies the mid-intertidal zone, which is occupied largely by barnacles, mussels, and seaweeds. Two common species of barnacles, the rock barnacle (*Balanus balanoides*) and the bay barnacle (*B. improvisus*), can be found glued to the rocks. Barnacles adhere to the rocks so strongly that the most powerful wave cannot dislodge them.

At high tide, barnacles are covered by water. During this time, they filter feed on plankton and organic debris by rhythmically whipping their feathery cirri to capture food. (You can observe filter feeding in the barnacle by doing the investigation at the end of this chapter.) When the tide goes out, the barnacles are exposed to air for several hours, so they shut their shells tight to keep from drying. The barnacle's sharp, overlapping shells also help to protect it from predators. However, a marine snail called the dog whelk (*Nucella lapillius*) can drill a hole through the shell and eat the barnacle. (The whelk produces an acidic secretion from its foot gland, which softens the barnacle shell before the whelk drills into it with its radula.) Along the rocky coast of the Pacific Northwest, the giant acorn barnacle (*B. nubilis*) grows up to 100 millimeters (mm) in size; this barnacle is traditionally harvested and eaten by Native Americans.

Below the layer of barnacles lie seaweeds and a densely packed bed of mussels. As mentioned above, mussels attach to the rocks by means of byssal threads. The threads prevent the mussels from being dislodged by waves. Like barnacles, mussels are filter feeders. The beating of the mussels' tiny cilia (on the surface of their gill membranes) creates currents that carry food and oxygen into, and wastes out of, their shells.

This bed of mussels attracts several predators. The dog whelk that eats barnacles also eats mussels. Sea stars raid the mussel beds and consume large numbers of these bivalves. The blackfish, or tautog (*Tautoga onitis*), has well-developed front teeth that can crush mussels. (A popular food fish caught by anglers from rock jetties and piers, it inhabits Atlantic waters from Nova Scotia to the Carolinas.) The blue mussel (*Mytilus edulis*) is found along the Atlantic and Pacific coasts and is harvested and eaten by humans.

The brown seaweed called rockweed (*Fucus*) also lives in the mid-intertidal zone. When the tide is low, thick mats of rockweed can be seen draped over the rocks. The seaweed clings to the rocks by means of its holdfast pad. Rockweed provides cover for a variety of marine animals that live in and among the rocks, such as snails, limpets, small crabs, and worms. When the tide is high, rockweed floats near the surface, buoyed up by its gas-filled bladders.

The Lower Intertidal Zone

Below the bed of mussels lies the lower intertidal zone, which is dominated by seaweeds. The red seaweed commonly known as Irish moss (*Chondrus crispus*) grows like a thick carpet over the rocks in this zone. When the tide is low, spaces between the rocks retain water, forming small habitats known as **tide pools**. (See Figure 3-9.) Tide pools are like natural aquarium tanks that contain algae, invertebrates (such as snails and crabs), and small fish.

The Subtidal Zone

Below the lower intertidal zone is the underwater subtidal zone. The rocky coast subtidal zone has an abundance of life. The best way to observe life in the subtidal zone is by snorkeling. You can observe sea urchins and snails grazing on algae. Sea urchins eat giant kelp,

Figure 3-9 Tide pools on a rocky coast contain a variety of plant and animal life.

the biggest seaweed in the ocean, by feeding on its holdfasts. Sea stars cling to the rocks. The movement of the sea stars is dictated by the tides. When the tide comes in, the sea stars move from the subtidal zone to the intertidal zone to feed on the mussels. There is also a variety of sea anemones, as well as crabs and lobsters, hiding in the rock crevices. Lobsters come out of their hiding places at night to feed. Predatory fish that swim in from the open ocean prey on the abundant invertebrates.

3.3 SECTION REVIEW

1. How do the rocky coast life zones differ from those of the sandy beach?

2. Compare the types of organisms found in the upper intertidal zone with those found in the lower intertidal zone.

3. Describe the organisms that are typical of the rocky coast mid-intertidal zone.

3.4 THE ESTUARY ENVIRONMENT

Along many coasts, freshwater rivers drain into the ocean. At the mouth of a river, where it enters the ocean, salt water and freshwater mix, forming the type of environment known as an **estuary**. Along the shores of an estuary, there are many inland bays and creeks. (See Figure 3-10.) The varied terrain of an estuary, along with its **brackish** water (the mixture of freshwater and seawater), provides diverse habitats for marine life. In fact, an estuary is one of

Figure 3-10 Cross section of an estuary and a barrier beach.

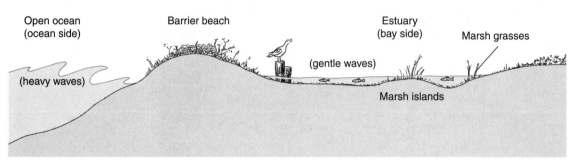

Open ocean (ocean side)   Barrier beach   Estuary (bay side)   Marsh grasses

(heavy waves)   (gentle waves)

Marsh islands

the most productive environments found along any coast. Many types of organisms lay their eggs in an estuary, and the young of many species develop in this nutrient-rich environment.

Many estuaries were formed at the end of the last Ice Age, about 10,000 to 15,000 years ago, when glaciers melted and the sea level rose. The ocean invaded low-lying coastal areas, flooding the mouths of rivers and streams. Sediments carried in by the ocean were deposited offshore, forming long ridges of sand called **barrier beaches**. On one side of a barrier beach is the bay and on the other side is the open ocean. The estuary lies on the bay side of a barrier beach. (See Figure 3-11.)

Estuaries have calm waters because they are protected from heavy wave impact by the barrier beaches. The estuaries have become natural sanctuaries for a wide variety of marine plants and animals. As a result, several different natural communities have been identified in the estuary environment.

The Salt Marsh Community

In many estuaries along the Atlantic, Pacific, and Gulf coasts, grasses grow abundantly in the shallow water, forming a **salt marsh community**, or **wetlands**. Look at a typical salt marsh community,

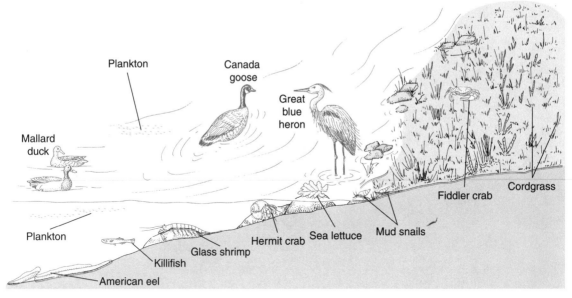

Labels on figure: Plankton, Canada goose, Great blue heron, Mallard duck, Cordgrass, Fiddler crab, Plankton, Mud snails, Sea lettuce, Hermit crab, Glass shrimp, Killifish, American eel

Figure 3-12 Characteristic organisms of a salt marsh community.

shown in Figure 3-12. Marsh grasses are the dominant species in the salt marsh. One type of grass, called cordgrass (*Spartina*), is tough, coarse, and resistant to the killing effects of salt; special glands in its leaves secrete salt crystals. Few animals eat cordgrass because it is tough and has a high salt content. However, since it is a producer, cordgrass is high in nutrients. After cordgrass dies, its dead organic matter is acted on by decay bacteria. Such decaying organic matter is referred to as *detritus*. During the process of decay, nutrients from cordgrass, and other forms of detritus, are released into the estuary's waters. (See Figure 3-13.)

The nutrients, which include phosphates (used for energy) and nitrates (used for growth), are taken up by plankton. The plankton flourish and become a food source for a variety of filter feeders. One of the filter feeders is the ribbed mussel (*Modiolus demissus*), which lives half buried in the mud. The water currents created by the mussel's cilia also benefit the cordgrass by increasing water circulation around its roots. Attached to the mussels are barnacles, which also filter feed. The tangle of cordgrass roots acts as a net to trap organic debris, which is consumed by another inhabitant of the salt marsh, the common shore shrimp (*Palaemonetes vulgaris*).

The lives of many inhabitants of the salt marsh are controlled by the tides. When the tide is low, small fiddler crabs (*Uca*) emerge from their holes in the sand to feed on bits of organic matter left

Figure 3-13 Nutrient-rich salt marshes are still found near large cities.

behind by the outgoing tide. If approached, they make a quick retreat into the nearest hole. Before the tide comes in, the fiddler crabs return to their tunnels and plug up the entrances.

Another crustacean that feeds on organic debris in the salt marsh is the hermit crab (*Pagarus*). You can see hermit crabs scurrying about in the shallow tidal creeks, feeding on food particles in the sand. Since they have soft, unprotected abdomens, hermit crabs live inside empty snail shells to protect themselves from predators. When hermit crabs outgrow their snail shells, they look for larger ones to inhabit.

The calm and nutrient-rich waters of the salt marsh provide an ideal environment for marine animals to produce offspring. In fact, these wetlands, which are home to a great variety of fish, are often described as the important "nurseries" for many species of ocean fish. Young flounders are among these fish. The flounders feed on killifish (*Fundulus*), which are regular inhabitants of the salt marsh. The killifish, in turn, feed on insect larvae.

The fish and invertebrates that inhabit the salt marsh are a food source for the many bird, reptile, and mammal species that live in and around the marsh. Many migratory birds also depend on the wetlands for food as they fly both north and south each year.

The Mud Flat Community

The part of the estuary environment that is characterized by dark, muddy sand and no marsh grasses is called the tidal **mud flat community**. The mud flat has a slightly sloping beach touched by gentle waves. (See Figure 3-14.) There is very little aeration of the muddy sands because of the slight wave impact. Flushing action by the outgoing tide is minimal. As a result, organic debris carried in by the incoming tide tends to accumulate in the sand. (See Figure 3-15.)

Just as the wetland is the estuary's nursery, the mud flat is its graveyard. Bacteria decompose the wastes and turn the sand into a dark mud. If you were to dig a hole in the sand, you would see that under the surface the sand is black and gives off a foul odor, like rotting eggs. The smell is caused by the presence of hydrogen sulfide (H_2S), a compound that is the product of decay and that accumulates in sediments deficient in oxygen. Microscopic organisms are abundant in the mud flats. Decomposers, like the decay bacteria

Figure 3-14 Characteristic organisms of a mud flat community.

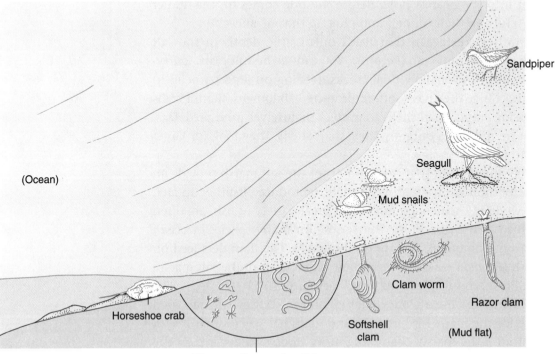

Figure 3-15 Gentle waves bring organic debris to the mud flat.

that live in the dark, moist sand and play an important role in the well-being of the mud flat community, convert wastes into useful nutrients. Tidal action transports the nutrients to other parts of the estuary and out to sea. Nutrients from the estuary are a major food source for the oceanic plankton.

Within the mud flat, a variety of invertebrates scavenge for food. At low tide, large numbers of mud snails (*Ilyanassa obsoleta*) feed on debris in shallow tide pools. These tiny "garbage eaters" do an efficient job of getting rid of excess wastes in the mud flat. Sand-worms and bloodworms tunnel through the sand, feeding on organic debris. Clams also live in the sand. The soft-shell clam (*Mya arenaria*) has a long siphon tube that it uses to take in water that contains plankton. When the tide is out, the clam retracts its siphon, leaving a hole in the sand. You can locate soft-shell clams by looking for these telltale holes in the sand. The slightest movement causes the soft-shell clams to squirt water up through the holes.

The mud flat is also home to another mollusk, the razor clam (*Ensis directus*). The razor clam has a well-developed muscular foot that it uses to dig quickly through the sand. The burrowing action of mollusks, worms, and other invertebrates in the mud flats helps to bring some much-needed air to the oxygen-deficient sediments.

The invertebrates also burrow in the mud and sand to escape predators, such as shorebirds. The bills of the various shorebirds are of different shapes and lengths, depending on the birds' feeding habits. The short bill of the plover and sandpiper is used to feed on invertebrates that live near the surface in the mud. Shorebirds with longer bills can reach invertebrates that burrow deeper into the mud.

The Mangrove Community

In regions with a tropical climate, such as Florida, the shores of some bays and inlets are covered by a thick growth of mangrove trees, and the coastal marsh takes the form of a **mangrove community**, or mangrove swamp. (See Figure 3-16.) The dominant plant species that grows in the water of the mangrove swamp is the red mangrove tree (*Rhizophora mangle*). At low tide, the arching roots, which anchor the mangrove trees in the muddy sand, are visible. At high tide, the water covers the roots, but the tree trunks and leaves remain above water. (See Figure 3-17.)

Figure 3-16 Characteristic organisms of a mangrove community.

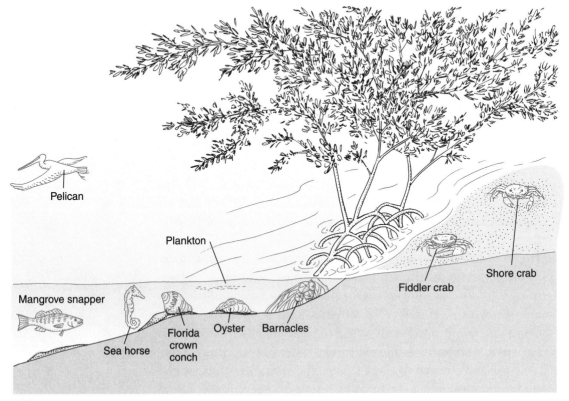

The incoming tide brings in organic debris that gets trapped in the tangle of mangrove roots. Dead leaves from the mangrove trees decay in the water. The products of decay enrich the mangrove swamp with nutrients, a food source for plankton. Filter-feeding animals such as barnacles and oysters consume the plankton. The outgoing tide flushes nutrients back out to sea, providing oceanic life with an important food source.

The mangrove swamp is like a natural wildlife sanctuary. Like the tangled roots of the mangrove trees, life in the mangrove swamp is a diverse web of interrelationships. A variety of marine animals, including snails, land crabs, and fiddler crabs, scavenge for food at low tide. In the evenings, raccoons search for crabs among the mangrove roots. When the tide is in, the crown conch (*Melongea corona*) preys on oysters; and fish, such as the mangrove snapper (*Lutjanus griseus*), feed on the smaller fish and invertebrates swimming among the roots. Nesting in the leafy trees are the brown pelican (*Pelecanus occidentalis*) and the osprey (*Pandion haliaetus*), which hunt and scavenge on the abundant marine life.

Mangrove swamps also protect the shore from erosion. The roots of the mangrove tree hold the sand and prevent it from being carried away by waves and currents. During storms, the mangrove swamp acts like a giant sponge, absorbing the water and impact of the storm. Mangrove communities are natural barriers that protect other habitats farther inland.

3.4 SECTION REVIEW

1. Why is an estuary such a productive aquatic environment?

2. How are organic wastes recycled in the mud flat community?

3. Briefly describe life around the tree roots in a mangrove swamp.

3.5 THE CORAL REEF ENVIRONMENT

The **coral reef** is among the most spectacular of marine environments. Look at the coral reef animals shown in Figure 3-18. As you can see, the coral reef contains a fantastic assortment of marine life. The foundation for this diverse community is the coral animal itself. The coral reef is a stony formation that is built up from the seafloor by a living organism called the coral polyp. (See Chapter 7 for more information on the coral polyp.)

The reef begins to form when microscopic coral larvae settle on a hard substrate in the sand and develop into coral polyps. The coral animals live in colonies, with each tiny polyp sitting in its own limestone home, which it builds. Each new generation of polyps lays down a new layer of limestone, causing the reef to expand upward (at about 2.5 cm per year) and outward. Some massive reefs are more than 40 meters high and more than 1000 km long. Corals can grow right up to the ocean surface but cannot grow out of the water.

Figure 3-18 Characteristic organisms of a coral reef community.

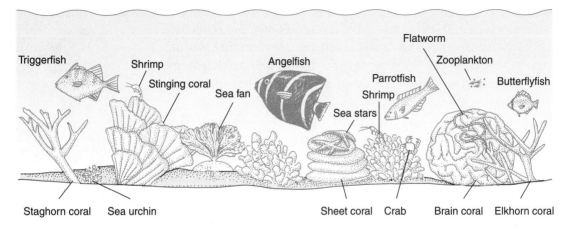

Coral Features

Coral reefs are found in the tropical and subtropical regions of the world, between 30 degrees north and 30 degrees south latitude. They are located in the Atlantic, Pacific, and Indian oceans. Why are coral reefs found only in the mid-latitudes? In these latitudes, the ocean's water is warm and clear, and there is plenty of sunlight. These are the conditions that are needed to promote the growth of symbiotic algae within the coral polyps that build the reefs.

Look at a world map and locate Australia's Great Barrier Reef, the longest reef system in the world; it is approximately 2000 km long. Large reefs are also found in the Caribbean, the Florida Keys, and off the coast of Belize in Central America. Living among the Great Barrier Reef's corals are more than 1500 species of fish, among other organisms.

Why do so many different organisms inhabit the coral reef? A reef consists of different types of coral growing together. The many kinds and shapes of corals create an irregular pattern of crevices, depressions, and caves in which organisms can live and hide. This great variety of life, or **biodiversity**, is typically found in habitats that provide many areas for hiding and surfaces for attachment.

Each type of coral has its own unique shape, size, color, and texture. Corals are classified into two types, hard corals and soft corals. Many corals are named after familiar objects, for example, hard corals such as elk horn (*Acropora palmata*), staghorn (*Acropora cervicornis*), and brain (*Diploria labyrinthiformis*), and soft corals such as fan (*Gorgonia ventalina*) and sea rod (*Plexaura flexuosa*), which are more flexible and sway in the currents. (See Figures 3-19 and 3-20.)

Figure 3-19 (left) The staghorn coral is an example of a hard coral.

Figure 3-20 (right) The sea fan is an example of a soft coral.

The coral reef is a very productive environment. However, coral reefs are fragile. Pieces of coral can be broken off easily; even touching coral can damage the thin membrane that protects its surface. Unfortunately, many coral reefs around the world are in danger. Development along coasts has clouded offshore waters and caused reefs to die. The unrestricted use of reefs for fishing and diving has contributed to their degradation.

Reef Inhabitants

When you dive along a coral reef, you encounter an oasis of colorful living things. Typical reef inhabitants include the butterfly fish with its elongated snout to feed on food particles on the reef surface, the barracuda with its needlelike teeth to grab and eat other reef fish, and the parrot fish with its beaklike mouth to nibble off chunks of reef. (See Figure 3-21.) The parrot fish eats the coral polyps and eliminates the indigestible limestone. (The egested particles of limestone trickle down to the seafloor to form coral sand.) Various sponges, worms, shrimps, anemones, sea stars, mollusks, and other fish also live in and around the reef.

Figure 3-21 The parrot fish uses its beaklike mouth to bite off chunks of coral; it eats the polyps and egests the limestone.

How do the inhabitants of the coral reef avoid predators? The reef provides ideal hiding places for its inhabitants, as mentioned above. In a coral crevice, you might see a spiny lobster (*Panulirus argas*), which retreats farther back into its hole if it detects danger. Predators also hide; for example, the moray eel (*Gymnothorax*) stays hidden in crevices and rarely ventures out during the day. In fact, many reef animals emerge only at night to feed.

Some fish depend on **camouflage**, the ability to match or blend in with their natural background, to avoid being detected. To hide, the sticklike trumpetfish (*Aulostomus*) floats motionless, with its head facing down, alongside branching corals and sponges. Such camouflage helps the trumpetfish escape detection by other fish, including those it may want to eat. In contrast, the spotted trunkfish (*Lactophrys bicaudalis*) lives at the bottom of the reef, where it is difficult to spot against the background of coral and speckled sand. (See Figure 3-22.) Fish can also gain some protection against predators by swimming in a large group, or school, of their own species. Schools of fish may stay in one location, or they may swim over long distances. Stationary schools of fish, such as that of the grunt

Figure 3-22 The spotted trunkfish is hard to see as it swims along the mottled reef bottom.

Figure 3-23 Some reef fish find protection from predators by swimming in a school.

(*Haemulon*), are common around coral reefs. There is security in numbers; the chances of any fish in particular being caught by a predator is reduced by its being within a school of hundreds of fish. (See Figure 3-23.)

Reef fish are well known for being very colorful. The black-and-yellow coloration of the rock beauty (*Holocanthus tricolor*) stands out in stark contrast to the background colors of the reef. The queen angelfish (*Holocanthus ciliaris*) and the butterfly fish (*Chaetodon*) also exhibit different, bold colors in a pattern known as **color contrast**. What is the adaptive value of color contrast in fish? Fish use color as a means of identifying members of their own species. This is a very important ability in the coral reef environment, where there are numerous species of fish, many of them closely related. Color is also used to confuse predators. For example, the four-eyed butterfly fish has two fake eyespots at the base of its tail fin, which trick predators into thinking that the back of the fish is its front. (See Figure 3-24.) Spots, bars, and stripes (such as on the banded butterfly fish), which obscure the outline of a fish, are called **disruptive coloration**. Such coloration makes it harder for a predator to see and catch the fish.

The reef is like an apartment building, with inhabitants living close to one another and at different levels. However, in the reef

Figure 3-24 The fake eye-spots near the tail fin of a butterfly fish can trick predators.

Figure 3-25 Reef sharks such as this one show territoriality when they patrol their home area.

there are no walls to separate one neighbor from another. Living so close together, the different types of fish need to establish a home area or turf. A home area with well-defined boundaries is called a *territory*. Within its territory, a fish may need to defend important resources such as food, a mate, or a nesting site. The damselfish (*Microspethodon chrysurus*), for example, will attack a much larger fish such as a parrot fish, or even a scuba diver, in an attempt to defend its territory. Such behavior by an organism in defending its home area is called **territoriality**. Larger fish, such as the barracuda and the black-tipped and white-tipped reef sharks, are also territorial. They swim along the reef and patrol their home area, threatening or even attacking perceived intruders, including divers. (See Figure 3-25.)

3.5 SECTION REVIEW

1. How is a coral reef built? What substance is it made up of?

2. What conditions are required for the growth of a coral reef?

3. What are some ways the inhabitants of a coral reef avoid detection?

Laboratory Investigation 3

Examining Beach Sand

PROBLEM: What are the physical properties of beach sand?

SKILLS: Using a dissecting microscope; measuring tiny sand grains.

MATERIALS: Transparent metric ruler, dissecting microscope (or compound microscope), sand samples, dissecting needle, petri dish, magnet.

PROCEDURE

1. Place the transparent metric ruler on the stage of the microscope.

2. Sprinkle some sand in the petri dish and put the dish on top of the ruler.

3. Observe the sand grains under the microscope. In your notebook, record the size of a single grain, in mm, on a copy of Table 3-1. If the grains vary widely in size, measure several and record the average size. Notice if there are any tiny shells or fragments that may be the remains of foraminifers.

4. Observe the physical properties of color, texture, luster, and shape in the grains. Use a dissecting needle to separate the grains into different piles, each representing a different property. Record your observations in the table. Make a sketch of a sand grain from each sample pile.

TABLE 3-1 **PHYSICAL PROPERTIES OF SAND SAMPLES**

Sample	Size	Color	Texture	Luster	Shape	Minerals
A						
B						
C						
D						
E						

TABLE 3-2   MINERALS AND THEIR PHYSICAL PROPERTIES

Mineral	Physical Properties
Quartz	Clear, glassy, resembles salt crystals, eroded from granite
Feldspar	Clear, tan or gray, usually square, eroded from granite
Hornblende	Dark gray to black, glassy
Mica	Thin shiny flakes, silver gray to black
Magnetite	Dark shiny triangles, clings to magnet, contains iron
Garnet	Purple to red, angular, abrasive, used in sandpaper
Olivine	Olive-green, glassy
Basalt	Gray to black, resembles lumps of coal
Pyrite	Pale orange to yellow, metallic luster (like gold)
Calcite	Opaque, glassy, composed of calcium carbonate, reacts with dilute HCl to produce bubbles of CO_2

5. Place a small magnet in contact with the sand. Do the sand grains cling to the magnet? If the sand clings, iron is present. Identify the mineral in Table 3-2 that contains the element iron. Use this table to identify the minerals that match the physical properties of the samples you observed. Record the information in your copy of Table 3-1.

OBSERVATIONS AND ANALYSES

1. Which type of sand would be a good source for magnetite—coral sand or volcanic sand? Explain.

2. What are the four physical properties of beach sand that are important for identification?

3. In what ways do you think the texture and shape of a sand grain are related to its source (where the sand is from; wave action; activities of organisms)?

Chapter 3 | Review

Answer the following questions on a separate sheet of paper.

Vocabulary

The following list contains all the boldface terms in this chapter.

adaptation, aphotic zone, barrier beaches, benthic zone, benthos, biodiversity, brackish, camouflage, color contrast, continental shelf, coral reef, disruptive coloration, estuary, intertidal zone, life zone, mangrove community, mud flat community, neritic zone, ocean basin, oceanic zone, pelagic zone, photic zone, photosynthesis, rocky coasts, salt marsh community, sandy beach, strandline, subtidal zone, supratidal zone, surf zone, territoriality, tide pools, wetlands, zonation

Fill In

1. The _____ is dominated by trees that live in salt water.

2. Organisms that live on the seafloor inhabit the _____.

3. The _____ includes the waters above the continental shelf.

4. The area between high tide and low tide is the _____.

5. Salt water and freshwater form an _____ at a river mouth.

Think and Write

Use the information in this chapter to respond to these items.

6. Compare the intertidal zone of a rocky beach with that of a sandy beach.

7. What are two important differences between the neritic and oceanic zones?

8. Compare the main traits of a salt marsh with those of a mud flat community.

Inquiry

Base your answers to questions 9 through 12 on Figure 3-3 on page 64, which shows a cross section of the major life zones of the ocean, and on your knowledge of marine science.

9. Identify, by name, the zone in which most of the world's commercial fishing takes place. Give two reasons why this zone is so biologically productive.

10. Identify, by name, the vast area of the oceanic zone that receives very little sunlight. Give an example of an organism that is adapted to live in this zone. Below which oceanic zone is it located (identify by name)?

11. Identify, by name, the largest marine life zone. What kinds of large marine animals swim freely here? What are the names of the two life zones that it includes?

12. Identify, by name, the zone in which marine organisms live that are specially adapted to survive alternating periods of high tides and low tides. How do barnacles survive there?

Multiple Choice

Choose the response that best completes the sentence or answers the question.

13. The arching roots of the mangrove tree serve all the following functions *except*   *a.* they anchor the trees in the muddy sand   *b.* they provide a habitat for small fish and invertebrates   *c.* they serve as a food source for the mangrove snapper   *d.* they hold the sand and prevent its erosion.

14. Which of the following organisms would be found in a rocky coast intertidal zone?   *a.* sea anemones, flounder, blue-green bacteria, barnacles   *b.* barnacles, snails, mussels, seaweeds   *c.* mussels, sea stars, cordgrass, seaweeds   *d.* sea stars, cordgrass, shore shrimp, fiddler crabs

15. Of the following marine organisms, which would probably *not* be found in a rocky tide pool?   *a.* crab   *b.* grazing snail   *c.* flounder   *d.* barnacle

16. A barrier beach is located between   *a.* a bay and an atoll
b. a bay and a river   *c.* a river and an atoll   *d.* a bay and
the ocean.

17. The marine environment that is characterized by a shifting,
unstable sediment is the   *a.* rocky coast   *b.* coral reef
c. sandy beach   *d.* tide pool.

18. The dominant plant life in the salt marsh is the   *a.* sea
lettuce   *b.* cordgrass   *c.* mangrove tree   *d.* red algae.

19. The salt marsh is more productive than the mud flat because
it   *a.* lies in deeper waters   *b.* has more salt   *c.* has more
producer organisms   *d.* lies in calmer waters.

20. In which geographic area would a mangrove swamp be
located?   *a.* Gulf of Maine   *b.* Puget Sound   *c.* San
Francisco   *d.* Florida Keys

21. Coral reefs are found in latitudes where the ocean's waters are
a. warm and clear   *b.* warm and murky   *c.* cold and clear
d. cold and murky.

22. The cordgrass *Spartina* has adapted to salt water by
a. growing deep roots   *b.* conserving freshwater in its leaves
c. secreting excess salt from its leaves   *d.* not taking in salt
through its roots.

23. The mud flat community has all of the following
characteristics *except*   *a.* dark, muddy sand
b. invertebrates that burrow   *c.* a high rate of
decomposition   *d.* turbulent wave action.

24. All the following factors contribute to the coral reef's
biodiversity *except*   *a.* the growth of algae within the coral
polyps   *b.* many crevices in which animals can live and hide
c. plenty of sunlight   *d.* coastal development.

25. The ability of a fish to blend in with its surroundings is called
a. color contrast   *b.* camouflage   *c.* disruptive coloration
d. territoriality.

Research/Activity

Many wetlands are being lost to coastal development. Report on
recent efforts to protect vanishing wetlands in the United States.

UNIT 2

KINGDOMS OF LIFE IN THE SEA

When you go to a supermarket, you see that there are thousands of food items from which to choose. How is it possible for you to find the particular one you want to buy? In a supermarket, the food is organized into sections, making it easier for you to find and buy the items you want.

Likewise, millions of different kinds of organisms live on our planet; and many thousands of these organisms live in the sea. Scientists have developed a system for organizing all living things into various categories. This classification system has several important benefits. It enables people to study organisms more easily. It shows the evolutionary relationships among organisms. And it lets you make certain assumptions about organisms within a particular group. In this unit, you will start to learn about the diversity of marine organisms.

4 Unicellular Marine Organisms

When you have completed this chapter, you should be able to:

UNDERSTAND and APPLY the rules of classification.

IDENTIFY and DESCRIBE the major groups of living things.

COMPARE and CONTRAST the monerans and the protists.

DISCUSS basic cell structure and different types of nutrition.

Look at the organism shown above. For many years, scientists disagreed on the classification of this tiny aquatic organism, called a *euglena.* Zoologists placed the euglena in the animal kingdom. After all, the euglena has animal-like traits, such as an eyespot that is sensitive to light, the ability to move from place to place, and the ability to ingest nutrients (by absorption).

However, botanists thought that the euglena should be placed in the plant kingdom. They argued that since this organism is able to make its own food, it should be included with the other organisms that are able to perform this function.

Actually, many living things are difficult to classify, because they possess characteristics of organisms that belong in two different groups. The argument about the euglena was finally settled when it was placed in a new kingdom, Protista, which contains mostly single-celled organisms.

Protists are widely distributed in the ocean. They play important parts in marine environments, so it is worthwhile to learn about them. However, we first will explore how all organisms are classified.

4.1 CLASSIFICATION

In 1758, the Swedish botanist Carolus Linnaeus (1707–1778) published a book describing his system of classification, which is the grouping of organisms according to similarities in structure. The science of classification that developed from this work is called **taxonomy**. Linnaeus classified all living things as belonging to either of two large taxonomic categories: the animal kingdom or the plant kingdom.

Linnaeus then divided the animal and plant kingdoms into smaller groups. The units he used were kingdom, phylum, class, order, family, genus, and species. The kingdom is the most inclusive group; it contains the largest variety of related organisms. The species is the smallest group; it contains only one kind of organism.

According to the system developed by Linnaeus, each organism is given a two-part scientific name that consists of a genus name and a specific name. For example, the scientific, or species, name of the blue whale is *Balaenoptera musculus*. *Balaenoptera* is the genus name and *musculus* is the specific name. The words used for scientific names are from the Latin and Greek languages.

You might wonder why scientific names are necessary. They sometimes seem difficult to read and even harder to pronounce. Since the different species of organisms number in the millions, it is necessary for scientists around the world to have a common language to be able to identify any organism with accuracy. As you know, people in different countries have different names for organisms. For example, the blue whale is known by different names in different languages. However, for scientists the world over—no matter what language they speak—the blue whale is always referred to by its scientific name, *Balaenoptera musculus*. This naming system prevents confusion.

The Five-Kingdom System

Since the time of Linnaeus, many new organisms have been discovered, and many classification systems have been proposed. Some of the systems have merit; all try to make sense of the great diversity of

TABLE 4-1 THE FIVE-KINGDOM CLASSIFICATION SYSTEM

Kingdom	Main Characteristics
Monera	Single-celled; lack nuclear membrane (bacteria, blue-green bacteria)
Protista	Mostly single-celled, some multicelled; have nuclear membrane (algae and protozoa)
Fungi	Single-celled and multicelled; have nuclei; absorb food from living and dead organisms
Plantae	Multicelled; have nuclei; make their own food through photosynthesis
Animalia	Multicelled; have nuclei; eat other organisms

life on our planet. However, most scientists now use the classification system that is composed of five kingdoms. (See Table 4-1.)

Monera: This group, commonly called the **monerans**, includes the bacteria and blue-green bacteria (formerly called blue-green algae). All **bacteria** are single-celled; all lack a nuclear membrane and thus their nuclear material is dispersed throughout the cell. Scientists call organisms that lack a nuclear membrane **prokaryotes**. These are probably among the most ancient organisms on Earth. The earliest fossil monerans are more than 3 billion years old. Even though they are composed of only a single cell, they are not such "simple" organisms. They are able to carry out all life functions and are remarkably successful to have survived on Earth for such a long time.

Protista: The kingdom of **protists** includes mostly single-celled, or **unicellular**, organisms, although some are composed of many cells or live together in small colonies. All protists have their nuclear material enclosed within a membrane; that is, they have a *nucleus*. Scientists call organisms with this feature **eukaryotes**. (All organisms on Earth other than those in the kingdom Monera are eukaryotes. See the section on diatoms for more information on eukaryotes.) The earliest fossil protists are about 1.5 billion years old. Tiny **protozoa** (animal-like organisms) and many kinds of **algae** (plantlike organisms) make up this group.

Fungi: The **fungi** include both unicellular and multicellular eukaryotic organisms that are not able to make their own food. Fungi absorb their nutrients from dead organic material and live tissues. The familiar mushroom is a type of fungus that you have probably seen. Certain diseases of the skin are caused by fungi, such as athlete's foot and ringworm. Like plant cells, fungal cells are enclosed by a rigid cell wall. Unlike plants, fungi do not contain the green pigment chlorophyll (see below). Fungi play an important role in breaking down dead organisms and recycling organic material. Hundreds of species of fungi exist, both on land and in the ocean.

Plantae: Plants are multicellular, eukaryotic organisms that are able to make their own food out of simple chemical substances. Plants contain **chlorophyll**, the green pigment that is able to capture the energy in light. (A **pigment** is a coloring matter found in the cells and tissues of plants and animals.) Plants use this energy during photosynthesis to make organic compounds out of water and carbon dioxide. During this process, plants give off oxygen as a by-product. Plant cells are surrounded by a rigid cell wall.

Animalia: Animals are multicellular, eukaryotic organisms. Unlike plant cells, animal cells lack cell walls. Also, unlike plants, animals are not able to make their own food, but must instead eat plants or other animals to obtain their nutrients. Most animals are capable of movement. We categorize animals in two main groups: those that lack a backbone and skull (the invertebrates) and those that have a backbone and skull (the vertebrates). Both groups are widely represented by animals that live in marine environments.

4.1 SECTION REVIEW

1. Why is it important to classify organisms?
2. How did Linnaeus classify organisms? Why is his system for naming organisms so useful?
3. How does the present system of classifying organisms differ from the system developed by Linnaeus?

4.2 BACTERIA

Bacteria are the most abundant organisms on Earth and are widely distributed in the ocean. Bacteria are microscopic single-celled organisms; they have a relatively thick outer cell wall that surrounds a thin cell membrane. If you examined some bacteria under a microscope, you would observe the three basic bacteria shapes.

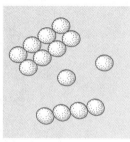

Ball-shaped
(cocci)

Rod-shaped
(bacilli)

Corkscrew-shaped
(spirilla)

Figure 4-1 The three shapes of bacteria: bacilli, cocci, and spirilla.

There are round bacteria called coccus (plural, cocci), rod-shaped bacteria called bacillus (plural, bacilli), and spiral-shaped bacteria called spirillum (plural, spirilla). (See Figure 4-1.) The first organisms that lived on Earth were prokaryotic cells that resembled these present-day bacteria. Since bacteria are structurally different from all other cells and organisms, they are classified in their own kingdom, Monera.

Bacterial cells can reproduce at a rapid rate, some every 20 minutes. All the instructions for reproduction are contained within threadlike structures called **chromosomes**. Because bacteria are prokaryotic (lack a nuclear membrane), this nuclear, or hereditary, material is dispersed throughout the cell's cytoplasm. The chromosomes are made up of molecules of DNA (deoxyribonucleic acid), which contain all the directions for a cell's structure and function within segments called genes. The total genetic make-up of an organism, that is, all its genes, is known as its **genome**. Bacteria are often used in modern recombinant DNA technology (genetic engineering) because they reproduce so rapidly.

Decay Bacteria

Bacteria play a very important part in the biological world. Bacteria are partly responsible for the decomposition, or breakdown, of dead organic matter. (Fungi also break down dead organic matter.) Dead matter is decomposed by a group of bacteria known as decay bacteria. In the ocean, decay bacteria break down organic matter into smaller molecules that are released into the water. These smaller molecules, such as phosphates, nitrates, and sulfates, are used as nutrients by different bacteria and other organisms. In this way, decay bacteria help recycle dead organic matter. Decay bacteria, and organisms like them, are called **decomposers**.

As you might suspect, decay bacteria are most abundant in bottom sediments where dead organic matter accumulates. There, the bacteria attach themselves to dead matter and secrete special chemicals that begin to break down organic matter in the sediments into nutrient molecules, some of which are taken in by the bacteria themselves. Feeding on organic matter in this way, the bacteria exhibit the type of nutrition normally found in fungi and animals.

Decay bacteria thrive in an environment that is warm, moist, dark, and rich in food. These conditions can be duplicated in the laboratory, where you can grow, or culture, marine bacteria. The food on which laboratory bacteria feed is a gel called nutrient agar (made from algae), which is put into a clear glass or plastic petri dish. An inoculating needle is used to transfer bacteria from water or from another sample to the nutrient agar. The inoculating needle is dipped into a sample that contains the bacteria to be cultured and then moved, or streaked, across the agar plate. The petri dish is covered and placed in an incubator (which keeps the dish warm) for a specified amount of time. During incubation, the bacterial cells along the streak will begin to multiply. They reproduce by dividing in two. After 24 hours, a single bacterial cell (bacterium) can grow into a colony, a bacterial population that contains millions of cells. Colonies on the agar may differ in color, form, and texture.

There are more than 5000 species of bacteria. One interesting group of bacteria is the so-called magnetic bacteria, found in some saltwater and freshwater marshes. Magnetic bacteria contain a string of magnetite (iron oxide) crystals that make the cell behave like a magnet. By clinging to the iron deposits in marshes, the magnetic bacteria can feed on dead matter in the sediments.

Some species of bacteria supply their energy needs in other ways. For example, the sulfur bacteria that live in marine mud use the compound hydrogen sulfide (H_2S), which is produced when organisms decay. (Hydrogen sulfide is the gas that smells like rotten eggs.) Oxidation of hydrogen sulfide provides energy that can be used by bacteria to form sugar from carbon dioxide and water. The hydrothermal vents on the deep ocean floor also contain bacteria that feed on sulfur compounds. The process by which organisms like sulfur bacteria derive energy from chemicals (that is, from inorganic raw materials) is called **chemosynthesis**.

Blue-Green Bacteria

Some microorganisms are not easily classified. The organism shown in Figure 4-2 on page 100 resembles an alga because it contains the green pigment **chlorophyll**. However, the organism also resembles a bacterium, because it lacks a membrane-bound nucleus. In fact, it

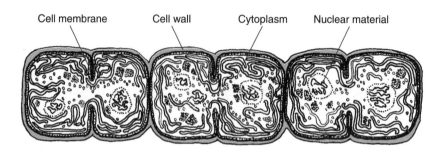

Figure 4-2 An example of a blue-green bacterium that is classified in a group with other cyanobacteria.

Cell membrane   Cell wall   Cytoplasm   Nuclear material

is classified in a group of organisms called **cyanobacteria**, also known as blue-green bacteria (formerly called *blue-green algae*). Most biologists classify cyanobacteria in the kingdom Monera with all other bacteria. Since they contain chlorophyll, cyanobacteria are able to make their own food. They are, in fact, the only moneran that is photosynthetic. In addition to chlorophyll, the cyanobacteria contain the blue pigment phycocyanin. The combination of green and blue pigments in the cyanobacteria produces their characteristic blue-green color. Some species of blue-green bacteria, such as the *Oscillatoria,* also contain the red pigment phycoerythrin. In shallow water, this species can produce a red color when a population bloom occurs. In fact, this cyanobacterium is responsible for periodically producing this effect in the Red Sea and may, in part, be responsible for giving this sea its name.

Cyanobacteria, which are found throughout the oceans, are very hardy organisms and can survive under a wide range of different environmental conditions. Some species of cyanobacteria live attached to rocks in the wave splash zone above the high tide mark. The cells are covered by a jellylike mass around their cell walls that prevents them from drying up when the tide is out. When these cells die, they stain the rocks black, a stain that resembles a recent oil spill. Other species of cyanobacteria secrete toxic chemicals that can produce a painful rash if they come into contact with skin. Cyanobacteria can also thrive in waters and sediments that are low, or lacking, in oxygen.

Scientists think that the first photosynthetic organisms to inhabit Earth were the cyanobacteria. The earliest cyanobacteria produced reeflike growths called stromatolites. Like coral reefs, the mushroom-shaped stromatolites built by cyanobacteria had a framework of calcium carbonate. Fossil stromatolites more than 3 billion years old have been found.

1. Why are cyanobacteria placed in the kingdom Monera?
2. How do most species of bacteria obtain their nutrients?
3. What is the evolutionary significance of the cyanobacteria?

4.3 DIATOMS

Diatoms are among the most common organisms found in the ocean. These single-celled protists usually float and drift near the ocean surface, though many diatoms also live in deeper waters. Diatoms are part of the ocean's community of **plankton** (meaning "wanderers" that drift, rather than swim) and are more accurately classified as **phytoplankton** ("plant wanderers"). Some diatoms (called encrusting diatoms) live attached to solid substrates, while others alternate between attached and free-floating forms.

Cell Structures of Diatoms

Like all eukaryotic cells, diatoms have a variety of structures called **organelles** that carry out important functions in the cell. The nucleus, which was mentioned in the discussion of classification, controls the growth and reproduction of the cell. In the nucleus are the *chromosomes,* coiled threads of nuclear material that carry genes, which determine a cell's characteristics. Most of the time, the chromosomes cannot be observed in the nucleus. However, with the use of special dyes, scientists can observe chromosomes, which become more visible during certain stages of cell division. A nuclear membrane encloses the nucleus, which floats in the fluid portion of the cell, the cytoplasm. A thin cell membrane, called the *plasma membrane,* surrounds all cells and regulates the entry and exit of materials. (See Figure 4-3 on page 102.)

Diatoms, like other algae and plants, contain the green pigment chlorophyll within special structures called *chloroplasts.* Because diatoms have a transparent cell wall, you can see the chloroplasts inside the cell. In fact, the cell wall of a diatom is made of silica, the main ingredient of glass. Why is it an advantage for a diatom to

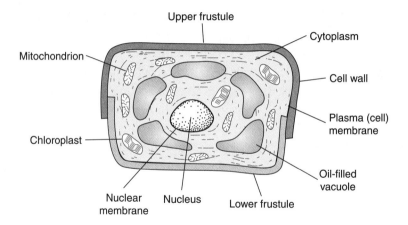

Figure 4-3 The cell structures of a diatom.

Upper frustule

Cytoplasm

Mitochondrion

Cell wall

Plasma (cell) membrane

Chloroplast

Oil-filled vacuole

Nuclear membrane

Nucleus

Lower frustule

have a transparent cell wall? Such a cell wall allows light to enter the cell. Inside the diatom, the light energy is trapped by chlorophyll in the chloroplasts. Notice the position of the chloroplasts. They lie next to the cell wall, where the intensity of light is greatest. Both the glassy cell wall and the position of the chloroplasts are adaptations that increase the rate of photosynthesis due to greater light absorption.

What happens to the oxygen produced by diatoms during photosynthesis? (Remember that oxygen is produced as a by-product of photosynthesis.) The cell wall of diatoms contains tiny holes, or pores. The pores allow carbon dioxide to enter and oxygen to leave the diatom. The pores are so small that you would need an electron microscope to see them! In fact, many of the other structures inside cells can be observed only by using an electron microscope.

The *endoplasmic reticulum* is a network of channels in the cytoplasm through which important chemicals are transported. Tiny particles called *ribosomes* are attached to the endoplasmic reticulum. Ribosomes are the places in the cell where proteins are assembled. Proteins are substances that are used by cells for growth and repair. The making of proteins is controlled by ribonucleic acid (RNA), a chemical present in the nucleus of a cell. Ribonucleic acid moves from the nucleus to the ribosome. In the ribosomes, RNA directs the manufacture of proteins. By controlling the kinds of proteins that are assembled in the cell, the nucleus acts as the control center for the cell's activities.

Cells need energy to do work. Located throughout the cell's cytoplasm are organelles called *mitochondria* (singular, mitochondrion). Mitochondria are the cell's energy factories in which sugar is broken down and chemical energy in the form of ATP (adenosine triphosphate) is produced. Food manufactured by a diatom is stored as an oil droplet in a structure called a *vacuole*. Small oval bodies called *lysosomes* are attached to the vacuole and produce chemicals that digest the food stored inside it. Other important chemicals needed by cells to carry out life functions are contained in stacks of flattened membranes called the *Golgi apparatus*. The Golgi apparatus releases, or secretes, these substances as needed.

Diatom Diversity and Life Functions

There are more than 25,000 species of diatoms, most of which inhabit the cold waters of the world. They are classified in phylum Chrysophyta, which means "golden algae." Diatoms exhibit great variety in their shapes. Some diatoms even appear to be strung together, like beads on a chain. (See Figure 4-4.)

Diatoms are, in fact, classified according to their shape. Round diatoms are called centric diatoms, and pen-shaped diatoms are called pennate diatoms. Some diatoms also have spines projecting from their cell wall, which help to prevent sinking. This feature

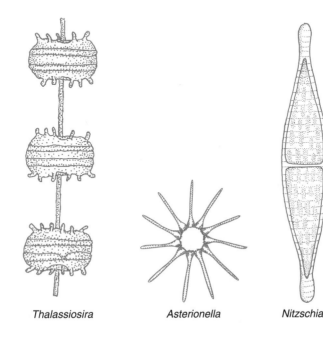

Figure 4-4 Three types of diatoms.

Thalassiosira　　　　*Asterionella*　　　　*Nitzschia*

(along with the stored droplets of food oil, which decrease cell density) aids survival, because diatoms that float close to the water's surface can absorb more energy from sunlight.

How does an organism with such a hard, glassy shell reproduce? Diatoms reproduce both sexually and asexually. During **asexual reproduction**, which is the production of offspring by one parent, a diatom divides to form two new cells. The two halves of a diatom's shell normally overlap, making the diatom resemble a box with a lid. Each half is called a frustule. When a diatom reproduces, its two frustules separate. Each half secretes a new frustule to complete the formation of two new diatoms.

During **sexual reproduction**, in which two parents are needed, a diatom develops into either a male or a female cell. A male cell produces sperm. The sperm swims to and enters the female diatom, where it unites with an egg nucleus. The fertilized egg cell develops into a mature diatom, completing the reproductive cycle.

When diatoms die, they fall to the ocean floor. The living material inside the diatom's shell decays, but the glassy cell walls remain. The shells accumulate on the ocean floor. Over time, these deposits form layers that may be hundreds of meters thick. These deposits of silica are known as diatomaceous earth. Since diatoms are porous, diatomaceous earth makes an excellent filtering material for aquariums and swimming pools. Diatomaceous earth is also used to purify drinking water.

Although diatoms are extremely small, they play an important part in the life of the ocean. Almost all animals in the sea ultimately depend on diatoms as a source of food. Tiny invertebrates, such as copepods, feed on diatoms. Shellfish such as mussels, clams, oysters, and scallops consume diatoms by filtering them from the seawater. Even humans depend on diatoms to some extent, when they eat organisms that have fed on diatoms either directly or indirectly.

Yet, important as they are, diatoms sometimes cause problems. A sudden increase in the diatom population, called an **algal bloom**, may occur from time to time in shallow coastal waters. During several summers in the late 1980s, the coastal waters of Long Island, New York, became so clouded with algae that the waters turned brown. Marine biologists analyzed water samples and found as many as 800,000 diatoms of one particular species in 1 milliliter of water. This kind of algal bloom is called a **brown tide**. The brown tide devastated the scallop industry in eastern Long Island. Biolo-

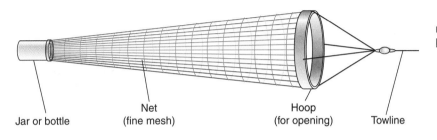

Figure 4-5 A plankton net, which is used to collect diatoms.

Jar or bottle

Net
(fine mesh)

Hoop
(for opening)

Towline

gists are still investigating the causes of these mysterious algal blooms.

Free-floating diatoms can be collected from seawater by use of a plankton net. (See Figure 4-5.) The plankton net can be pulled through shallow water, alongside a pier, or towed behind a moving boat. The plankton get caught in the mesh of the nylon net, and then fall into the collecting jar at the bottom of the net. A few drops from this sample can be observed under the microscope. In addition, encrusting diatoms can be scraped off the walls of a saltwater aquarium tank for observation in the classroom. You can view diatoms under the microscope by doing the lab investigation at the end of this chapter.

4.3 SECTION REVIEW

1. What adaptations do diatoms have for photosynthesis?

2. Explain how marine animals are dependent upon diatoms.

3. How can an algal bloom be detrimental to people?

4.4 DINOFLAGELLATES

Members of another protist group often found near the ocean's surface are the **dinoflagellates**. Three types of dinoflagellates are shown in Figure 4-6 on page 106. They are classified in phylum Pyrrophyta, which means "red (or fire) algae." How do dinoflagellates compare with diatoms? Dinoflagellates have two flagella. A flagellum (singular) is a microscopic hairlike structure. Each flagellum whips back and forth, helping to move the dinoflagellate along, although it still floats with the currents. In contrast, diatoms are not

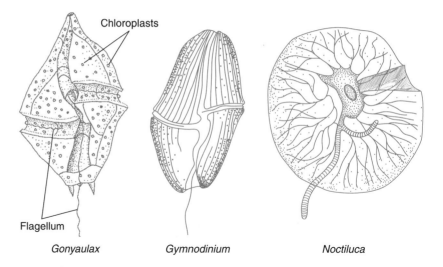

Figure 4-6 Three types of dinoflagellates.

Chloroplasts

Flagellum

Gonyaulax   *Gymnodinium*   *Noctiluca*

able to propel themselves at all; instead, they are just pushed along by the movement of water.

Dinoflagellates also possess chloroplasts and, like diatoms, are able to make and store their own food. Many dinoflagellate species have an eyespot that is sensitive to light. They use the eyespot to move toward the light, thus increasing their ability to make food. Unlike diatoms, dinoflagellates are also able to take in food. In this way they resemble the euglena, a freshwater protist that is able both to make food and to ingest it.

The cell walls of dinoflagellates and diatoms differ in structure and composition. Notice the plates and grooves of the dinoflagellates shown in Figure 4-6. Dinoflagellate cell walls are made of cellulose like those of plants, not of silica like those of diatoms. A cellulose cell wall is not as transparent as a diatom's glassy cell wall.

Effects of Dinoflagellates

Some dinoflagellates, like *Noctiluca* (meaning "night light"), are rather spectacular for so small an organism. Have you ever run your fingers through the ocean water at night and seen it sparkle? In places where many *Noctiluca* are present, the water will glow in the dark when it is disturbed mechanically. The movement of a boat propeller or the splashing of fish can cause the *Noctiluca* to emit a greenish-blue light. This ability of an organism to produce light,

ENVIRONMENT
Red Tides and Muddy Waters

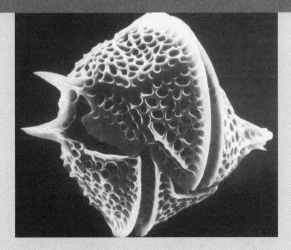

Each year, various coastal regions around the world suddenly become "killing fields" as large numbers of fish and shellfish die. The only warning is the appearance of a reddish tint in the water, the result of a swift and unexpected increase in the population of dinoflagellates. These one-celled algae contain pigments that are responsible for producing the so-called "red tide" in the ocean. They also harbor toxic chemicals that cause death when ingested by marine animals. On windy days, a mass of toxic dinoflagellates on the ocean surface can even cause wheezing and eye irritation in people who are exposed to the sea spray.

To protect their countries' billion-dollar fishing industries, marine scientists in South Korea and Japan developed a new way to diminish the destructive effects of the red tide. What was their solution? Spray the ocean surface with muddy water! The muddy spray contains clay particles that cling to the toxic cells, causing them to sink to the bottom. Using this method, fisheries in Korea reduced their losses from $100 million in 1995 to just $1 million in 1996.

The National Oceanic and Atmospheric Administration (NOAA) estimates that the red tide will cost the U.S. economy approximately $1 billion over the next decade. So, American scientists are testing this new method in small-scale experiments. In one such experiment, seawater taken from a red tide site in the Gulf of Mexico was transferred to a large outdoor tank and sprayed with a fine mist of clay. Within two-and-a-half hours, the clay had removed about 70 percent of the toxic cells.

Despite the initial success, some researchers are worried that introducing clay particles that fall to the seafloor may harm bottom-dwelling organisms. In addition, other scientists think that some toxic dinoflagellates may still survive on the sea bottom, where they could cause another kind of killing field. The research continues in order to determine if muddying the waters is an ecologically safe and sound method for minimizing the negative effects of the red tide.

QUESTIONS

1. What natural occurrence is responsible for producing the red tide?

2. How does the red tide affect the economy of various countries?

3. Describe the new method used to reduce the effects of the red tide.

4. Why are some scientists worried about using mud to control the red tide?

called **bioluminescence**, is also seen in a few other species of phytoplankton and in some deep-sea fishes as well.

Another interesting, although unpleasant, phenomenon associated with dinoflagellates is the **red tide**. Suddenly, with no warning, some shallow coastal waters turn red during the summer. At the same time, many hundreds of fish die; this is called a fish kill. When the water is analyzed, marine biologists find large numbers of a dinoflagellate that belongs to the genus *Gymnodinium*. This dinoflagellate contains a pigment that produces the red color in the water. Powerful toxins (poisonous substances produced by living things) made by these organisms accumulate in shellfish such as clams and mussels, which eat the algae and then poison the other organisms that eat them, such as fish, marine birds, and even humans. The algal bloom also reduces oxygen levels in the water, which further contributes to the fish kill.

Another dinoflagellate that produces a red tide belongs to the genus *Gonyaulax*. This organism causes paralytic shellfish poisoning, which leads to illness and death in fish and in humans. *Gonyaulax* contains a toxic substance called saxotoxin. This toxin interferes with the functioning of the nervous system in vertebrates. Saxotoxin is transferred from one organism to another during feeding. For example, mussels feed on *Gonyaulax* in the water. The mussels are not affected by the saxotoxin, but it accumulates in their body tissues. People who eat mussels that are contaminated by saxotoxin become sick and may even die.

4.4 SECTION REVIEW

1. Why are dinoflagellates classified as protists?

2. How can some dinoflagellates harm humans?

3. How does a dinoflagellate bloom cause a fish kill?

4.5 NUTRITION IN ALGAE

The food that is made by **producers**, such as diatoms and other algae, is used by them for all their growth and energy needs. Utilization of food by living things for growth and energy is called nutrition. Producers are also called **autotrophs**, a word that means "self-feeders." Their method of obtaining food is called autotrophic nutrition. The food that is made by producers is also used by other organisms; that is, it is used by the **consumers**, which are unable to make their own food. Organisms that live on the food made by other organisms—by eating them—are called **heterotrophs**, a word that means "other-feeders." This type of food-getting is known as heterotrophic nutrition. Since dinoflagellates both make and ingest food, they are considered to be autotrophs as well as heterotrophs.

Photosynthesis

Diatoms and dinoflagellates have chloroplasts—food factories in which photosynthesis takes place. The word *photo* means "light," and *synthesis* refers to the process of manufacturing large molecules from smaller ones. When you combine the two words, you have *photosynthesis,* the process by which autotrophs can, in the presence of sunlight, make food (glucose, or sugar) from simple raw materials (carbon dioxide and water). During this reaction, oxygen gas is produced as a by-product. The following chemical equation summarizes the process of photosynthesis.

$$6\ CO_2 + 12\ H_2O \xrightarrow[chlorophyll]{sunlight} C_6H_{12}O_6 + 6\ H_2O + 6\ O_2$$

The chemical equation can be put into words as follows: six molecules of carbon dioxide plus twelve molecules of water, in the presence of sunlight and chlorophyll, yields one molecule of glucose plus six molecules of water plus six molecules of oxygen.

After the oxygen is released by the protists or algae, some of its molecules dissolve in the water. The rest of the oxygen enters the atmosphere at the water's surface (some of which goes back into the sea). The oxygen you take in with every breath probably includes some molecules that come from marine algae!

How do producers use light to make food? Look again at the structure of the diatom shown in Figure 4-3. Since the diatom lives in water, the surrounding water (H_2O) moves directly into its cell. And since carbon dioxide gas is dissolved in water, the CO_2 also moves directly into the diatom. After entering the cell, carbon dioxide and water are taken up by the chloroplasts. The chloroplasts contain the green pigment chlorophyll. It is this green pigment that is able to absorb energy present in light. Chloroplasts use this energy to put together the molecules of water and carbon dioxide to produce or synthesize sugar.

Without photosynthesis, life as we know it could not exist on Earth. Why is photosynthesis so important for living things? During photosynthesis, plants are able to turn the light energy in sunlight into the chemical energy in the sugar glucose. Glucose is used by plants for their own energy needs. Since animals are unable to make their own food from simple compounds, they use the energy present in the compounds made by plants instead. When an animal eats a plant, the chemical energy present in the plant is transferred into the animal, that is, from the producer into the consumer. The next time you eat a filet of fish, remember that the energy in the food you are eating came originally from the energy that is present in sunlight.

Where on the planet does most photosynthesis occur, on land or in the ocean? Many people would probably say on land. After all, people are most familiar with the plants found on farms and in fields and forests. However, land makes up only about 29 percent of Earth's surface, and a large percentage of this land does not even support plant life. So this guess is incorrect. The ocean covers most of Earth's surface. It can support algae and plant life in the areas where light penetrates. So that is where most photosynthesis occurs.

The Cell Theory

So far, you have learned three important facts about living things. First, all living things are composed of one or more cells. Second, all cells perform the same basic life functions; for example, they make or obtain their food, they get rid of wastes, and they reproduce. And third, all cells (and, therefore, all organisms) come from preexisting cells. These three fundamental facts are part of what is called the **cell theory**. In the following chapters, you will see how the cell theory applies not only to unicellular forms of marine life but to all other marine organisms as well.

4.5 SECTION REVIEW

1. How is the nutrition of dinoflagellates more complex than the nutrition of diatoms?

2. Explain why protists that obtain food by photosynthesis are so important for other organisms.

3. Why does more photosynthesis take place in the ocean than on the land?

Laboratory Investigation **4**

How Diatoms Perform Their Life Functions

PROBLEM: How are diatoms adapted for carrying out their life functions?

SKILL: Using a microscope to observe unicellular organisms.

MATERIALS: Slides, medicine droppers, live diatoms, microscope, coverslips.

PROCEDURE

1. Place a drop of water that contains diatoms on a clean slide. Cover the sample with a clean coverslip. (*Note*: If fresh diatoms are not available, use a prepared slide of diatoms.)

2. Place the slide on your microscope stage. Move the low-power objective into position. Focus the lens. Move the slide until you observe cells that contain green, yellow, or orange pigments. These cells are diatoms.

3. Move the high-power objective into position. Focus on a single diatom. Notice the pigment color in the diatom you are viewing. The green pigment is chlorophyll, the yellow pigment is xanthophyll, and the orange pigment is carotene. All are involved in nutrition.

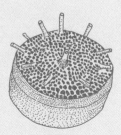

Figure 4-7
An example of a diatom.

4. Make a sketch of the diatom you are observing. If possible, color it in appropriately and label any parts you can identify. Check to see if you recognize any of the diatoms shown in Figure 4-4 or in Figure 4-7.

5. Move the slide to locate other types of diatoms. Sketch each one you observe.

OBSERVATIONS AND ANALYSES

1. Why are the pigments in the diatom visible?

2. Identify two life functions carried out by diatoms.

3. How are diatoms adapted for making their food?

Chapter 4 Review

Answer the following questions on a separate sheet of paper.

Vocabulary

The following list contains all the boldface terms in this chapter.

algae, algal bloom, asexual reproduction, autotrophs, bacteria, bioluminescence, brown tide, cell theory, chemosynthesis, chlorophyll, chromosomes, consumers, cyanobacteria, decomposers, diatoms, dinoflagellates, eukaryotes, fungi, genome, heterotrophs, monerans, nutrition, organelles, phytoplankton, pigment, plankton, producers, prokaryotes, protists, protozoa, red tide, sexual reproduction, unicellular

Fill In

Use one of the vocabulary terms listed above to complete each sentence.

1. The ability of an organism to produce light is called _____.

2. The plantlike marine protists are known as _____.

3. The animal-like marine protists are known as _____.

4. The glassy-shelled algae that cause brown tides are the

   _____.

5. Phytoplankton that have two flagella and cause red tides are

   _____.

Think and Write

Use the information in this chapter to respond to these items.

6. Why is it important to use Greek or Latin words for scientific names?

7. What very important part do bacteria play in the biological world?

8. Name two ways diatoms and dinoflagellates are similar and two ways they are different.

Inquiry

Base your answers to questions 9 through 12 on the diagram below, which shows a device used by marine biologists, and on your knowledge of marine science.

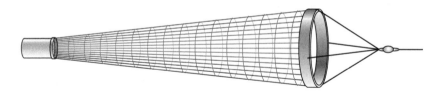

9. (a) Identify the item that is illustrated. (b) What is its main function? (c) Name two of its important parts.

10. Which of the following kinds of organisms are intentionally caught using this device? *a.* monerans *b.* red tide algae *c.* free-floating diatoms *d.* crustaceans

11. Of the following methods listed, which one is *not* used with this device? *a.* pulling alongside a pier *b.* towing behind a moving boat *c.* pulling through shallow water *d.* dragging along the seafloor

12. The best way to observe the organisms that are obtained with this device is to *a.* snorkel underwater with a sample of them *b.* observe them under the microscope *c.* place them in a saltwater aquarium *d.* place them in a petri dish.

Multiple Choice

Choose the response that best completes the sentence.

13. The utilization of chemicals from food for growth and energy is called *a.* photosynthesis *b.* nutrition *c.* reproduction *d.* bioluminescence.

14. Marine phytoplankton are not found at great depths in the ocean because of the lack of *a.* animal life *b.* warm temperatures *c.* sunlight *d.* nutrients.

15. The monerans include the unicellular organisms known as
    a. seaweeds   *b.* fungi   *c.* bacteria   *d.* diatoms.

16. If you scoop seawater from the ocean surface and examine a
    drop under the microscope, you may not see any plankton.
    The most reasonable explanation is that   *a.* the cells were
    not stained   *b.* the plankton were too widely dispersed
    c. a red tide wiped them out   *d.* the water was too rough.

17. An important factor for classifying organisms together in a
    taxonomic group is their similarity in   *a.* color   *b.* habitat
    c. body structure   *d.* body size.

18. Organisms whose cells lack a nuclear membrane are placed in
    the kingdom   *a.* Monera   *b.* Protista   *c.* Plantae
    d. Fungi.

19. Which of the following statements is *not* true?   *a.* Diatoms
    and dinoflagellates both carry out photosynthesis.
    b. Diatoms have cells walls made of silica.   *c.* Dinoflagellates
    have cells walls made of cellulose.   *d.* Diatoms and
    dinoflagellates both have flagella for locomotion.

20. Organisms that live on food that is made by other organisms
    are called   *a.* producers   *b.* heterotrophs   *c.* autotrophs
    d. cyanobacteria.

21. Of all the pigments involved in photosynthesis, the most
    important one found in all marine algae is   *a.* carotene
    b. xanthophyll   *c.* chlorophyll   *d.* phycocyanin.

22. A protist that can both make and take in its food is the
    a. diatom   *b.* dinoflagellate   *c.* cyanobacterium   *d.* ameba.

23. An algal bloom of dinoflagellates that causes fish kills is called
    the   *a.* saxotoxin   *b.* brown tide   *c.* bioluminescence
    d. red tide.

24. Organisms such as decay bacteria that help recycle dead
    organic matter are called   *a.* autotrophic   *b.* decomposers
    c. phytoplankton   *d.* eukaryotes.

Base your answers to questions 25 and 26 on the following chemical equation and on your knowledge of marine science.

$$6CO_2 + 12H_2O \xrightarrow[chlorophyll]{sunlight} C_6H_{12}O_6 + 6H_2O + 6O_2$$

25. This chemical equation represents the process known as
    a. respiration   *b.* decomposition   *c.* photosynthesis
    d. chemosynthesis.

26. In which of the following organisms is this reaction carried
    out?   *a.* decay bacteria   *b.* diatoms   *c.* blue whales
    d. fungi

27. By what means does the organism
    shown here satisfy its nutritional
    needs?
    a. heterotrophic only
    b. autotrophic only
    c. heterotrophic and autotrophic
    d. chemosynthetic only

Chloroplasts

Flagellum

Research/Activity

Construct a plankton net from household materials. Show the net
to your classmates and explain how you built it. If possible, use
the net to obtain plankton from the seashore. Bring live plankton
back to the classroom to be viewed.

5 Marine Algae and Plants

When you have completed this chapter, you should be able to:

DESCRIBE the three main groups of macroscopic algae.

DISCUSS adaptations of beach plants and marine grasses.

EXPLAIN the importance of mangrove trees to marine life.

The ocean, like the land, provides food for its inhabitants because it contains chlorophyll-bearing organisms, called *producers*. One important group of producers is the algae, plantlike aquatic organisms that vary in size from microscopic to macroscopic (multicellular), with some species more than 60 meters long. Unlike most plants, macroscopic algae do not have true stems, roots, or leaves. Although multicellular, these algae are classified in the kingdom Protista.

Where the land meets the sea, plants are usually found in great abundance. Plants that live in the ocean or along the shore are called marine plants. Like algae and land plants, marine plants produce food and oxygen through the process of photosynthesis.

Some marine plants may be familiar to you. On tropical beaches, you might see palm trees or perhaps a cluster of mangrove trees in shallow water near the shore. Along temperate coasts, tall grasses grow in and near bay waters. Marine plants have true stems, leaves, and roots; they are classified in the kingdom Plantae. In this chapter, you will learn how marine algae and plants are adapted to marine environments and why they are important to other marine organisms.

5.1 MARINE ALGAE

What do ice cream and toothpaste have in common? It may surprise you to learn that both products contain substances that come originally from seaweed. **Seaweeds** are multicellular algae that live in the sunlit waters of the ocean. Organisms that are **multicellular** are made up of more than one cell. You can find seaweeds deposited along the strandline that runs the length of a beach, forming the boundary between the intertidal and supratidal zones. The strandline is composed of the seaweed and debris (flotsam and jetsam) that are washed up on the beach by the incoming tide. (Refer to Figure 3-2 on page 63.)

One common seaweed you may come across in the subtidal zone is the tissue-thin sea lettuce *Ulva*. (See Figure 5-1.) The bright green color of *Ulva* is due to the chloroplasts in its cells. As you know, chloroplasts are food factories, the places where the sugar glucose is manufactured. Some marine animals eat this alga for the nutrients it provides. When sea lettuce dies, it decomposes. Then the glucose and other nutrients inside it are released into the water. Microscopic animallike organisms that are part of the plankton feed on these nutrients. Larger organisms, such as barnacles and

Figure 5-1 *Ulva*, or sea lettuce, is a common green alga.

mollusks, filter the nutrients from the water. Clearly, marine animals are dependent on algae as a food source, both directly and indirectly.

Sea lettuce is also an oxygen producer. An experiment that tests for photosynthesis in this alga can be performed in the classroom. The seaweed is exposed to light for several hours. A gas is collected in a test tube placed over the opening of the container the seaweed is in. A student can then test the gas in the tube to determine that it is oxygen and confirm that it is being produced as a by-product of photosynthesis.

More than 500 species of macroscopic algae live in the ocean. These algae are classified according to the color of the pigments in their cells. The three main groups classified in this way are the green algae, brown algae, and red algae.

Reproductive Cycle of Algae

At first glance, algae may appear to be simple organisms. However, they often exhibit a very complicated reproductive cycle. We can examine the reproductive process in the green alga *Ulva,* as shown in Figure 5-2 on page 120. This life cycle is typical of many species of both green algae and brown algae.

The leafy part of a seaweed is called the **thallus**. When *Ulva* reaches maturity, it is ready for asexual reproduction. Specialized cells at the edge of the thallus produce spores. The thallus that produces spores is called the sporophyte thallus, which is diploid. However, the **spore** is a reproductive cell that contains the organism's haploid number of chromosomes. (Diploid refers to the normal number of chromosomes, found in the body cells of an organism. Haploid refers to one-half the normal number of chromosomes, found in the reproductive cells. Chromosomes carry the hereditary material for the cell. This material directs special structures in the cell to assemble proteins.)

Spores have flagella that beat back and forth, thus moving them along. Eventually, the spores reach the ocean bottom and, if they fall on a suitable substrate, each develops into a leafy thallus that produces gametes. This thallus is called the gametophyte thallus, and it is haploid. The **gametes** it produces are reproductive cells that

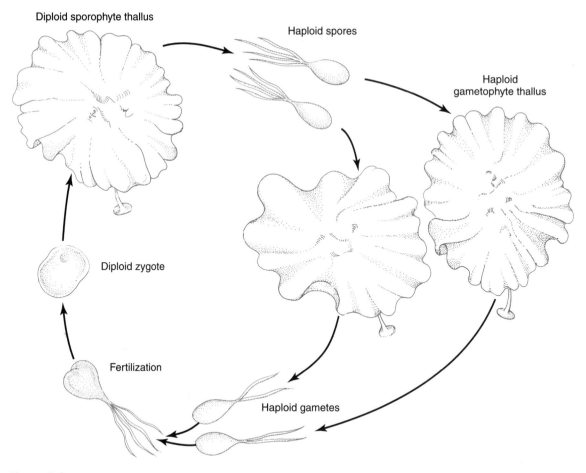

Diploid sporophyte thallus

Haploid spores

Haploid gametophyte thallus

Diploid zygote

Fertilization

Haploid gametes

Figure 5-2 The reproductive cycle of *Ulva* exhibits alternation of generations.

contain the haploid number of chromosomes necessary for sexual reproduction. The union of a sperm cell with an egg cell is called *fertilization*. The two gametes fuse to produce a **zygote**, a fertilized egg cell that contains the species' normal diploid number of chromosomes. (Each half of the zygote's chromosomes comes from one of the gametes that fused.) The zygote divides and develops into the next leafy sporophyte (diploid) thallus.

The life cycle of *Ulva* is composed of two separate stages or generations—the sporophyte generation and the gametophyte generation—where one generation follows the other. This succession of two types of generations (sporophyte/asexual and gametophyte/sexual) is called *alternation of generations*. This type of reproductive cycle is also found in some land plants.

Green Algae

The green algae are classified in the phylum Chlorophyta. They are thought to be the algae most closely related to plants, due to the similarity of their pigments. Three types of ocean-dwelling green algae are shown in Figure 5-3. Many species of green algae grow attached to rocky substrates on or near the ocean's surface. In general, because they are attached to a substrate, they are not tossed up on the beach by waves. However, some green algae may be torn from their substrates during storms and by heavy wave action.

Green algae lack the typical roots, stems, and leaves that are found in most land plants. In such land plants, roots and stems transport water from the ground to the leaves through specialized water-conducting cells. Land plants that have water-conducting cells are called vascular plants. Plants that do not have special water-conducting cells are called nonvascular plants. Even without such water-conducting cells, algae are quite successful. Since they live in an aquatic environment, algae have no need for specialized tissues that conduct water. Water passes directly into the algae's cells from their surroundings.

Figure 5-3 Three types of green algae.

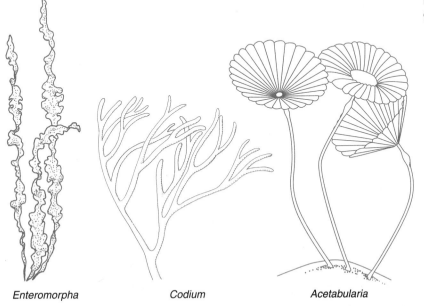

Enteromorpha   *Codium*   *Acetabularia*

ENVIRONMENT
Caulerpa–The Plantlike Pest

Looks can be deceiving. What appears to be a vascular plant growing on the seafloor of some coastal areas is actually a very long unicellular alga, *Caulerpa prolifera*, the largest one-celled organism in the world. Not only is *Caulerpa* a giant cell, but it has become a giant pest as well. In recent years, some species of *Caulerpa* have invaded the coastal waters of California and Florida, far from their original habitats in the Caribbean and Southeast Asia.

Marine biologists have started to document the effects of this exotic species on native habitats. Growing as much as 1 cm per day, *Caulerpa* has already colonized the eel grass beds in southern California, where it threatens to disrupt the habitat of small bottom-dwelling fish, bivalve mollusks, the spiny lobster, and other invertebrates. In southern Florida, the alga has spread to the reefs, forming thick mats that cover parts of the coral. Some marine scientists have reported that this seaweed also chokes sponges and disrupts the habitat for other marine animals that live on the reef.

In an attempt to prevent further damage, the State of California has gone ahead and declared *Caulerpa* a pest and banned its importation. Aquarium owners in California will no longer be able to purchase this attractive, bright-green seaweed; they are thought to have accidentally introduced *Caulerpa* to the marine environment in the first place, when their aquarium water was dumped into local waterways.

Other measures are being contemplated to stop the infestation of this wandering alga, including smothering it with a tarp. The management of this pest, however, may cause more damage to the environment than does the alga itself. More research is needed to determine if *Caulerpa* is, in fact, a dangerous pest or just a strange seaweed that needs time to adapt to new environments in a nondestructive way.

QUESTIONS

1. Why is *Caulerpa* classified as an alga rather than as a true (vascular) plant?

2. What are some effects *Caulerpa* has had on marine environments in Florida and California?

3. Describe two measures that may help limit the spread of *Caulerpa* in U.S. waters.

4. Explain why research on this "pest" should continue before a conclusion is reached.

Some species of green algae are very hardy. For example, *Entero-morpha*, a filamentous alga, thrives under environmental conditions that are unsuitable for most kinds of algae and plants. *Enteromorpha* grows abundantly in shallow coastal waters, where it carpets rocks and other hard substrates in the upper intertidal zone. Often mistaken for moss, a land plant, *Enteromorpha* can tolerate temperatures that vary widely from summer to winter as well as alternating periods of wetness and dryness. When the tide comes in, *Enteromorpha* is covered with water; when the tide goes out, this green alga can survive in the dry air. And when it rains heavily, *Enteromorpha* is even able to adapt to a temporary freshwater environment.

The seaweeds *Codium* and *Acetabularia* live in the more stable subtidal zone. *Acetabularia* is interesting because it is actually a very large single cell that grows to about 8 cm in length, large enough to be visible with the unaided eye. This green alga, shaped like a miniature umbrella, grows in the warm waters of the Gulf of Mexico and off the coast of Florida.

Codium is a spongy green alga with a branching structure. In tropical waters, *Codium magma* can grow to more than 6 meters in length. A smaller species, *Codium fragile,* lives in temperate waters and grows to about 1 meter in length. *Codium* attaches itself to hard substrates, such as rocks and shells. In recent years, this seaweed has invaded scallop, mussel, and oyster beds in the Northeast. When *Codium* attaches itself to the mature shellfish, it affects their survival and makes it difficult to harvest them.

Brown Algae

The brown algae are classified in the phylum Phaeophyta. Three types of brown algae are shown in Figure 5-4 on page 124. These algae have a brown or olive-green color. This color results from the mixture of pigments in the cells of the algae—particularly the green pigment chlorophyll and the yellow pigment xanthophyll. The variable blending of these pigments gives brown algae their characteristic range of colors.

Brown algae are important members of various marine ecosystems, providing shelter or nutrients for other organisms. Likewise, they provide materials that people find valuable. For example, the sea palm, which grows on rocks and resembles a tiny palm tree, can

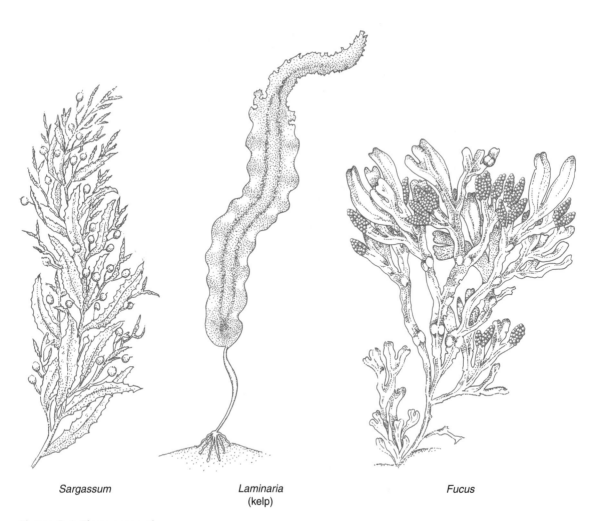

Sargassum

Laminaria
(kelp)

Fucus

Figure 5-4 Three types of brown algae.

be eaten raw or cooked. Some brown algae provide substances that people use in the preparation of foods and other products. (See the discussion of kelp on the following page.)

A common brown seaweed is the marine alga *Fucus,* also called rockweed. *Fucus* attaches to rocks in the intertidal zones along the Atlantic, Pacific, and Gulf coasts. A tough, fibrous pad of tissue, called a **holdfast,** anchors *Fucus* to rocks. The holdfast prevents an alga from being dislodged from its substrate by strong currents and waves. However, during severe storms, heavy waves can tear *Fucus* from the rocks.

If you look closely at a piece of *Fucus,* you will notice some air-filled bladders. (The air bladders act like a life preserver.) If you place a piece of *Fucus* in a bowl of water, it will stay afloat. Air bladders help keep *Fucus* upright in the water, so that it can absorb more of the sun's energy to carry out photosynthesis. (*Note*: Not all species of *Fucus* have air-filled bladders.)

A rocky coast at low tide is the best place and time to find *Fucus*. When the tide is out, you can see thick mats of *Fucus*. If you turn this seaweed over, you will see snails, small crabs, barnacles, worms, and other small creatures underneath it. These organisms hide under *Fucus* when the tide is out. If you do not live near a beach, visit a store that sells fish and ask for some rockweed to bring back to class for study—*Fucus* is often used as a packing material for lobsters. By doing the laboratory investigation at the end of this chapter, you will be able to observe ways *Fucus* is adapted to life in a marine environment.

The largest seaweeds in the ocean are the brown algae known as **kelp**. One species of kelp, *Laminaria,* thrives in the colder waters of the temperate zone, especially along the coasts of Maine and California. This brown alga lives in the subtidal zone, where it attaches to rocks with a large and very sturdy holdfast. Kelp grows rapidly from the seafloor to the surface. The giant kelps, such as *Nereocystis* and *Macrocystis,* can reach a length of more than 60 meters. Marine biologists have measured the growth rate and found it to be about a third of a meter per day. Numerous creatures, such as fish, shellfish, sea urchins, sea lions, sea otters, and sharks, live in and around these giant kelp forests.

A chemical in kelp called **algin** is used in many different industries. Algin is an important ingredient in various prepared foods, medicines, paints, and paper products. Large aquatic mowing machines attached to barges are used to cut and harvest kelp in some areas along the Pacific coast. Then algin is extracted from the kelp for commercial purposes.

Not all species of brown algae are anchored to a substrate. One type, *Sargassum,* floats on the water's surface in the South Atlantic Ocean and in some seas off the coasts of Asia. It is believed that pieces of the alga break off from rocky shores and drift out to sea, where they form floating mats. *Sargassum* flourishes in one area of the Atlantic where both water and weather are calm. The alga grows in such abundance that the area is known as the Sargasso Sea. The

Sargasso Sea supports a rich community of organisms, including fish, shellfish, and young sea turtles that live within the protective covering of the seaweed.

Red Algae

Red algae are the most abundant, and commercially valuable, of the marine algae. They are classified in the phylum Rhodophyta. Three types of red algae are shown in Figure 5-5. Red algae are found on rocky shores from the intertidal to the subtidal zones. Some species are found at much greater depths than either brown or green algae. The red pigment phycoerythrin and the blue pigment phycocyanin enable red algae to use the limited light that penetrates these deeper waters to carry out photosynthesis. Phycoerythrin masks the green pigment chlorophyll, which is also present in red algae.

Many species of red algae are thin and delicate. The thin, sheet-like alga *Porphyra*, also called nori, grows attached to rocks in the lower intertidal zone, from the Carolinas northward. *Porphyra* is a tasty seaweed that is cultivated in Japan.

Irish moss *(Chondrus crispus)* is a short, bushy seaweed found in the lower intertidal and subtidal zones. Irish moss carpets rocks with a dense, spongy growth. This seaweed is harvested for use as a food item. It also contains a chemical called **carrageenan**, which is used

Figure 5-5 Three types of red algae.

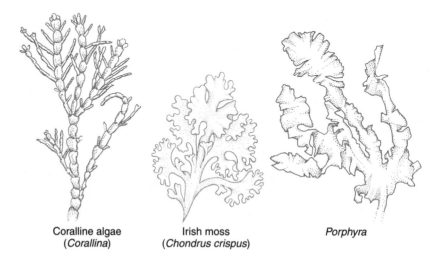

Coralline algae
(*Corallina*)

Irish moss
(*Chondrus crispus*)

Porphyra

as a binding agent in ice cream, puddings, and toothpaste. Other soft red algae supply a chemical called **agar**, which is also used to make food and medicinal products, and as a medium for growing bacteria.

A few species of red algae are hard and brittle. The coralline seaweeds *(Corallina)* have calcium carbonate in their cell walls, the chalky substance found in shells and corals. This branching alga can be found attached to rocks in the lower intertidal zone from Canada to Long Island. Another group of red algae, known as encrusting stony red algae *(Lithothamnion),* grows on rocks and on the shells of hermit crabs, adorning their surfaces with a bright violet color.

5.1 SECTION REVIEW

1. Describe the importance of seaweeds in marine ecosystems.

2. How are marine algae adapted for carrying out photosynthesis?

3. In what ways are seaweeds important to people?

5.2 BEACH PLANTS

A great variety of plants grow along sandy beaches. These beach plants are found in an area called the upper beach. The area closer to the water, called the lower beach, does not have any plants. Why? High tides and heavy surf make it very difficult for plants to take root in the sand along the lower beach. Also, the salty conditions in the sand, and in the misty air that blows off the ocean, make this area very inhospitable for most plant species.

Conditions in the upper beach are much more suitable for the growth of plants. Here, winds move the sand into small hills called **dunes**. The dunes are held in place by the roots of beach plants. (See Figure 5-6 on page 128.) For many species of plants and animals, dunes are very important. Yet these small hills of sand are also very delicate. Even a footstep can begin the process that leads to a dune's destruction. That is why you often see signs asking people not to walk on dunes. The beach grass *Ammophila* has long underground

Figure 5-6 Beach plants grow on sand dunes; the plants' stems and roots help hold the dunes in place.

stems and deep roots that help hold the sand in place and thus stabilize the dunes. Beach grasses are also widely spaced to minimize competition with one another.

Adaptations to the Upper Beach

In many ways, the upper beach resembles the kinds of conditions you would find in a desert. During the summer months, the temperature can often exceed 37°C. A typical desert plant, the prickly pear cactus *Opuntia compressa*, grows in this region of the beach. This plant has a thick waxy covering to minimize water loss from evaporation. Another common dune plant, the seaside goldenrod, stores water in its stem. (See Figure 5-7.)

At the summit of the dunes, woody shrubs and trees such as the beach plum and pitch pine are often found. Many plants also grow on the side of the dune that faces away from the ocean. Here they are sheltered by the dunes from the drying effects of the winds that blow off the ocean. An interesting phenomenon can be observed here. Trees often grow as tall as the dunes, but no taller. Can you offer a reason why this is so? Again the answer is found in the winds that blow off the ocean. The dunes offer some protection, but only for plants whose growth is no taller than the height of the dune. Once the plants grow taller than the dunes, the drying effects of

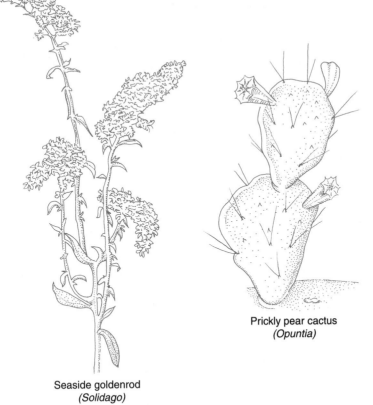

Figure 5-7 Two common beach plants.

Prickly pear cactus
(*Opuntia*)

Seaside goldenrod
(*Solidago*)

ocean winds act like giant pruning shears, taking off their top growth and preventing them from growing taller.

Most beach plants are vascular plants, since they have the specialized tissues in their roots, stems, and leaves that conduct food and water throughout the plant's body. As such, beach plants are included in the phylum Tracheophyta, along with all other vascular plants. Beach plants also produce flowers and seeds.

5.2 Section Review

1. Why do beach plants grow in the upper beach and not in the lower beach?

2. How are plants in the upper beach adapted to survive in a desertlike environment?

3. How are the beach plants different from the marine algae?

The cry of sea gulls and the smell of salty air are sure signs that the ocean is not far away. When you begin to see tall grasses waving gently in the breeze, you know that you are within walking distance of the water. A variety of marine grasses are typically found on the shores of protected bays and inlets along the Atlantic, Gulf, and Pacific coasts. Let's see what kinds of grasses can adapt to this marine environment.

Marsh Grasses

A variety of plants called **marsh grasses** grow along the sandy beaches of calm bays. One type is the tall **reed grass** called *Phragmites*. This marsh grass can be easily identified by its fluffy brown tassels. Along the water's edge, in the intertidal zone, you often find two species of **cordgrass** (*Spartina*). (See Figure 5-8.) A tall, coarse species of cordgrass, *Spartina alterniflora,* grows in the lower intertidal zone, where it is covered by water during periods of high tides; it can tolerate changes in salinity and temperature. This cordgrass is a very important member of salt marsh communities; its survival is linked to that of the fiddler crabs and mussels that live on and around its roots. In addition, cordgrass has the ability to break down industrial pollutants that flow into marshes, releasing the chemicals as harmless gases. A shorter, more delicate cordgrass, *Spartina patens,* is found in the upper intertidal zone, where it gets flooded only during periods of very high tides.

Cordgrass species have adaptations that enable them to survive in water that is salty. Special glands located in the leaves are able to excrete excess salt. (If you study these grasses in nature, you will notice that light is reflected by salt crystals on the leaves.) Since marsh grasses have a short life cycle, much of the salt marsh contains dead and decaying grass. When cordgrass dies, the decay products from the plants enrich the water with important nutrients. Plankton feed on these nutrients. Due to the high level of nutrients in the water, great numbers of plankton can thrive in marshes. The plankton are a major food source for other marine organisms. In

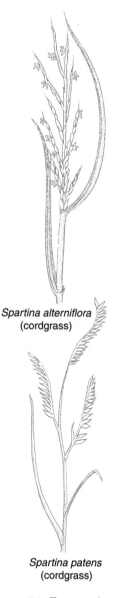

Spartina alterniflora
(cordgrass)

Spartina patens
(cordgrass)

Figure 5-8 Two species of marsh grass.

fact, marshes are among the most biologically productive ecosystems in the world.

Another salt-tolerant marsh grass is the glasswort *Salicornia*. Glasswort (also called pickle weed) grows in the upper intertidal zone, in areas from Massachusetts to the Gulf Coast. The short, thick waxy stems of the glasswort store the fresh water that the plant needs to survive.

Sea Grasses

Have you ever seen grass growing underwater? In the shallow subtidal zones along many shores, different types of **sea grass** can be found. Look at the two species of sea grass shown in Figure 5-9. In the cooler waters along the Atlantic and Pacific coasts, you can find the **eel grass** *Zostera marina*. Eel grass lives in the protected bays and inlets of the subtidal zone. The tufts of eel grass grow close together, forming beds that provide hiding places for mollusks, arthropods (invertebrates such as crabs), and fish.

In the bays and inlets of warmer waters, along the coasts of Florida and the Gulf of Mexico, large beds of the **turtle grass**

Figure 5-9 Two species of sea grass.

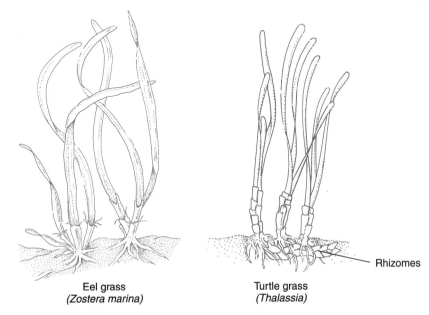

Eel grass
(*Zostera marina*)

Turtle grass
(*Thalassia*)

Rhizomes

Thalassia grow. Turtle grass has underground stems called **rhizomes**, which form an interlocking mat that helps stabilize the sandy seafloor. Turtle grass is home to a variety of sea animals. Fish hide in the grasses, and invertebrates attach to the blades of grass. Also, as you might guess from the name, turtle grass is an important food source for sea turtles.

How does a plant like sea grass reproduce underwater? Like many land plants, sea grass produces flowers. The flowers are small structures located at the base of the plant. Pollen from the flowers is dispersed in long threads in the water. When sea grass egg cells are fertilized by pollen, seeds are produced and shed into the water. If they settle on a suitable substrate, the seeds will germinate.

5.3 SECTION REVIEW

1. How are marsh grasses adapted to survive in salt water?
2. Of what importance are sea grasses to aquatic communities?
3. Describe how sea grasses reproduce underwater.

5.4 MANGROVE TREES

Certain trees are able to grow in salt water. Along tropical shores around the world, including the bays and inlets of Florida, the **mangrove trees** grow so close together they form a thick jungle of vegetation, called a mangrove swamp or community (see Chapter 3). When the tide is low, you can see the arching **prop roots** of the red mangrove tree *(Rhizophora mangle)*. (There are black mangrove and white mangrove trees, too.) Prop roots anchor the mangrove trees into the muddy sand. The tangle of mangrove roots also acts as a net to trap organic debris brought in by the tides. (See Figure 5-10.)

The mangrove roots are covered at high tide, while the stems and leaves remain above water. Seedpods dangle from the branches of red mangroves. These pods are 10 to 12 cm in length and look like small pencils. When they are ripe, the pods fall into the water. They float vertically and are carried by ocean currents to other loca-

Figure 5-10 A mangrove tree with its prop roots showing at low tide.

tions. When a seedpod makes contact with a suitable muddy bottom, it begins to grow into a mangrove seedling. When several seedlings take root, a new red mangrove community is established.

Mangroves, Marshes, and Wildlife

Like salt marshes, mangrove communities are biologically productive areas. Many different species of animals find safety and sustenance living in the communities formed by these marine plants. Salt marshes and mangroves are enriched with nutrients carried in by every movement of the tides. Bacteria, plankton, and the decaying remains of many kinds of marine organisms that are trapped by the mangrove roots provide nutrients for the mangrove community. These nutrients serve as food for plankton, and for every other organism that feeds on them, and as fertilizers that help marine plants grow.

In return, the plants provide food and hiding places for many animals. In fact, salt marshes and mangrove swamps are often considered to be the "nurseries" of the sea. Young fish and other small animals survive by hiding in the grasses or within the tangled network of mangrove roots, where larger animals cannot pursue them. Many birds and mammals survive by eating the food provided by these communities. In North America, birds such as rails, herons, egrets, and terns, and mammals such as raccoons, muskrats, deer, and foxes, may be found living in and around salt marshes. In South Asia, the unusual proboscis monkey lives almost exclusively in the trees of the mangrove swamp, relying on the leaves for its sustenance.

5.4 SECTION REVIEW

1. Why do mangrove swamps contain such a rich abundance of organisms?

2. How does a ripe mangrove seedpod find a suitable place to grow?

3. Explain how the roots of mangrove trees help other organisms survive.

Laboratory Investigation 5

Adaptations of a Marine Alga

PROBLEM: How is a marine alga adapted to live in the ocean?

SKILL: Observing the external structure of a seaweed.

MATERIALS: *Fucus,* pan, seawater, scissors.

PROCEDURE

1. Put a piece of *Fucus* in a pan of seawater. Notice the alga's brown-green color, which results from the mixture of green and yellow pigments in its cells.

2. Observe the flattened shape of the seaweed. More specifically, look at the shape of its stem. (Refer also to Figure 5-11.)

3. In your notebook, make a drawing of the whole specimen that you are examining.

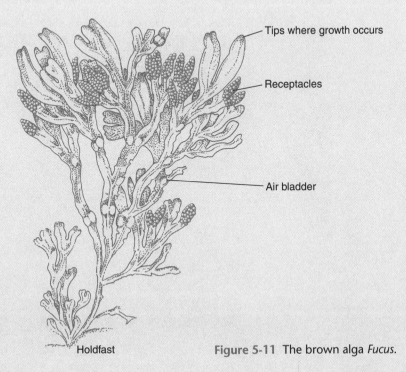

Tips where growth occurs

Receptacles

Air bladder

Holdfast

Figure 5-11 The brown alga *Fucus.*

4. Use your fingers to locate a small sac along one of the stems. Cut open the sac with your scissors. Make a drawing of what you observe.

5. Growth occurs from the tips of the alga. Find some forked stem tips that are flat. Draw what you observe.

6. You may also find some tips that are swollen. The swollen tips are the receptacles. These are the reproductive organs that contain the sperm cells and egg cells.

OBSERVATIONS AND ANALYSES

1. What are three adaptations *Fucus* shows for carrying out photosynthesis?

2. What structure prevents *Fucus* from being washed away?

3. Why do you think *Fucus* lives close to shore and not out in the open sea?

Chapter 5 | Review

Answer the following questions on a separate sheet of paper.

Vocabulary

The following list contains all the boldface terms in this chapter.

agar, algin, carrageenan, cordgrass, dunes, eel grass, gametes, holdfast, kelp, mangrove trees, marsh grasses, multicellular, prop roots, reed grass, rhizomes, sea grass, seaweeds, spore, thallus, turtle grass, zygote

Fill In

Use one of the vocabulary terms listed above to complete each sentence.

1. *Spartina* is a type of _____ that grows in a salt marsh.
2. The leafy part of a seaweed is called the _____.
3. A _____ is a tough pad of tissue that anchors a seaweed.
4. The largest seaweed in the ocean is the brown alga _____.
5. The _____ of mangrove trees anchor them in the sand.

Think and Write

Use the information in this chapter to respond to these items.

6. Why are the algae and beach grasses both called producers?
7. Describe the alternation of generations seen in some marine algae, such as *Ulva.*
8. Discuss two ways that mats of algae, sea grasses, and mangrove roots serve a similar function for marine life.

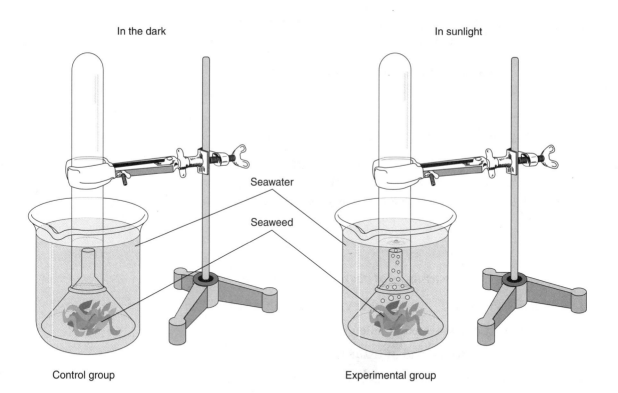

In the dark   In sunlight

Seawater

Seaweed

Control group   Experimental group

Inquiry

Base your answers to questions 9 through 12 on the diagram above, which shows an experimental setup, and on your knowledge of marine science.

9. What life activity is being studied in this experiment?
   a. chemosynthesis   *b.* photosynthesis   *c.* respiration
   d. reproduction

10. The organism being studied in this experiment is a type of
    a. vascular plant   *b.* alga   *c.* decomposer   *d.* spore.

11. Which statement is accurate regarding the results of this
    experiment?   *a.* The gas produced in the experimental group
    is carbon dioxide.   *b.* The gas produced in the experimental
    group is oxygen.   *c.* The seaweed in the experimental group
    is decaying.   *d.* The seaweed in the experimental group is
    excreting salt crystals.

12. Which is a valid conclusion that can be drawn from this experiment? *a.* Seaweeds can produce oxygen at night. *b.* Plants can make food during the day. *c.* Seaweeds do not contain chlorophyll. *d.* Light is necessary for oxygen production.

Multiple Choice

13. All of the following are characteristics found in seaside plants *except* *a.* long underground stems *b.* vascular tissue *c.* holdfast attachments *d.* chloroplasts.

14. The waxy covering of the prickly pear plant is an adaptation for *a.* deterring animals *b.* minimizing water loss *c.* stabilizing dunes *d.* secreting poison to trap insects.

15. What do all three organisms shown below have in common with one another? *a.* They are consumers. *b.* They are photosynthetic. *c.* They are heterotrophic. *d.* They reproduce asexually.

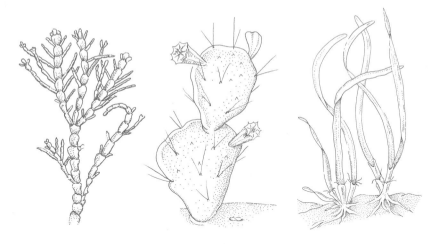

16. Cordgrasses are adapted to live in salt water because they have *a.* special glands in their leaves to excrete excess salt *b.* thick leaves that store water and pump salt out *c.* roots that absorb excess salt *d.* waxy stems that store fresh water.

17. The organic debris brought in by tides and trapped by the roots of mangrove trees   *a.* serves as a source of food for marine organisms   *b.* reduces the force of the tides   *c.* serves as a depository for salt   *d.* provides a surface on which plankton can grow.

18. The mangrove tree is well adapted to a coastal environment because   *a.* its prop roots anchor the tree into the sand   *b.* it can carry out photosynthesis underwater   *c.* its prop roots are used like a net to catch fish   *d.* its seedpods are dispersed through the air.

19. Which of the following marine plants or algae is least related to the other ones?   *a. Spartina*   *b. Thalassia*   *c. Fucus*   *d. Phragmites*

20. Haploid cells necessary for sexual reproduction are produced by a seaweed's   *a.* sporophyte thallus   *b.* holdfast   *c.* gametophyte thallus   *d.* flagellated spore.

21. The largest and fastest growing producer in the ocean is the   *a.* sea lettuce   *b. Fucus*   *c.* kelp   *d. Sargassum.*

22. Seaweeds produce oxygen through the process of   *a.* respiration   *b.* photosynthesis   *c.* chemosynthesis   *d.* alternation of generations.

Research/Activity

If you live near a beach, collect and display seaweeds found on the shore. (Ask your teacher how to preserve the specimens.) Classify the seaweeds into their appropriate groups and describe their major distinguishing characteristics.

6 Simple Marine Animals

When you have completed this chapter, you should be able to:

DISCUSS nutrients and the digestion of food by animals.

EXPLAIN the importance of zooplankton in marine food chains.

DESCRIBE the three groups of protozoans and their life functions.

COMPARE and CONTRAST the sponges, rotifers, and bryozoans.

Look at the ocean scene shown above. Here, corals, sponges, sea urchins, fish, lobsters, and feather worms all live in a small area of the ocean. What do these organisms have in common? They are all animals. What is an animal? In general, an animal is a multicellular organism that consumes food and is able to move. There are at least 1 million species of animals on Earth. More than 100,000 of these animal species are known to inhabit the world's oceans.

Along with the animals, there are marine creatures so small that you need a microscope to see them. Many of these microscopic organisms are unicellular. At one time, they were placed in the animal kingdom, along with the multicellular animals. Today, scientists place most unicellular animal-like organisms in the protist kingdom. In this chapter, you will continue your study of life in the sea as you meet the unicellular protists and some multicellular "simple" animals.

6.1 NUTRITION IN ANIMALS

Fish swim, snails crawl, worms burrow, and whales dive. These are only a few of the many activities that marine animals perform in their daily struggle for survival. Animals use energy to carry out these tasks, and they get the energy they need from food. The process by which organisms use food to perform their life activities is called **nutrition.**

Animals, unlike green plants and algae, cannot make their own food. Therefore, animals must take in food in order to satisfy their energy needs. Food contains useful chemical compounds called **nutrients.** The basic nutrients needed by animals to survive are sugars, starches, proteins, fats, minerals, vitamins, and water. The process by which animals break down and utilize these nutrients is called **metabolism.** As discussed in Chapter 4, the consumption of food by organisms is called heterotrophic nutrition; thus, animals are heterotrophs.

Carbohydrates

Two nutrients that animals are able to derive energy from quickly are sugars and starches. Sugars and starches comprise the carbohydrates, compounds that contain the elements carbon, hydrogen, and oxygen in definite proportions. The molecular formula for the simple sugar glucose is $C_6H_{12}O_6$. That means there are 6 carbon atoms, 12 hydrogen atoms, and 6 oxygen atoms in a molecule of glucose. Notice that the ratio of hydrogen atoms to oxygen atoms is the same as the ratio of hydrogen to oxygen found in water, that is, 2:1. You can see how carbohydrates got their name, for they can be considered "hydrated carbons." The energy in glucose is found in the carbon–hydrogen bonds. A compound like glucose that contains the element carbon is classified as an organic compound.

When glucose is not being used in the body, it is changed into and stored as a starch. The chemical reaction by which glucose is changed into starch is called dehydration synthesis. During dehydration synthesis, one oxygen and two hydrogen atoms are removed from two glucose molecules, thus forming one molecule of water and one molecule of maltose. This reaction is as follows:

$$C_6H_{12}O_6 + C_6H_{12}O_6 \longrightarrow C_{12}H_{22}O_{11} + H_2O$$

Glucose   Glucose   Maltose   Water

Glucose, a single sugar, is a monosaccharide. Maltose contains two glucose units, so it is a double sugar, or disaccharide. Double sugars have the formula $C_{12}H_{22}O_{11}$. Starch is produced when maltose combines with other glucose molecules. Thus, starch is a polysaccharide (meaning "many sugars"), because it contains a long chain of glucose units. The simplest formula for starch is $(C_6H_{10}O_5)_n$, where n represents the number of glucose units.

Starches can be changed back into molecules of glucose when an animal needs energy. The chemical reaction that changes starch back into glucose is called hydrolysis. During hydrolysis, the larger starch molecule is changed into smaller glucose molecules by the addition of water. Hydrolysis, a breaking-down process, occurs when food is digested. Energy comes from the breaking of chemical bonds. Dehydration synthesis, a building-up process, occurs during growth. Living cells carry out these important chemical reactions to satisfy their growth and energy needs. Notice that dehydration synthesis and hydrolysis are opposite chemical reactions.

Chemical reactions in living things cannot occur without enzymes. An **enzyme** is a protein (see section below) that regulates the speed of a chemical reaction without itself being changed. For example, the enzyme maltase aids the breakdown (hydrolysis) of the double sugar maltose into two units of the simple sugar glucose. The reaction is as follows:

$$C_{12}H_{22}O_{11} + H_2O \xrightarrow{\text{maltase}} C_6H_{12}O_6 + C_6H_{12}O_6$$

Maltose   Water   Glucose   Glucose

During hydrolysis, large molecules are acted on by enzymes and changed into small molecules, as follows: proteins, via protease, to amino acids; lipids, via lipase, to fatty acids and glycerol; and starch, via amylase, to glucose. Enzymes are also called organic catalysts.

Lipids

The high-energy nutrients known as fats and oils are called lipids. Unlike the 2:1 ratio in carbohydrates, the ratio of hydrogen atoms to oxygen atoms in lipids varies. Compare the formula for a lipid molecule such as castor oil, $C_{18}H_{34}O_3$, with that of a carbohydrate molecule such as glucose, $C_6H_{12}O_6$. Which has more energy? Recall that energy is found in the carbon–hydrogen bonds. Notice that there are many more carbon–hydrogen bonds in a fat molecule than in a carbohydrate molecule. Because of its greater number of carbon–hydrogen bonds, a molecule of fat contains more energy than does a carbohydrate molecule. During hydrolysis, the carbon–hydrogen bonds in fats are broken and energy is released.

Proteins

Living things need proteins for the growth and repair of their cells. Proteins are composed of smaller "building blocks" called amino acids. There are 20 different amino acids. Each amino acid molecule has an amino group (NH_2) at one end and a carboxyl group (COOH) at the other end. Amino acids differ from one another based on structural differences in their "R (radical) groups." Growth occurs by the process of dehydration synthesis, when amino acids join together inside the cell to make the proteins. When two amino acids join, they form a dipeptide; when more amino acids join, they form a polypeptide. (*Note*: Enzymes aid in dehydration synthesis, just as they do in hydrolysis.) The reaction that forms a dipeptide from two amino acids is as follows:

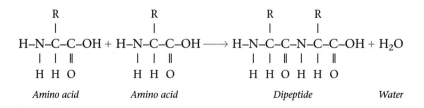

| Amino acid | Amino acid | Dipeptide | Water |

Minerals

Minerals are elements and compounds found in water and soil that do not contain the element carbon. Chemical substances that do not contain carbon are considered to be inorganic. Thus, the mineral known as table salt (NaCl) is a type of inorganic compound.

All living things require minerals for their normal growth and health. Marine plants absorb the minerals they need from water.

Animals that eat plants incorporate the plants' minerals into their body tissues. The mineral calcium is needed for the growth of bones and teeth. Iron is found in red blood cells and is used to carry oxygen to, and carbon dioxide away from, cells. The muscles and nerves of animals require sodium and chloride ions to function. Phosphorus is an essential element in ATP, the energy compound, and in DNA, the genetic material. Another mineral, silica, the main ingredient in glass, is found in the cell walls of diatoms. Seafoods are rich in iodine, a mineral found in thyroxin, the hormone that regulates growth and metabolism in vertebrates.

Vitamins

Vitamins are organic compounds that are needed, in small amounts, to maintain good health. They aid in the functioning of enzymes. Vitamin C, found in fresh fruits and vegetables, is needed to prevent the vitamin deficiency disease called scurvy. Vitamin D, necessary for healthy bone growth, is produced in small amounts in marine mammals when ultraviolet light reacts with the fat located just under their skin. Marine plants are a rich source of other vitamins, including vitamins A, E, K, and the B vitamins.

Water

Water (H_2O), an inorganic compound, is the most abundant nutrient in the body. About 80 percent of an organism's body weight is water. However, the exact amount of water in different organisms varies. For example, the human body is about 67 percent water, while jellyfish are about 95 percent water. Body fluids such as blood, lymph, sweat, and tears are made up almost entirely of water. Water takes part in important chemical reactions such as photosynthesis and hydrolysis. In addition, water contains and transports many dissolved and suspended substances within the bodies of organisms.

6.1 SECTION REVIEW

1. Why are dehydration synthesis and hydrolysis necessary processes for living things?

2. Discuss the importance of minerals for animal nutrition.

3. Describe the role of water in living things.

6.2 ZOOPLANKTON IN THE SEA

Animals and animal-like organisms that float and drift on the ocean surface are considered to be part of the plankton population. (Remember that *plankton* means "wanderers." The name is used because large masses of plankton are carried great distances by ocean currents.) Plankton include the unicellular protists and multicellular organisms such as the jellyfish. The plantlike plankton, which contain chlorophyll, are referred to as phytoplankton. The animal and animal-like plankton are called **zooplankton**. This category describes the organisms' lifestyle; it does not represent a true taxonomic grouping.

Zooplankton vary in size. The jellyfish are the largest of the plankton species (see Chapter 7). However, most plankton species are microscopic. You can catch plankton by dragging a plankton net through the water. The best time for a field trip to net plankton is in the spring and early summer, when the longer days cause a rapid increase in plankton populations. By doing the lab investigation at the end of this chapter, you will learn that there are many different kinds of plankton that can be discovered from a plankton tow at the shore.

Zooplankton Diversity

Zooplankton are so varied that they are divided into two groups: the temporary zooplankton and the permanent zooplankton. The temporary zooplankton (also called meroplankton) are the embryos or larvae of fish, crabs, sponges, lobsters, clams, and other invertebrates. (See Figure 6-1.) These animals spend the early part of their life cycle floating and drifting near the surface of the ocean. When they mature, the temporary zooplankton settle to the bottom, where they develop into adults. As adults, they are no longer considered to be part of the plankton population.

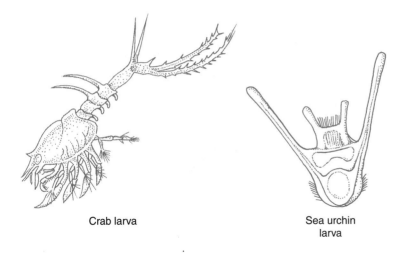

Crab larva

Sea urchin
larva

Clam larva

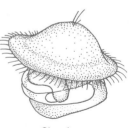

Figure 6-1 Three types of temporary zooplankton.

The permanent zooplankton are those species that remain in the plankton population throughout their entire life cycle. (See Figure 6-2.) The **foraminiferan** (meaning "hole-bearing"), or foram for short, is a unicellular protist. Forams are encased in a shell, or test, made up of calcium carbonate ($CaCO_3$). Parts of a foram's cytoplasm flow out through holes, or pores, in the shell and form a sticky surface for catching food. When a foram dies, its shell falls to the seafloor. Over many years, sediments of these shells accumulate, forming thick chalk deposits. Sometimes these seafloor deposits are lifted up to Earth's surface. Chalk deposits in Georgia and Mississippi, and the white cliffs of Dover, in England, are examples of sediments formed mostly from the shells of forams.

Another type of permanent zooplankton is the **radiolarian**. The

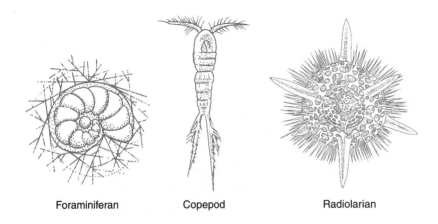

Figure 6-2 Three types of permanent zooplankton.

Foraminiferan   Copepod   Radiolarian

unicellular radiolarian is transparent because its cell wall (like that of a diatom) is composed of silica. Notice the shapes of the radiolarians. Long spines branch out from a radiolarian's body, like the spokes of a wheel, for added buoyancy and protection.

Of all the permanent zooplankton, the copepod is the most numerous. This tiny shrimplike animal, the size of a grain of sand, feeds on phytoplankton such as diatoms. In turn, the copepod is eaten by larger zooplankton, small fish, and whales. Thus, the copepod is an important link in many marine food chains.

Sea Soup

As noted above, for sea creatures ranging from the smallest of fish to the largest of whales, plankton is an important food source. In some ways, the ocean can be thought of as a thin soup, with plankton as the food particles suspended in it. Several species of whales feed by plowing through the water with their mouths open. The plankton are filtered out as water is forced from the whale's mouth through huge fringed plates. Animals that strain their food from the water are called **filter feeders**. In the Antarctic, whales feed on large schools of shrimplike zooplankton called krill. The krill, which are only 4 to 5 cm long, are being considered as a potential food source for humans.

Bottom-dwelling mollusks living in shallow waters, such as mussels, clams, oysters, and scallops, depend on plankton as their main food source. These animals also filter plankton from the water. In the case of clams (and other bivalves), cells on the inside of the animal contain microscopic hairs called **cilia**. The cilia beat back and forth, causing currents of water to enter and leave the clam. In this way, the clam filters food out of the water as it passes through its body. Other invertebrates, such as shrimp, also feed on plankton.

Some newly hatched small fish that have used up the food supply in their yolk sac feed directly on plankton until they are large enough to eat other organisms. Larger fish and other animals, including humans, depend indirectly on plankton as a food source. In a marine food chain, each organism serves as food for another. However, the plankton form the foundation of it all. (See Chapter 21 for more information on marine food chains and food webs.)

1. How does a whale feed on zooplankton?

2. Describe how filter-feeding mollusks obtain their food.

3. Discuss and illustrate the role of plankton in a marine food chain.

6.3 PROTOZOANS

Marine creatures can be found living anywhere from the ocean's surface to its bottom. You learned that plankton are usually found near the surface. Zooplankton such as forams and radiolarians are both members of the larger group of unicellular animal-like organisms called protozoa. Thousands of species of protozoans are found living on the surface of marine substrates and in the bottom sediments. These one-celled organisms are classified within the protist kingdom, along with the algae. The protozoa are subdivided into three major groups: the Ciliophora, Zoomastigina, and Sarcodina.

The Ciliophora are the largest group of protozoa, composed of thousands of freshwater and marine species, all having cilia. Most members of this phylum, such as the *Spirostomum*, are free-swimming and use their cilia for locomotion. Others, such as the *Stylonychia*, use their cilia like tiny feet to crawl on substrates. A few, such as the *Vorticella*, live attached to a substrate, where they use their cilia for feeding rather than locomotion. (See Figure 6-3 on page 150.)

The Zoomastigina consist of a group of animal-like protists that move through the water by means of whiplike flagella. Members of this group live in freshwater, salt water, and also as parasites within the bodies of other organisms. Included in this group are the euglena and the dinoflagellates, which take in (as well as make their own) food. (See Chapter 4.)

The Sarcodina are the protozoan group that includes the forams and radiolarians, as well as the amebas. They live on the surface of substrates and move by means of cytoplasmic extensions called **pseudopods** (meaning "false feet"). This kind of movement, typical of the ameba, is called *ameboid movement*. The moving pseudopods are also used to surround and engulf food particles and living

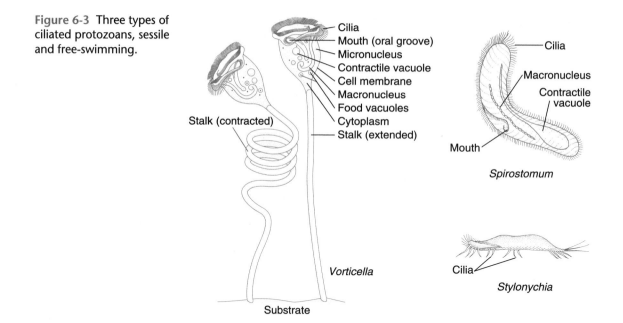

Figure 6-3 Three types of ciliated protozoans, sessile and free-swimming.

Cilia
Mouth (oral groove)
Micronucleus
Contractile vacuole
Cell membrane
Macronucleus
Food vacuoles
Cytoplasm
Stalk (extended)

Stalk (contracted)

Vorticella

Substrate

Cilia
Macronucleus
Contractile vacuole
Mouth

Spirostomum

Cilia

Stylonychia

prey. Interestingly, there are cells in the human body that also show ameboid movement. For example, some kinds of white blood cells show ameboid movement when they engulf bacteria.

Life Functions of Protozoans

The protozoa are an amazing group of organisms. They carry out all necessary life functions within a single cell. Multicellular animals use large numbers of cells to perform their life functions. How, then, are protozoans able to carry out all the life functions—ingestion and digestion, respiration, transport, water balance and excretion, sensitivity, and reproduction—within a single cell?

Ingestion and Digestion in Protozoa

Look at the *Vorticella,* a ciliated protozoan shown in Figure 6-3. The *Vorticella* can be easily observed, because it is often found attached to a substrate. Organisms that live attached to substrates are called **sessile** organisms. How does the *Vorticella* carry out the life function of *ingestion*, the taking in of food? Food is swept toward the *Vorticella*'s "mouth" (oral groove) by movements of its ring of cilia. (Protozoa that are motile actively catch and ingest their food.)

The large food particles ingested by the *Vorticella* enter its food vacuoles. There, they are broken down into smaller particles through the process called *digestion*. Food vacuoles are located in the cytoplasm, the fluidlike material that makes up much of the cell. Digestion that takes place inside a cell is called intracellular digestion.

Respiration in Protozoa

How do protozoa obtain energy for their cell's activities? Like other organisms, protozoa take in oxygen and combine it with glucose to produce chemical energy, a process called respiration. The energy in glucose that cannot be directly used by a cell is transformed, in a series of enzyme-controlled reactions, to the usable form of chemical energy called adenosine triphosphate, or ATP. This process, which occurs inside the cell, is called cellular respiration (or aerobic respiration) and can be summarized by the following chemical equation:

$$C_6H_{12}O_6 + 6\ O_2 \xrightarrow{\text{enzymes}} 6\ CO_2 + 6\ H_2O + 36\ ATP$$

This equation can be put into words as follows: one molecule of glucose plus six molecules of oxygen (in the presence of enzymes) yields six molecules of carbon dioxide plus six molecules of water plus 36 molecules of adenosine triphosphate (ATP).

Oxygen enters the mitochondria, the energy factories of the cell, where it combines with glucose. For every molecule of glucose that is burned, or oxidized, in the cell, 36 molecules of ATP are produced. The ATP both stores and releases the chemical energy that is used by a cell to do work. For example, *Vorticella* uses the energy stored in ATP for contraction of its stalk and movement of its cilia.

In a unicellular organism, what structures are required to help carry out the function of respiration? Notice that *Vorticella* has a cell membrane, which permits oxygen to enter the cell from the surrounding water and allows the waste product of respiration, carbon dioxide, to exit the cell.

Transport in Protozoa

What causes oxygen and carbon dioxide to enter and leave the cell? The movement of substances into, out of, and within a cell is called *transport.* The concentration, or number, of oxygen molecules per unit area is greater outside the *Vorticella* cell (in the water) than inside the cell. Since the concentration of oxygen molecules is greater outside the *Vorticella* than inside it, oxygen molecules will move from outside to inside the cell. The movement of molecules from an area of higher concentration to an area of lower concentration is called **diffusion**. Diffusion is an example of passive transport, meaning no energy from ATP is used in the process.

Diffusion also explains the transport of carbon dioxide from the inside to the outside of a cell. As a result of cellular respiration, carbon dioxide accumulates as a waste product inside the cell. Therefore, the concentration of carbon dioxide is greater inside than outside the cell. The concentration of CO_2 is lower outside the *Vorticella* because it is diluted by the large volume of ocean water. Carbon dioxide diffuses from inside to outside the cell because of this concentration difference. Transport also occurs within the cell. Food vacuoles move about and distribute nutrients inside the cell as a result of the flowing of cytoplasm, a process called cyclosis.

Water Balance and Excretion in Protozoa

Being aquatic, *Vorticella* and other protozoans live in an environment where the concentration of water molecules is greater outside the cell than inside the cell. There is a lower concentration of water molecules inside the *Vorticella,* because dissolved substances inside its cell take the place of water molecules. This unequal concentration of water molecules causes them to move from outside the cell to inside the cell. The movement of water molecules from an area of higher concentration to an area of lower concentration across a cell membrane is called **osmosis**. Like diffusion, osmosis is an example of passive transport.

What prevents a protozoan from swelling up and bursting as a result of inward osmosis? Excess water is pumped out of the cell through a structure called the **contractile vacuole**. The forceful closing of the contractile vacuole requires energy from ATP, so this process is an example of active transport.

Liquid wastes are also eliminated from the cell in a process called *excretion*. A contractile vacuole called the excretory vacuole carries out the life function of excretion. The elimination of excess water and liquid wastes from inside the cell enables the protozoan to maintain a proper water balance and a stable internal environment. In general, the ability of an organism to maintain a stable internal environment is called **homeostasis**.

Sensitivity in Protozoa

You blink when there is dust in your eye, sneeze when there is pollen in your nose, and jump when you are startled by a loud noise. These movements and reactions are called responses. To stay alive, organisms must be able to respond to changes in their environment. These changes, known as stimuli (singular, stimulus), are what cause an organism to respond. In the above examples, the dust, pollen, and noise are the stimuli.

The ability of an organism to respond to environmental stimuli is called *sensitivity*. How does a unicellular organism like the *Vorticella* carry out the life function of sensitivity? The *Vorticella* responds to the stimulus of touch by contracting its elongated stalk into a tight coil. Contraction occurs when other protozoa make contact with the *Vorticella*. Following contraction, the stalk rapidly uncoils. What is the adaptive value of this response? The contraction is an avoidance reaction that the *Vorticella* makes in response to a stimulus that may be harmful. Interestingly, the *Vorticella* also contracts spontaneously in the apparent absence of a mechanical stimulus, a behavior that is not fully understood at this time.

Reproduction in Protozoa

All life comes from existing life-forms. The production of offspring by living organisms is the life function called *reproduction*. In the *Vorticella,* one cell divides to form two cells, called the daughter cells. The production of offspring by a single parent is called asexual reproduction. Many unicellular organisms reproduce asexually.

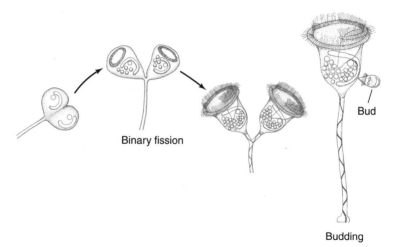

Figure 6-4 Asexual reproduction in the *Vorticella*—by binary fission and by budding.

Binary fission

Bud

Budding

Reproduction is controlled by the nucleus, which contains the hereditary material of the cell, called DNA (deoxyribonucleic acid). Before the cell divides, the nucleus duplicates itself to form two nuclei. During cell division, one nucleus moves into each of the newly formed cells. Division of a cell's cytoplasm into two daughter cells of equal size is called **binary fission.** The offspring that result from binary fission are identical. *Vorticella* and other protozoans can also reproduce by **budding**. In budding, the division of the cytoplasm is uneven (but the DNA is the same). As a result, the new daughter cell, or bud, is smaller than the parent cell. Eventually, the bud breaks away and forms its own stalk. Asexual reproduction ensures the existence of another generation and also ensures genetic continuity from one generation to the next. (See Figure 6-4.)

Protozoa, such as the *Vorticella*, can also reproduce sexually. In sexual reproduction, two parents are needed to produce offspring. Among protozoans, two "parent" cells come into contact to exchange hereditary material, a process known as **conjugation**. As a result of the exchange of genetic material, the offspring of sexually reproducing parents are not identical to either parent. In protozoa, conjugation is not, strictly speaking, a reproductive process since there are still only two organisms after conjugation occurs. However, genetic material is exchanged between the two conjugating cells. This exchanged material adds genetic variability to a population that eventually increases in number through binary fission.

6.3 SECTION REVIEW

1. In what kingdom are protozoa classified? Identify three main groups of protozoa.

2. How does a protozoan carry out the process of respiration?

3. Describe two ways that protozoa can reproduce.

6.4 SPONGES, ROTIFERS, AND BRYOZOANS

In this section, you will learn about three distinct phylums of marine invertebrates—the sponges, rotifers, and bryozoans. Although each group of species is unique enough to be placed in its own phylum, they do share some characteristics. Rotifers and bryozoans are both microscopic, but sponges can grow quite large. Sponges and bryozoans are both sedentary, but rotifers are capable of movement. However, the animals found in all three groups are multicellular, bottom-dwelling invertebrates.

Sponges

Animals that are composed of more than one cell are called multicellular. The sponge is a multicellular, primarily marine animal that has few specialized structures. Sponges have two layers of mostly undifferentiated cells: an inner layer called the endoderm and an outer layer called the ectoderm. Between the two layers is a jellylike material called the mesenchyme. (See Figure 6-5 on page 156.)

Sponges are classified in the phylum Porifera (meaning "pore bearing"). Since they inhabit the seafloor, from the intertidal zone down to the depths of the ocean, sponges are considered **benthic**, or bottom-dwelling, organisms.

Life Functions of the Sponge

How does the sponge feed if it can't move? A sponge's body contains many holes, or pores. Tiny food particles and plankton enter

Figure 6-5 Structure of the sponge, a multicellular benthic animal.

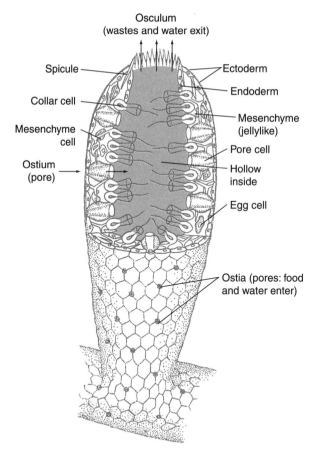

Osculum (wastes and water exit)

Spicule

Ectoderm

Endoderm

Collar cell

Mesenchyme (jellylike)

Mesenchyme cell

Pore cell

Ostium (pore)

Hollow inside

Egg cell

Ostia (pores: food and water enter)

through the small pores, called **ostia** (singular, ostium), which are surrounded by pore cells. Water and wastes exit through the large hole, called the **osculum** (plural, oscula), usually located at the top of the sponge. Inside the sponge are special **collar cells** that contain flagella, which beat back and forth. The coordinated movements of these flagella produce the currents that pump water into and out of the sponge.

Collar cells are also involved in food getting. Plankton, bacteria, and other tiny particles of food brought in on currents of water are trapped, ingested, and digested by the collar cells. Other cells, called amebocytes, which are found in the mesenchyme, also ingest and digest food. Digestion in the sponge occurs within food vacuoles inside the individual cells; as in the protozoa, it is intracellular.

How does a sponge take in oxygen and get rid of carbon dioxide? Since the cells of a sponge are in direct contact with water, gas exchange occurs across cell membranes. The water that comes in through the ostia contains dissolved oxygen, which diffuses into the sponge's cells. Carbon dioxide, the waste product of respiration, diffuses out of the cells and is expelled through the sponge's oscula into the water.

Sponges are not very responsive. If you touch a living sponge, it will not move. Since the sponge lacks a nervous system, rapid reflex movements do not occur. However, the sponge has muscle-like cells called myocytes, located near the ostia and oscula. When the myocytes contract, the ostia close, preventing water from entering the pores. This ability to close the ostia is probably a defensive reaction that protects the sponge from taking in any toxic substances in the water.

Life Cycle of the Sponge

Sponges will attach to a variety of substrates, including rocks, mollusk shells, and the hulls of ships. How can a sponge find, and attach to, a substrate if it doesn't swim? You can find the answer to this question by examining Figure 6-6. The life cycle of a sponge

Figure 6-6. Development of the sponge, from zygote to adult stage.

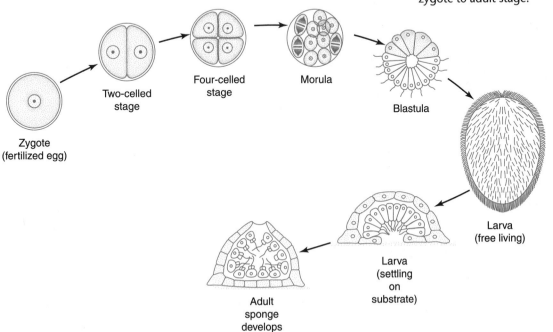

Zygote
(fertilized egg)

Two-celled
stage

Four-celled
stage

Morula

Blastula

Larva
(free living)

Larva
(settling
on
substrate)

Adult
sponge
develops

DISCOVERY

A Challenge to Taxonomy

Every year, scientists discover new species of organisms. After careful study of a new organism's important characteristics, the scientists classify it in the appropriate taxonomic categories. Recently, however, researchers in Denmark discovered a unique, tiny marine animal that presented a challenge to taxonomy. "What is it?" the two Danish scientists asked themselves when they first saw the organism on the lip of a lobster. This tiny creature (about the size of a dot above the letter "i") lives attached to the mouthparts of a lobster by means of an adhesive disk, where it feeds off scraps of food. The organism also has an unusual life cycle—it can be either male or female, and it can take part in sexual or asexual reproduction.

The scientists went about the job of trying to classify this organism. First, they knew it belonged to the animal kingdom, since it was multicellular, had the ability to move, and consumed food. Next, when they examined its body structure, they saw that it was an invertebrate. However, when they tried to determine which of the 35 known phylums it might belong to, they discovered that it didn't resemble any of the organisms in these groups. So, they created a new phylum for it, called Cycliophora, which refers to its circular mouth.

So far, this exotic creature is the only member of phylum Cycliophora. But what is its name? The last step in classification of an organism is to identify it scientifically by genus and species. The Danish scientists named it *Symbion pandora* because it lives symbiotically (in close association) with another organism (the lobster) and because it exhibited an unexpected and complex life cycle (likened to the strange surprises found in the mythological Pandora's box).

QUESTIONS

1. Using one or more complete sentences, explain how *Symbion pandora* was discovered.

2. Describe how the two scientists went about classifying this organism.

3. Why did the scientists feel it necessary to create a new phylum for this organism?

begins with adult sponges releasing eggs and sperm into the water. Typical of sexual reproduction in most animals, a single sperm unites with a single egg to produce a fertilized egg cell. This zygote represents the first stage in the development of the sponge. Next, the zygote divides to form two cells. The cells then divide again to form four cells. Cell division continues until a solid ball of cells is formed, called the morula. (*Morula* is the Latin word for "raspberry," which is what the ball of cells looks like.) The rapid division of cells in this early stage of development is called cleavage, and the organism at this point is referred to as an **embryo**.

In time, a hollow area develops inside the embryo. This hollow ball of cells is now called a blastula. The cells of the blastula develop whiplike flagella, which enable the embryo to swim. At this stage in its development, the embryo is called a **larva** (plural, *larvae*), which refers to any free-living stage in the early development of an animal. The swimming larva becomes part of the plankton population. After the sponge larva makes contact with a hard substrate, such as the hull of a ship, it attaches to it and starts to develop into an adult sponge.

Individual sponges can be either male or female, or can have both male and female reproductive organs within them. Animals that possess both ovaries and testes are called hermaphrodites. Sponges that are hermaphrodites produce eggs and sperm at different times, thus ensuring that self-fertilization does not occur. Sponges can also reproduce asexually; pieces of a sponge may break off and then grow into a whole new sponge. This mode of reproduction, in which a whole body can be regrown from parts of the parent body, is called **regeneration**, and it occurs in some other invertebrates.

Sponge Diversity

Sponges are hardy creatures that are found in a variety of marine and freshwater habitats, ranging from warm tropical seas to cold polar waters. Some representative sponges are shown in Figure 6-7 on page 160. The bath sponge *(Euspongia)*, which lives in warm tropical waters, may be most familiar to you. At one time, thousands of people were employed harvesting this sponge for commercial use in America and Europe. Today, sponge collecting is largely a tourist

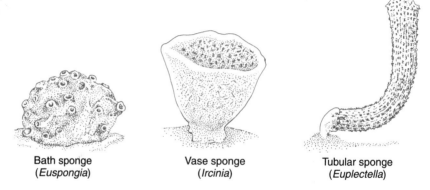

Figure 6-7 Three representative types of sponges.

Bath sponge
(*Euspongia*)

Vase sponge
(*Ircinia*)

Tubular sponge
(*Euplectella*)

industry. The market for sponges has declined sharply because synthetic sponges have largely replaced the use of natural ones.

While natural sponges now have limited commercial value, they play a very important role in the marine environment. Sponges are very efficient filter feeders. It is estimated that a single bath sponge can filter 100 liters of seawater in an hour. Sponges are also important in recycling minerals back into the water.

Sponges such as the yellow boring sponge *(Cliona)* grow on the shells of clams and other shellfish. The boring sponge is so named because it uses an acid to bore holes into shells in order to attach to them. You may have seen clamshells on the beach that were pockmarked with holes caused by *Cliona*. The sponge also recycles calcium carbonate, the mineral found in seashells, back into the water. Living things such as sponges that grow over the surfaces of substrates are called **encrusting organisms**.

Some sea stars, snails, and fish eat sponges, particularly the young sponges that have begun to colonize substrates. However, few animals eat sponges, because they often are composed of hard mineral matter, such as calcium carbonate or silica. Such sponges have a rigid structure, due to their skeleton of chalk or silicon spines called **spicules**. In contrast, some sponges, such as the natural bath sponge, have an elastic framework of protein fibers called **spongin**. When you squeeze one of these sponges, you can feel its elasticity.

Sponges show a variety of interesting shapes, sizes, and colors. The beautiful Venus's flower basket (*Euplectella*), which lives at great depths, is a tubular sponge composed of a delicate network of glassy spicules. The vase sponge *(Ircinia),* found on sandy bottoms near coral reefs, grows vertically. And some basket sponges, found in tropical waters, grow so large that a person could sit inside of them.

Rotifers

Many species of microscopic organisms are composed of more than one cell. Look at the multicellular organism called the **rotifer** (meaning "wheel bearer"), shown in Figure 6-8. Dozens of rotifer species live in the moist sands along the shore and in the gravel of aquarium tanks. (Rotifers are also common in freshwater.) Rotifers, which are in their own unique phylum Rotifera, are able to change the shape of their body. For example, when they swim by means of their cilia, rotifers can telescope their bodies to facilitate movement.

Some rotifers are predatory, while others scavenge on debris. When attached to a substrate, the rotifer's crown of beating cilia in its head region gives the animal the appearance of a miniature spinning wheel. The moving cilia create a water current that pulls floating food toward the rotifer's mouth. Food enters the mouth and then passes through a food tube, or one-way digestive tract,

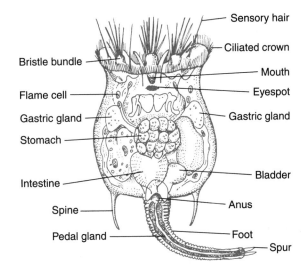

Figure 6-8 Structure of the rotifer, a microscopic multicellular animal.

Rotifer

consisting of an esophagus, a stomach, and intestines. Food is digested in the stomach and intestines, and waste products are eliminated through the anus. How is digestion in rotifers and humans similar? In both, digestion occurs in a one-way tube, in which food enters at one end of the tube and wastes come out at the other end.

Reproduction in the Rotifer

How does the rotifer reproduce? The male rotifer produces sperm in its testes, the male reproductive organs. The female produces eggs in its ovary, the female reproductive organ. Sperm and egg cells (gametes) contain the hereditary material from each of the parents that is necessary for sexual reproduction. The gametes are released into the water, where they unite.

Sperm cells are able to move through water because each one possesses a taillike flagellum. The sperm cell unites with an egg cell to produce a zygote. After fertilization, the rotifer zygote gradually develops into an adult. Rotifers have external fertilization and external development, meaning both of these events take place outside the body of the female.

Rotifers can also reproduce by an asexual process called parthenogenesis. In this process, the female produces an egg, without fertilization, that has a complete (double) set of chromosomes. The egg develops into a female rotifer that can reproduce either sexually or parthenogenetically.

Bryozoans

Another benthic organism, which is sometimes mistaken for a sponge, is the **bryozoan** (meaning "moss animal"). The bryozoan, classified in its own phylum Bryozoa, is a microscopic multicellular animal that lives within a box- or vase-shaped compartment made of calcium carbonate or chitin (the material found in crab and lobster shells). Branching colonies composed of hundreds of individual bryozoans cover the surfaces of rocks, seaweeds, and shells. Colonies of different types of bryozoans have different shapes; the

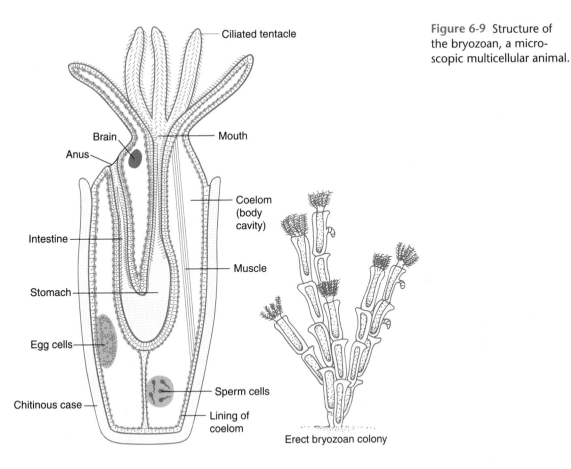

Ciliated tentacle

Brain

Anus

Mouth

Coelom (body cavity)

Intestine

Muscle

Stomach

Egg cells

Sperm cells

Chitinous case

Lining of coelom

Erect bryozoan colony

Figure 6-9 Structure of the bryozoan, a microscopic multicellular animal.

two main types are encrusting bryozoans and erect bryozoans. (See Figure 6-9.)

Life Functions of the Bryozoan

The bryozoan feeds on plankton and organic debris, which it captures with its ciliated tentacles. Food is digested in a one-way digestive tract that consists of a mouth, stomach, and intestines. Wastes are eliminated through the anus. Bryozoans can reproduce asexually by budding, the process by which a smaller individual develops on, and then separates from, the larger parent body. This is a fast way to increase the bryozoan colony's size. They can also reproduce sexually, by producing sperm and egg cells. Most bryozoans are hermaphrodites; that is, they contain both ovaries and testes within the same individual. As a result, in some species of bryozoans, self-fertilization actually occurs.

6.4 SECTION REVIEW

1. Draw and describe how a sponge filter feeds.
2. Draw and label the life cycle of a sponge.
3. Compare the structures of a bryozoan and a rotifer.

Laboratory Investigation **6**

Observing Diverse Zooplankton

PROBLEM: What kinds of zooplankton can be found in seawater?

SKILL: Observing tiny organisms under the microscope.

MATERIALS: Compound microscope, zooplankton samples, medicine droppers, slides, coverslips.

PROCEDURE

1. Prepare a wet mount slide of plankton. Use a medicine dropper to put 1 or 2 drops of your plankton sample on a glass slide. Place a coverslip over the slide. Make sure the bottom of your slide is dry.

2. Place the slide on the stage of a microscope and observe under low power.

3. Focus with the coarse adjustment. Move the slide around until you see something that moves.

4. In your notebook, make sketches of each of the zooplankton you observe. Locate and sketch as many zooplankton as you can in the allotted time. You may want to ask your teacher to help you try to identify them.

OBSERVATIONS AND ANALYSES

1. Why might you have difficulty in finding zooplankton in a bucket of seawater?

2. How would you know that the organism you observed is a type of zooplankton or phytoplankton?

3. Why is it more difficult to find zooplankton under high power than under low power?

Chapter 6 Review

Answer the following questions on a separate sheet of paper.

Vocabulary

The following list contains all the boldface terms in this chapter.

benthic, binary fission, bryozoan, budding, cilia, collar cells, conjugation, contractile vacuole, diffusion, embryo, encrusting organisms, enzyme, filter feeders, foraminiferan, homeostasis, larva, osculum, osmosis, ostia, pseudopods, radiolarian, regeneration, rotifer, sessile, spicules, spongin, zooplankton

Fill In

Use one of the vocabulary terms listed above to complete each sentence.

1. Sponges often have a rigid structure due to their _____.

2. An animal's early, free-living stage is called a _____.

3. Animals that strain tiny food from the water are _____.

4. Water is pumped out of a protozoan by its _____.

5. Bottom-dwelling organisms are referred to as _____.

Think and Write

Use the information in this chapter to respond to these items.

6. Compare the structure and function of the *Vorticella's* cilia and a sponge's flagellated collar cells. How are they similar?

7. Explain the difference between temporary and permanent zooplankton.

8. What is the main difference between binary fission and budding? For each method, name an organism that uses it.

Inquiry

Base your answers to questions 9 through 12 on the information below and on your knowledge of marine science.

A student performed an experiment to test the effect of salinity variations on the contraction rate of the stalked marine ciliate *Vorticella*, measured in contractions per minute. All factors except salinity (such as water temperature, light, and so on) were kept constant for the control and experimental groups. The control group had a salinity of 3.5 percent (like that of seawater); the experimental groups had different salinities. Results are shown in the table below.

THE EFFECT OF SALINITY VARIATIONS ON THE CONTRACTION RATE OF THE *VORTICELLA* (CONTRACTIONS PER MINUTE)

Number of Readings (Minutes)	3.5% Salinity (Control Group)	1.75% Salinity (Experimental Group 1)	0.87% Salinity (Experimental Group 2)
1	3	7	8
2	5	5	9
3	5	6	6
4	4	7	7
5	4	8	8
6	5	5	8
Average	4.5	6.3	7.6

9. What activity was measured in this experiment?   *a.* salinity   *b.* speed of the *Vorticella*   *c.* feeding rate of the *Vorticella*   *d.* contraction rate of the *Vorticella*

10. At which salinity were the *Vorticella* the least active?   *a.* 5 percent   *b.* 3.5 percent   *c.* 1.75 percent   *d.* 0.87 percent

11. Which is an accurate statement regarding the data in the table?   *a.* The average rate of the control group was higher than the average rate of either experimental group.   *b.* The average rate of each experimental group was higher than the average rate of the control group.   *c.* The highest contraction rate occurred at a salinity of 1.75 percent.   *d.* The lowest contraction rate occurred at a salinity of 1.75 percent.

12. What conclusion can be drawn based on the data in the table?   *a.* As salinity increases, the contraction rate also increases.   *b.* The contraction rate is not affected by a change in salinity.   *c.* As salinity decreases from 3.5 to 0.87 percent, the contraction rate increases.   *d.* The contraction rate will decrease if the salinity decreases below 0.87 percent.

Multiple Choice

Choose the response that best completes the sentence.

13. The ameboid movement of protozoans is carried out by their
    a. spicules *b.* cilia *c.* pseudopods *d.* flagella.

14. The sponge is considered a "simple" animal because it
    a. lives on the bottom of the ocean *b.* lacks a high degree
    of specialization *c.* filter feeds *d.* is generally small.

15. The group of marine protozoans that lacks microscopic hairs
    is the *a.* ciliates *b.* flagellates *c.* amebas *d.* sponges.

16. A life function carried out by protozoans but not by sponges is
    a. ingestion *b.* respiration *c.* locomotion *d.* excretion.

17. A good place to find sponges locally is *a.* floating on the
    ocean surface *b.* under the sand *c.* in the upper intertidal
    zone *d.* attached to the bottom of a wharf piling.

18. A marine biology student looking for protozoans would most
    likely find them in *a.* gravel from a marine aquarium
    b. a water sample from the ocean surface *c.* a dried sponge
    d. a water sample from the surface of a marine aquarium.

19. A marine biology student added a drop of dilute HCl to a
    sponge and observed effervescence. The student should con-
    clude that *a.* the internal skeleton is composed of calcium
    carbonate *b.* the spicules contain silica *c.* spongin was
    present *d.* the sponge prefers an acid environment.

20. The inability of a sponge to make rapid, reflex movements is
    due to its lack of *a.* contractile tissue *b.* a nervous system
    c. a mechanism for respiration *d.* a method of ingestion.

21. Protozoans are classified into three groups based on their
    being either ameboid, ciliated, or *a.* larval *b.* sessile
    c. flagellated *d.* planktonic.

22. Cilia and flagella are mainly used by protozoans for
    a. locomotion *b.* reproduction *c.* digestion *d.* fission.

23. Water and wastes exit a sponge's body through its
    a. ostia *b.* amebocytes *c.* osculum *d.* myocytes.

Research/Activity

- Compare one of the body systems of a sponge or protozoan with that of a human. Describe how they are similar and/or different.

- Prepare a wet mount of marine aquarium gravel to observe living protozoans. Identify and draw the different types; then do library or Internet research to learn more about the species you observed. Write a report that you can present to your class or your school science fair.

UNIT 3
MARINE INVERTEBRATES

If you take a walk along the shore at low tide, you may observe mussels attached to rocks, crabs scurrying about in shallow water, or snails moving slowly over rocks, grazing on the algae-coated surfaces. If you dig in the sand, you may unearth some sandworms. When you look into a tide pool, you may find colorful sea anemones with tentacles outstretched, waiting to catch a meal. What do all these animals have in common? All of them are invertebrates—that is, they have no backbones.

Marine invertebrates are divided into several phylums. Your study of marine invertebrates actually began in Chapter 6, with an examination of the rotifer, sponge, and bryozoan. In this unit, you will learn more about invertebrates and the adaptations they have that enable them to survive in the marine environment.

171

7 Cnidarians

When you have completed this chapter, you should be able to:

DISCUSS the characteristics and life cycle of the jellyfish.

COMPARE and CONTRAST the corals and sea anemones.

EXPLAIN how coral reefs are built up over time.

DESCRIBE some important features of the hydroids.

7.1
Jellyfish

7.2
Sea Anemones

7.3
Corals

7.4
Hydroids

A swimming race in the ocean is canceled when the competitors complain of painful stings; a boat sinks into the waters off the Florida Keys when its hull is ripped open by an underwater hazard. What do these two mishaps have in common? The swimmers in the race were stung by jellyfish; and the ship sank after it hit a coral reef. Both incidents were caused by animals that have stinging tentacles.

Animals with stinging tentacles are classified in the phylum Cnidaria. Besides jellyfish and corals, the phylum Cnidaria includes sea anemones (shown above), hydras, and other similar animals. There are more than 9000 species in this phylum, divided into three classes—Scyphozoa, Anthozoa, and Hydrozoa. In this chapter, you will find out how these unusual animals carry out their life functions.

7.1 JELLYFISH

Jellyfish are typical members of the phylum Cnidaria. Members of this phylum, called **cnidarians**, are characterized by two cell layers, a saclike digestive tract, and tentacles. In addition, cnidarians' tentacles, which are long, flexible appendages, are arranged in a ring around a central mouth, thereby giving these animals **radial symmetry**. Cnidarians have no brain, so there is no central coordination of movement. Instead, a network of nerve cells and receptor cells make up the **nerve net**, which is a simple nervous system; when one part of the body is stimulated, the whole animal responds. (See Figure 7-1.)

The jellyfish is not, in fact, a fish. Jellyfish differ from true fish in many ways; the most fundamental difference is that the jellyfish, as an invertebrate, does not have a backbone. Now you can understand the meaning of the expression "You're as spineless as a jellyfish."

Jellyfish are members of the class Scyphozoa. All members of this class have an umbrella-shaped structure called the **medusa**, with tentacles hanging down from it. The medusa is composed of two membranes: an epidermis, or outer membrane, and a gastrodermis (meaning "stomach skin"), or inner membrane. Lying between the two membranes is a jellylike mass called the **mesoglea**. The shape and thinness of the membranes, and the low density of

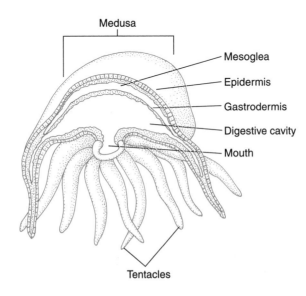

Medusa

Mesoglea

Epidermis

Gastrodermis

Digestive cavity

Mouth

Tentacles

Figure 7-1 Structure of the jellyfish, a typical cnidarian.

Figure 7-2 A jellyfish stranded on the shore by waves.

the layer of mesoglea between them, enable a jellyfish to float with ease. You may recall that jellyfish are considered part of the plankton population. However, they do have limited abilities of locomotion. Jellyfish use muscles to contract their medusa in a rhythmic fashion, causing them to pulsate gently through the water. Since jellyfish are such weak swimmers, they are often deposited on sandy beaches by waves and tides. (See Figure 7-2.)

How does the jellyfish carry out respiration? Since the membranes of a jellyfish are so thin, oxygen diffuses directly from the water into the animal's cells. The waste gas, carbon dioxide, diffuses in the opposite direction, from the cells into the water. This gas exchange occurs over the entire surface of the jellyfish.

Feeding in the Jellyfish

"Are there any jellyfish in the water?" Bathers will often ask a lifeguard this question before diving into the ocean. Swimmers are concerned because some species of jellyfish, such as the lion's mane jellyfish *(Cyanea),* can inflict painful stings. The sting of some jellyfish, such as the sea wasp, has even been known to cause death. Jellyfish use their stinging tentacles for defense and for getting food. The tentacles contain stinging cells, called **cnidoblasts**. Inside each cnidoblast there is a coiled thread with a barb at the end, called a **nematocyst**. (See Figure 7-3.) In most species, the barb contains a paralyzing toxin.

The nematocysts can be discharged in response to either a mechanical or chemical stimulus. Just touching a tentacle (an example of a mechanical stimulus) is sufficient to cause a nematocyst to

Figure 7-3 Cnidoblasts with the nematocysts coiled and discharged.

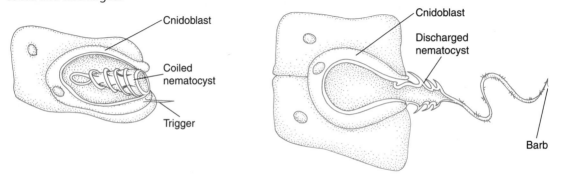

discharge. (Note the "trigger" in the cnidoblast.) However, once a nematocyst is discharged, it cannot be used again. New nematocysts are constantly produced to replace discharged ones. When the nematocysts are discharged into prey, such as a fish, they paralyze it. Contracting tentacles bring the fish up to the mouth, located in the center of the medusa, where it is ingested (taken in). Most of the food is digested in the saclike digestive cavity by enzymes secreted by its lining. Digestion in the jellyfish is both extracellular (in the digestive sac) and intracellular (in food vacuoles). Waste products are egested, or expelled, through the mouth.

The jellyfish can also catch food with its medusa. The surface of its medusa produces a sticky mucus. As a jellyfish moves through the water, plankton get stuck on this surface. Tracts of ciliated cells move the food from the medusa to the mouth, where it is ingested.

Life Cycle of the Jellyfish

How do jellyfish reproduce? Jellyfish have separate sexes, and their life cycle includes both sexual and asexual reproduction phases. You can see the stages in the life cycle of the moon jelly, *Aurelia*, illustrated in Figure 7-4 on page 176. *Aurelia* is a saucer-shaped jellyfish that has a weak sting; it often washes up on the beach during the summer. In the *Aurelia,* as in many other jellyfish, the ovaries and testes are organized into a four-leaf-clover pattern in the medusa. During the sexual phase, the testes produce sperm, which swim out of the male's mouth and into the female's mouth and digestive sac. The sperm fertilize the eggs in the ovary. Some species of jellyfish release eggs and sperm directly into the water, where external fertilization then occurs.

The fertilized egg cell, or *zygote,* passes through a series of developmental stages before becoming an adult. As in sponges and other animals, organisms in the early stages of development are called *embryos.* The first embryonic stage occurs when the zygote divides in two. These cells continue to divide until the solid ball of cells called a *morula* is produced. Following the morula stage, the hollow ball of cells called a *blastula* is formed. (Refer to Figure 6-6.)

Shortly thereafter, the surface cells of the growing blastula become ciliated, forming a swimming larva called the **planula**. In some cases, the planula larva attaches to a hard substrate and then

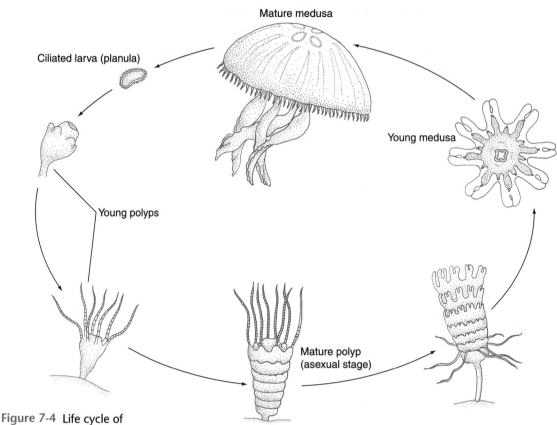

Figure 7-4 Life cycle of the moon jelly *Aurelia*.

develops into a structure called a polyp. (In other cases, there is no polyp stage.) The **polyp** has a mouth, tentacles, and digestive cavity, and its shape resembles an upside-down jellyfish medusa. Immature jellyfish, in the form of small umbrella-shaped medusas, then develop and break off from the polyp by budding, which is the asexual phase in jellyfish reproduction. The life cycle is completed when each medusa develops into an adult jellyfish.

Comb Jellies

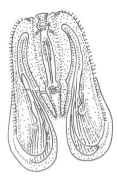

Figure 7-5 The comb jelly is a ctenophore (not a cnidarian).

An animal that is often mistaken for a jellyfish, but is not a cnidarian, is the *comb jelly*. The comb jellies are classified in their own phylum, Ctenophora. They all have eight rows of long, fused cilia that look like combs, hence the group's name. (See Figure 7-5.) The beating of the cilia helps move the animal through the water.

ENVIRONMENT
Floating Beauties, Stinging Beasts

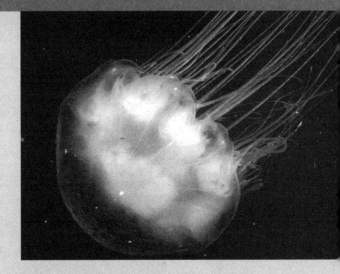

Have you ever seen, or been stung by, a jellyfish? Beware, jellyfish are found everywhere-from the Arctic to the Antarctic seas, and from the ocean surface down to the great depths. These largely transparent, often beautiful creatures look very delicate and vulnerable as they gently pulsate through the water. But don't be fooled. Jellyfish are among the invertebrates that possess deadly stinging cells for feeding and for protection. Most marine animals have learned to steer clear of jellyfish, with the exception of the loggerhead sea turtle, which regularly preys on them. Humans, too, have learned the hard way just how dangerous the "jellies" can be for people who enter the marine environment. There are more than 20,000 species of jellyfish, but only about 70 are potentially harmful to humans. Which jellies are the most dangerous? Marine biologists agree that, of the dozens of jellyfish species that can deliver a painful and toxic sting, the following three top the "most dangerous jellyfish" list. (Being largely transparent, jellies are difficult to spot in the water. Check on jellyfish conditions before swimming. If you do get stung, wash your skin with vinegar or rubbing alcohol and immediately notify a lifeguard.)

Lion's Mane Jellyfish
(*Cyanea capillata*)

Description: Largest of all the jellyfish, grows up to 2 meters wide (the medusa) and 40 meters long (the tentacles).

Danger: The sting causes painful reactions; it can be deadly.

Location: Atlantic Ocean, from the Arctic Circle to Florida and the Gulf of Mexico; Pacific Ocean, from Alaska to Southern California.

Note: See photograph above.

Sea Nettle
(*Chrysara quinquecirrha*)

Description: Grows up to 1 meter wide and 4 meters long; has reddish-maroon markings.

Danger: The sting causes painful reactions; it can be deadly.

Location: Atlantic Ocean, from Massachusetts to Florida and the Gulf of Mexico; Pacific Ocean, from Alaska to Southern California. It is particularly common in Chesapeake Bay, Virginia.

Sea Wasp [Box Jelly]
(*Chironex fleckeri*)

Description: Very transparent, boxlike medusa, can grow to size of a basketball; has approximately 60 tentacles, up to 5 meters long.

Danger: The sting causes painful reactions; most dangerous of all jellies, it can be deadly. Kills dozens of swimmers in Australia every year. Its toxins attack cardiac, respiratory, and nervous systems.

Location: Pacific Ocean, near Australia and the Philippines.

QUESTIONS

1. Which jellyfish is considered the most dangerous? Why?
2. How do the stinging cells of a jellyfish aid their survival?
3. Why are most victims unaware of the presence of jellyfish?

Cnidarians **177**

Along with short tentacles, the cilia also bring food, such as cope-pods, to the mouth. A nerve net controls movement of the cilia. Most species also have long tentacles to help capture food. One species of ctenophore has nematocysts in its tentacles. Comb jellies usually float on or near the surface of the water. They are bioluminescent, and light up when disturbed. Comb jellies release their gametes into the water for external fertilization.

7.1 SECTION REVIEW

1. Explain how the jellyfish is adapted for floating.
2. How do jellyfish capture and digest their prey?
3. Describe how a typical jellyfish reproduces.

7.2 SEA ANEMONES

Sea anemones, which look like colorful underwater flowers, are members of the class Anthozoa (meaning "flower animal"). How are they related to the jellyfish? Like jellyfish, **sea anemones** possess stinging tentacles, radial symmetry, and a nerve net. Unlike jellyfish, the adult sea anemone lives as a polyp, attached to a substrate by means of a muscular foot. (See Figure 7-6.) As such, the sea anemone is a stationary, or sessile, animal capable of only limited movement. How, then, does the sea anemone respond to its environment? The sea anemone has a very simple nervous system; if you touch part of the sea anemone, the entire animal contracts. For example, if you touch a tentacle, the anemone quickly withdraws all tentacles into its fleshy polyp.

Figure 7-6 The sea anemone is a sessile cnidarian; the adult lives as a polyp.

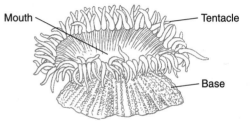

Mouth

Tentacle

Base

Feeding in the Sea Anemone

How does the sea anemone capture its food? Typical of the cnidarians, the sea anemone uses its stinging tentacles to get food. Tiny organisms and small fish that swim into the tentacles are paralyzed by the discharging nematocysts. You can observe discharging nematocysts in a sea anemone by doing the laboratory investigation at the end of this chapter. One type of sea anemone *(Metridium)* captures its food as jellyfish do, by trapping it in the mucus and cilia on its surface.

Interestingly, some large sea anemones have little shrimp and clownfish that live unharmed among their tentacles. The clownfish have a protective skin coating, so they are not injured by the stinging tentacles. By living in the midst of the tentacles, clownfish gain protection from predatory fish; at the same time, they protect the anemone's tentacles from being bitten by other fish. The shrimp help keep the anemone clean and may also help protect it from predators. This kind of mutually beneficial relationship between different species is known as **symbiosis**. The participants are referred to as **symbionts**. (See Chapter 21 for more on symbiosis.)

After an anemone captures its prey, it brings the food to its mouth. Food is digested inside the digestive sac by enzymes. As in the jellyfish, digestion in the sea anemone is mostly extracellular. Vacuoles in specialized cells in the digestive tract's lining engulf and digest food particles, so there is intracellular digestion as well. Undigested food particles and other wastes are eliminated through the mouth. Because of the saclike structure of the digestive cavity, both ingestion and elimination occur through the mouth. Therefore, digestion in sea anemones and jellyfish occurs in what is called a two-way digestive tract.

Life Cycle of the Sea Anemone

The pattern of sexual reproduction in the sea anemone is very similar to that of the jellyfish. However, the medusa stage, typical of the jellyfish, is not present in the sea anemone life cycle. The dominant structure in the sea anemone is the polyp. Some sea anemones can reproduce asexually by splitting in half, or they can regenerate from pieces of the polyp's base that are broken off as an adult anemone moves along the substrate. They can also reproduce sexually. Anemones produce embryos that develop into planula larvae, which settle and develop into adult polyps.

7.2 SECTION REVIEW

1. How does the adult body form of a sea anemone differ from that of a jellyfish?
2. Describe how the sea anemone obtains its food.
3. How does the sea anemone respond to stimuli?

7.3 CORALS

One of the most spectacular structures in nature is a coral reef. Some are huge, like the Great Barrier Reef off the coast of Australia, which is about 2000 km long and 80 km wide. However, the organism responsible for the formation of reefs is so small that it might escape notice outside of its stony home. In fact, the coral animal is so tiny that some can be seen only through a microscope. Yet, different species of coral animals are capable of building a variety of structures that range in size from small and gemlike to the massive reefs that can pose a danger to ships. Corals come in a variety of shapes and sizes, often resembling familiar objects for which they are named.

Coral Types and Structure

There are two types of corals: stony (hard) corals and soft corals. The stony corals are made up of limestone (calcium carbonate), and they can form massive stony structures. Examples of hard corals are the brain coral, staghorn coral, and star coral. The soft corals are composed of a fibrous protein, which gives them flexibility. Underwater, the soft corals look more like plants than animals, as they sway back and forth with the waves and currents. Examples of soft corals

Figure 7-7 A variety of hard corals and soft corals can be found together on a reef.

are the sea fan, sea whip, and sea plume. (See Figure 7-7, and refer to Figures 3-19 and 3-20.)

The basic structure of the coral animal is the **coral polyp**, shown in Figure 7-8 on page 182. As you can see, the coral polyp resembles a very small sea anemone. Life activities such as ingestion, digestion, sensitivity, exchange of gases, and reproduction are similar in both anemones and coral polyps. Due to their similarities in structure and function, corals are placed in the class Anthozoa along with the sea anemones. However, unlike sea anemones, which live alone or in small groups, coral polyps live in large groups as **colonial animals**, attached to one another by a thin membrane. The membrane connects the polyps' digestive systems, so there is nutritional sharing among them.

The Coral Polyp

The coral polyp is a tiny mound of tissue that contains a saclike digestive tract and a mouth surrounded by stinging tentacles. In hard corals, each polyp sits in a cup-shaped depression that it forms on the surface of the reef. The stony depression is composed of limestone, which is added to—layer by layer over many years—by the polyps that live on the surface.

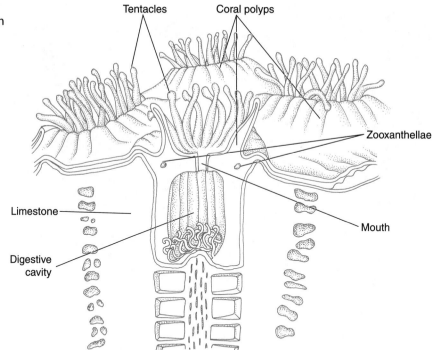

Figure 7-8 Structure of the coral polyp, shown in cross section.

Tentacles

Coral polyps

Zooxanthellae

Limestone

Mouth

Digestive cavity

How does a coral polyp build a reef of stone? The polyp needs a great deal of energy to build a stony reef. It gets its energy from the food that it eats (tiny plankton trapped by its tentacles at night) and from the food that is made during photosynthesis. The photosynthesis is actually carried out by tiny symbiotic algae called **zooxanthellae**, which live inside the tissue of each polyp. The zooxanthellae are types of dinoflagellates that are captured by the coral polyp from the marine environment. As in all algae, the zooxanthellae contain chloroplasts, which use the energy of sunlight to make energy-rich sugar (glucose) from other simple compounds. The coral animals use the glucose to supplement the energy they get from captured food and to build their limestone homes.

The zooxanthellae benefit by living safely within the coral polyps' cells and receiving nutrients from the polyps. The polyps also benefit by receiving oxygen as a by-product of the algae's photosynthesis. This is another example of symbiosis in marine organisms. The life activities of the coral polyps and the zooxanthellae benefit one another.

Coral Reefs

As mentioned in Chapter 3, a reef forms when coral larvae settle on a substrate in the sand and develop into polyps. What kind of building materials do the coral polyps need to construct the reef? A coral reef is a massive limestone structure composed of the compound calcium carbonate ($CaCO_3$), the same material that makes up mollusk shells. To build the reef, the coral polyp needs a source of calcium and carbon. Cells within the polyp absorb calcium (Ca) from the seawater; and the zooxanthellae take up carbon dioxide (CO_2) that is produced by the polyp during respiration. When these two substances combine inside the tissue of the polyp, the $CaCO_3$ needed to build the limestone reef is produced.

Interestingly, the process of reef building is similar in principle to the production of bones and teeth in your own body. Teeth and bones are composed of the mineral calcium. Special cells in your skeletal system remove calcium from the blood to build your teeth and bones.

Coral reefs are found in tropical areas, where the waters are warm and clear. Since the zooxanthellae require energy from sunlight, the waters that corals live in must be clear and free from sediments. And coral polyps need warm temperatures in order to secrete their calcium carbonate skeletons. Each new generation of polyps lays down another layer of stone as the reef grows upward at a rate of about 2.5 cm per year. The reefs provide homes and habitats for many other animal species. Large reefs also protect nearby coastlines from the harsh effects of ocean waves.

At low tide, the top of the reef may lie just beneath the water's surface. Ships that come too close may hit a reef and damage their hulls. As a result, coasts with coral reefs are graveyards for many ships. Likewise, many reefs are damaged when ships run aground on them. A large ship, or its anchor, can break off chunks of living coral and damage the protective mucus coating on the coral polyps. Coral reefs are also damaged by water pollutants, severe storms, freshwater runoff, sediments, and unusually low or high water temperatures (which can disturb the algae living within the polyps). When this happens, a condition known as "coral bleaching" occurs, in which the living polyps die off, and only the white coral

Figure 7-9 An example of coral bleaching: the polyps have died, leaving behind the stony white skeleton.

skeletons remain. This problem has been noticed in several places around the world. (See Figure 7-9.)

7.3 SECTION REVIEW

1. Why are coral reefs found only in clear tropical waters?
2. Compare the coral polyp with a sea anemone. What are the similarities and differences?
3. Why is the coral polyp considered a colonial animal?

7.4 HYDROIDS

Another type of cnidarian that may resemble a plant is the **hydroid**. Hydroids, which are members of the class Hydrozoa, live in the intertidal and subtidal zones. Hydroids are actually colonial animals, made up of many individual polyps that function together as a single organism. The snail fur (*Hydractinia*) is a hydroid colony whose pink and reddish growth coats the shells of many kinds of ocean animals, including hermit crabs, There are some exceptions to the rule; for example, the hydroid *Tubularia* is not colonial.

One common hydroid, *Obelia*, forms bushy colonies attached to rocks, seaweeds, and other substrates in the intertidal zone. *Obelia* is made up of two types of polyps: feeding polyps (part of the sessile asexual phase) that have nematocysts, and reproductive polyps that form the brief medusa (free-swimming sexual) phase. (See Figure 7-10.) The life cycle of *Obelia* is similar to that of both jellyfish and sea anemones. Like the jellyfish, hydroids have a medusa and a planula larva. However, like the sea anemone (which has no medusa phase), the hydroid's dominant phase is the asexual polyp.

Portuguese Man-of-War

An interesting and much-feared hydroid colony is the Portuguese man-of-war (*Physalia*), so named by British sailors in the eighteenth century as an insult to the ships of the Portuguese navy. *Physalia* is classified with the hydrozoans, not with the true jellyfish which it resembles, because it is a colony made up of different types of

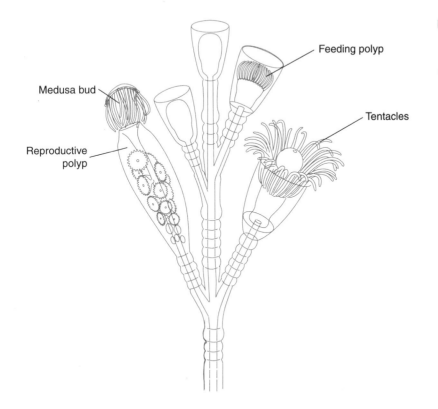

Figure 7-10 Structure of *Obelia*, a colonial hydroid; its dominant phase is the sessile asexual polyp.

Medusa bud

Feeding polyp

Reproductive polyp

Tentacles

Figure 7-11 The Portuguese man-of-war, *Physalia*, a colonial hydroid.

polyps. Like other cnidarians, it has batteries of nematocysts on its tentacles. The sting of the Portuguese man-of-war is very painful. People become ill from the sting; some even die.

The different polyps that make up *Physalia* are adapted for a variety of functions. One kind of polyp makes up the gas-filled bag that keeps the colony afloat. This gas-filled float, composed of many individual buoyant polyps, resembles a sail. Winds push against the sail, thus moving the colony along in the water. (*Physalia* does not actively chase its food.) Another kind of polyp makes up the stinging tentacles, which are several meters long and hang like a deadly curtain beneath the float. Other polyps digest food that is caught and killed in the tentacles. And, as in other hydroids, some polyps serve a reproductive function. All in all, the Portuguese man-of-war is quite a remarkable organism—yet one that you would not want to meet up close and personal! (See Figure 7-11.)

7.4 SECTION REVIEW

1. Why are hydroids referred to as colonial animals?
2. List the different functions of hydroid polyps.
3. Why is the Portuguese man-of-war classified as a hydroid?

Laboratory Investigation 7

Observing Stinging Tentacles

PROBLEM: How can you observe the discharging nematocysts of a stinging tentacle?

SKILL: Using the compound microscope to observe tiny cell structures in action.

MATERIALS: Tiny sea anemone or jellyfish tentacles, forceps, slides, coverslip, compound microscope, absorbent paper, medicine dropper, dilute acetic acid (or clear vinegar).

PROCEDURE

1. Cut a small piece of a fresh jellyfish tentacle or, using your forceps, remove a tentacle from a sea anemone. (*CAUTION:* Do not touch the tentacles with your hands!)

2. Place the tentacle on a glass slide. Put 1 or 2 drops of seawater on the tentacle. Carefully apply the coverslip.

3. Place the wet mount slide on the microscope stage under the low-power objective.

4. Look for the tentacle under low power. Focus along the edge of the tentacle. In your notebook, draw a section of the tentacle.

5. Now place a piece of absorbent paper on the glass slide to the left of the coverslip. Fill the medicine dropper with clear vinegar or dilute acetic acid.

6. While looking through the microscope at the edge of the tentacle, squeeze 2 or 3 drops of acid on the slide so that the acid makes contact with the right side of the coverslip. Then place the absorbent paper in contact with the left

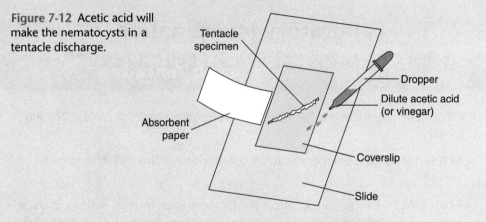

Figure 7-12 Acetic acid will make the nematocysts in a tentacle discharge.

Tentacle specimen

Dropper

Dilute acetic acid (or vinegar)

Absorbent paper

Coverslip

Slide

side of the coverslip. (See Figure 7-12.) The paper will absorb the seawater and draw the acid into contact with the tentacle.

7. Look for many fine threads shooting out from the tentacle. These threads are the discharging nematocysts. Sketch your observation. (Refer to Figure 7-3 on page 174.)

8. Examine the nematocyst under high power. Draw and label the structure.

OBSERVATIONS AND ANALYSES

1. What is the adaptive value of stinging tentacles?

2. Why do you think the nematocysts reacted to the acetic acid?

3. Why do some fish avoid swimming near sea anemones?

Answer the following questions on a separate sheet of paper.

Vocabulary

The following list contains all the boldface terms in this chapter.

cnidarians, cnidoblasts, colonial animals, coral polyp, hydroid, medusa, mesoglea, nematocyst, nerve net, planula, polyp, radial symmetry, sea anemones, symbionts, symbiosis, zooxanthellae

Fill In

Use one of the vocabulary terms listed above to complete each sentence.

1. A jellyfish's umbrella-shaped structure is the _____.
2. The stinging cells inside tentacles are called _____.
3. An adult sea anemone lives as an attached _____.
4. The name for the algae that live inside coral polyps is

   _____.
5. The Portuguese man-of-war is a dangerous colonial _____.

Think and Write

Use the information in this chapter to respond to these items.

6. Diagram the life cycle and reproduction of a jellyfish.
7. Describe the main difference between a coral colony and a hydroid colony.
8. Name two similarities and two differences between corals and sea anemones.

Inquiry

Base your answers to questions 9 through 11 on the following diagram, which illustrates the life cycle of the moon jelly Aurelia, *and on your knowledge of marine biology.*

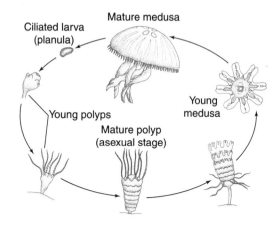

9. What kind of reproduction is illustrated in this life cycle?
   a. sexual only   *b.* asexual only   *c.* both asexual and sexual
   d. parthenogenesis

10. Which statement is *not* correct regarding the stages of this life cycle?   *a.* The polyp stage lives attached to a substrate.
   b. The medusa drifts among other plankton.   *c.* All stages of development are free-swimming.   *d.* The free-swimming and sessile stages alternate.

11. During which two stages does embryonic development take place?   *a.* young medusa to mature medusa   *b.* ciliated larva to young polyp   *c.* young polyp to mature polyp
   d. mature polyp to young medusa

Multiple Choice

Choose the response that best completes the sentence or answers the question.

12. Inside each stinging cell of an anemone is a coiled thread with a barb at the end, called a   *a.* cnidoblast   *b.* cilia
   c. nematocyst   *d.* zooxanthellae.

13. A jellyfish responds to external stimuli through use of its
    a. nematocysts   *b.* mesoglea   *c.* nerve net
    d. zooxanthellae.

14. Corals and jellyfish are classified together mainly because
    they   *a.* share the same habitat   *b.* have stinging tentacles
    c. are tropical   *d.* are planktonic.

15. The large reef-building corals grow only in tropical waters
    because of the warm temperatures and   *a.* plentiful sunlight
    b. room for more growth   *c.* greater water pressure
    d. fewer predators.

16. Which of the following organisms with tentacles is *least*
    related to the others?   *a.* sea anemone   *b.* brain coral
    c. octopus   *d.* Portuguese man-of-war

17. All of the following are functions of the jellyfish medusa
    except   *a.* reproduction   *b.* locomotion   *c.* protection
    d. ingestion.

18. The cnidarians are characterized by the presence of stinging
    tentacles and a   *a.* streamlined shape   *b.* saclike digestive
    tract   *c.* tube-shaped digestive tract   *d.* ventral nerve cord.

19. The basic structure of the coral animal is the   *a.* polyp
    b. medusa   *c.* nematocyst   *d.* cnidoblast.

20. The beneficial living arrangement between the coral polyps
    and the tiny algae within them is known as   *a.* radial
    symmetry   *b.* symbiosis   *c.* zooxanthellae   *d.* cnidarian.

21. Which of the following is planktonic in its adult stage?
    a. sea anemone   *b.* elkhorn coral   *c.* hydroid   *d.* jellyfish

22. In the cnidarians, oxygen and carbon dioxide gas exchange
    occurs through the   *a.* gills   *b.* lungs   *c.* mouth   *d.* cell
    membranes.

23. What is the main function
    of the structures labeled *A*
    in the animal shown here?
    a. reproduction
    b. food-getting
    c. locomotion
    d. respiration

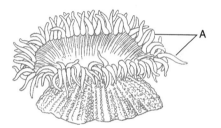

24. The cnidarians have tentacles that grow in a ring around a central mouth, an arrangement known as   *a.* symbiotic   *b.* planktonic   *c.* colonial   *d.* radial symmetry.

Research/Activity

- Do a report on a species of jellyfish that lives in your area. If possible, photograph jellyfish that have washed up on your beach or that can be seen floating near a pier. If you live inland, try to report on a freshwater cnidarian, such as the *Hydra*. You can do additional research in the library or on the Internet.

- The Portuguese man-of-war is really many different organisms that function together as one. Report on how *Physalia* accomplishes this task. Include a labeled drawing.

- Participate in a NOAA-sponsored jellyfish sighting survey by going on line with Sea Nettle Homepage.

8 Marine Worms

When you have completed this chapter, you should be able to:

IDENTIFY the basic characteristics of the various worm phylums.

DESCRIBE the life functions of the different groups of worms.

DISCUSS some adaptations of worms to the marine environment.

In their general shape, worms tend to resemble one another. You are probably most familiar with earthworms—the soft, moist, slow-moving creatures that live in the soil. Most worms move through substrates by wriggling their bodies from side to side. Appendages are missing or greatly reduced in size. But not all worms are alike in how they look and behave.

The delicate-looking feather duster worms shown above actually live in stony tubes, which offer them some protection against other animals. The worms' colorful featherlike structures are thrust out of their tubes to catch plankton and to take in oxygen.

Worms vary in size from microscopic to over several meters in length. In fact, there are enough differences among the groups of worm species to justify their being placed into several phylums. In this chapter, you will learn about some similarities and some differences among the worms that live in marine environments.

8.1
Flatworms and Ribbon Worms

8.2
Roundworms and Segmented Worms

8.3
Giant Tube Worms and Arrow Worms

The vertical zone of water that extends from the top of the ocean to its bottom is called the **water column**. There are few places in the water column where worms cannot be found. They inhabit the seas from the surface down to the seafloor. There are significant differences among the various groups of worms that live at the different depths. Most of the worms you will learn about in this chapter live at the bottom of the water column, near or in the ocean's sediments. Some are found closer to the surface. Yet others are parasitic, living within the bodies of different marine organisms.

Flatworms

Look at the group of worms shown in Figure 8-1. As you can see, the bodies of these worms are flat. The flat-body form is the distinguishing trait used to classify these worms in the phylum Platyhelminthes (meaning "flatworms"). There are both freshwater and saltwater species of **flatworms**. Although some flatworms are microscopic, many others can be seen without a microscope. In fact, some species can reach nearly 20 meters in length! There are more than 10,000 species of flatworms.

Figure 8-1 The three types of flatworms.

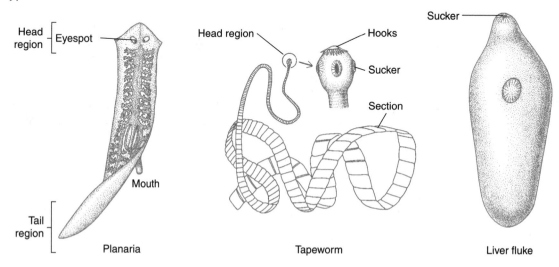

The Planarian

Look at a macroscopic flatworm, the **planarian** *(Planaria)*, shown in Figure 8-1. Planarians, which feed on small organisms and organic debris, are found in freshwater and marine habitats. These worms have a mouth on their lower, or ventral, surface through which they take in food as they move along rocks and other substrates. The mouth connects to a branched digestive cavity. Unlike other worms, planarians have a two-way digestive tract. Food enters the mouth and is digested in the intestine; the nutrients diffuse throughout the body (there are no circulatory or respiratory systems). The undigested materials are discharged from the mouth (there is no anus).

The planarian glides along the bottom, moving its head from side to side. Locomotion results from the contraction of body muscles and the action of cilia, which are attached to cells on the worm's ventral side. When the cilia beat backward, the planarian moves forward along the surface of a substrate. Planaria have two eyespots in their head region, which are surrounded by the **ganglia** (nerve cell clusters). The ganglia act like a simple brain, sending nerve impulses along two ventral nerve cords to the rest of the body. This lets the flatworm respond to stimuli (such as light) from its environment.

Since the structures on the right side of its body are the same as the structures on the left side of its body, the planarian is said to have **bilateral symmetry**. In fact, bilateral symmetry, which first appears in the flatworm, is a characteristic feature of all other worms and more structurally complex animals. This type of body plan is associated with the development of a head region, such as that seen in the flatworm. Another feature first seen in the flatworm is the possession of three cell layers: ectoderm, endoderm, and mesoderm (the middle layer). This is important for the development of organ systems, also first seen in the flatworms.

Planaria are capable of both asexual and sexual reproduction. During asexual reproduction, a planarian can attach to a substrate, stretch its body, and break in two. Each half can then regenerate the parts needed to form a whole new organism.

The planarian, like most flatworms, is a hermaphrodite—it contains both male and female reproductive organs. However, self-fertilization does not occur. During mating, two flatworms exchange sperm, so that sperm from each of the flatworms fertilizes the eggs of the other flatworm internally. Development of fertilized eggs occurs externally in tiny capsules or cocoons. Offspring hatch from the cocoons in two to three weeks.

In the same class as the planaria are other marine flatworms. These worms are also bilaterally symmetrical, although the head region is not obvious as in the planaria. They are characterized by very beautiful colors and patterns, and frilly edges that seem to flutter as they swim.

Tapeworms and Trematodes

Some flatworm species live as parasites. A **parasite** is an organism that obtains its food by living in or on the body of another organism. The organism that is fed on is called the **host**. The **tapeworm**, shown in Figure 8-1, is sometimes found in the intestines of fish and other animals, including humans. The tapeworm attaches itself to the intestinal lining of its host and absorbs nutrients directly through its thin body wall. As a result, it has no need for a digestive system. This feeding method works well for the tapeworm—some can grow to more than 18 meters in length!

Another flatworm that is a parasite is the **trematode**. The trematode, or fluke, lives in the bodies of mollusks, fish, birds, and other animals. The liver fluke, which lives as a parasite in mammals, is shown in Figure 8-1. The blood fluke is a parasite that attaches to a fish's skin and then forms a cyst in which it lives within the fish's tissues. Trematode parasites that are accidentally eaten in raw fish may then reproduce in the digestive tracts of people. Human wastes that are discharged into bodies of water may contain the trematode's eggs. These eggs develop into swimming larvae, which penetrate the soft tissue of their first host—a snail. The larvae develop inside the snail and, in time, are discharged into the water, where they seek out their second host—a fish or aquatic bird—thus completing their life cycle. If you happen to be in the water when the trematode larvae are looking for a host, they may attach to your skin, causing what is called swimmer's itch.

Ribbon Worms

You have just learned that some flatworms (the tapeworms) can grow nearly 20 meters in length. Interestingly, there are other types of worms that grow very large. The largest free-living worm in the sea is the **ribbon worm** *(Cerebratulus lacteus),* shown in Figure 8-2. The nearly 1000 species of ribbon worms are classified in the phylum Nemertea. Although some are relatively small, the average length of a ribbon worm, or nemertean, is about 1 meter. However, ribbon worms 10 to 12 meters long have been observed. All ribbon worms are unsegmented and ciliated, have a milky color, and are thin and flat like ribbons, hence their common name. Although they resemble flatworms in their general shape, ribbon worms have more "highly" developed circulatory and digestive systems.

Ribbon worms live in the intertidal zone. Most are free-living, although some live inside the shells of clams and oysters. Ribbon worms burrow in the sand and also swim with a gently undulating motion among animals that live encrusted on rocks. Prey such as small fish and polychaete worms (see Section 8.2) are speared by a sharp, sticky extension called a **proboscis**, which the ribbon worm extends from a recessed area in its head region. Food is digested in a one-way digestive tract (that has a stomach and intestine) and wastes exit through the anus. Ribbon worms have a simple, closed circulatory system for the distribution of nutrients and oxygen.

Ribbon worms can reproduce asexually by breaking into pieces; each piece is capable of regenerating into a whole new worm. They can also reproduce sexually, having external fertilization and development. Like many other worms, the ribbon worm has a nervous system that consists of ganglia (more concentrated than those of the flatworm), which send impulses to its muscles by way of two nerve cords.

Proboscis

Figure 8-2 The ribbon worm.

8.1 SECTION REVIEW

1. What two features are first seen in the planarians?

2. Describe asexual and sexual reproduction in the planarian.

3. How do the planarian's and the ribbon worm's digestive and circulatory systems differ?

8.2 ROUNDWORMS AND SEGMENTED WORMS

Roundworms

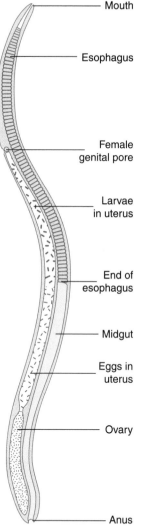

Mouth

Esophagus

Female genital pore

Larvae in uterus

End of esophagus

Midgut

Eggs in uterus

Ovary

Anus

Figure 8-3 A marine roundworm.

The most numerous of all worms in the sea—in numbers of individuals and of species—are the **roundworms**. There are more than 10,000 species of roundworms. Very tiny roundworms live in the sand and mud at the bottom of the water column. When the tide goes out, roundworms can be found in the moist substrate of the intertidal zone. Gravel samples taken from an aquarium tank and examined under a microscope will also reveal a number of roundworms. (See Figure 8-3.)

Typical whipping movements of the roundworm's body propel it through the spaces between sand grains. Some roundworms can also swim through the water. In the roundworm's head region are the ganglia, which connect to nerves that run through the body. These nerves control the muscles that enable the worm to move and actively respond.

Roundworms are classified in the phylum Nematoda; they are often referred to as **nematodes**. As their name implies, roundworms are characterized by a cylindrical body shape, which is tapered at both ends. Most roundworms are small, but some reach up to a meter in length. The sexes are usually separate (a few nematode species are hermaphrodites), and reproduction is sexual. Fertilization is internal; development of eggs is external. Males and females tend to differ in their size and shape.

Most marine nematodes live freely in bottom sediments and feed on organic debris. Like the ribbon worm, roundworms have a one-way digestive tract. Food enters the mouth, is digested in the stomach and intestine, and undigested wastes pass out through the anus, located at the opposite end. Like the flatworm, the roundworm has no circulatory or respiratory systems. Nutrients diffuse into its cells; respiratory gases and cellular wastes pass through its skin.

Segmented Worms

Of all the worms, the earthworm is probably the most familiar to you. Earthworms are typically found in soil and other moist

DISCOVERY

The Cells of *C. Elegans*

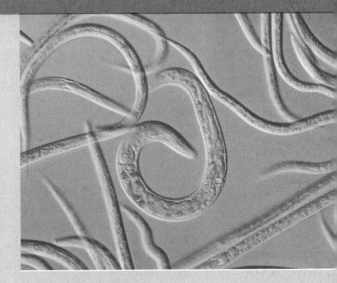

Dozens of scientists around the world are focusing their efforts, and their microscopes, on *Caenorhabditis elegans*, a transparent roundworm just 1 millimeter in length. This common worm—found scavenging on debris in moist sediments everywhere—offers scientists an uncommon chance to investigate biological processes such as cell division, embryology, and disease.

Many scientists choose to study *C. elegans* because the adult worm contains a total of only 959 cells—compared with millions of cells in other multicellular animals. The worm's transparent skin enables scientists to observe its muscle cells and nerve fibers at work. And it allows them to trace the pathway that the worm's cells take during cell division, from the fertilized egg cell to the 959 cells of the adult just three days later. That means that the researchers have observed differentiation, the process by which embryonic cells develop into adult muscle, nerve, and skin cells.

Scientists also found that some cells in the embryo are predestined to die rather than differentiate. It is not known why the cells die off during development. However, researchers think that the study of these cells may help them find out which genes control cell death. This, in turn, may lead to a better understanding of what happens during diseases such as Alzheimer's,

in which a person's nerve cells inexplicably die, resulting in severe memory loss.

One of the most spectacular achievements in the last few years was the deciphering of the complete genome of *C. elegans*—the first multicellular organism for which the entire DNA was mapped! Scientists discovered that *C. elegans* has more than 19,000 genes, which code for over 19,000 different proteins.

Deciphering an organism's genome is the first step in learning which genes are responsible for a cell's fate. It is hoped that, in the years to come, further research on *C. elegans* will continue to throw more light on our understanding of cell processes and diseases in humans.

QUESTIONS

1. State three advantages of using *C. elegans* over other animals for cell research.

2. How might studying *C. elegans* help us understand Alzheimer's disease in humans?

3. Why was deciphering of the roundworm's genome a significant achievement?

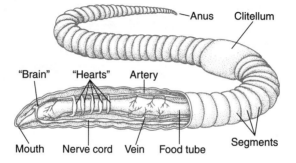

Figure 8-4 An earthworm (partly cutaway view).

sediments. Some of the internal and external features of an earthworm are shown in Figure 8-4. Notice that the worm's body is divided into compartments, or segments. Referred to as the **segmented worms**, or **annelids**, earthworms and all other annelids are classified in the phylum Annelida (meaning "little rings"). There are three classes of annelids, comprising more than 10,000 species. Earthworms, and their aquatic relatives that feed on organic matter in sediments, belong to the class Oligochaeta (meaning "few bristles").

The Sandworm

A common marine annelid is the clamworm, or **sandworm** *(Nereis).* (See Figure 8-5.) Sandworms live in muddy sands in the intertidal and subtidal zones of inland bays and marshes. When the tide comes in, sandworms crawl around on the bottom and prey on tiny invertebrates. When the tide moves out, sandworms burrow in the sand and scavenge on organic debris. Some live within tubes.

How do sandworms carry out their life functions? *Nereis* captures its prey with two sharp hooks located in its mouth. The sandworm also has a proboscis, which it can extend to catch prey or bits of seaweed. Food is digested in a one-way digestive tube; waste products of digestion are eliminated through the anus. The digestive tract is separated from the skin by a fluid-filled space called the **coelom**. (All annelids have a coelom.)

Sandworms belong to the class Polychaeta (meaning "many bristles"). They wriggle through the wet sand by using their paddle-like appendages called **parapodia** (singular, parapodium). The parapodia are located on each segment. Each parapodium also has

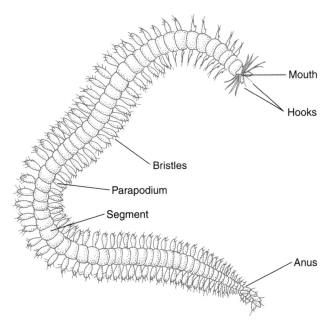

Figure 8-5 The sandworm *Nereis*.

Mouth

Hooks

Bristles

Parapodium

Segment

Anus

hairlike bristles called **setae** sticking out from it. Movement results when longitudinal and circular muscles located under the worm's skin are contracted.

The sandworm has a nervous system to coordinate the movements of its body. Eyelike receptors that receive light stimuli, and touch receptors in the skin that pick up other external stimuli, send impulses to a ventral nerve cord. The impulses travel to the brain-like ganglia. Movement occurs when the "brain" sends the impulses via nerve cords to muscles, causing them to contract.

Like earthworms, sandworms breathe through their skin—in this case, through their parapodia. Because the worm's skin is thin and moist, oxygen diffuses from the surrounding water into the body, and carbon dioxide diffuses from inside the body to the outer environment. Nutrients and respiratory gases are transported around the body in a closed circulatory system. This consists of a dorsal blood vessel, arteries, veins, and capillaries. The dorsal blood vessel contracts and pumps the blood. Arteries and veins carry the blood back and forth; capillaries connect arteries with veins.

Sandworms have a well-developed excretory system. Waste products of metabolism are excreted from a pair of coiled tubes located in each body segment. These organs, which carry out excretion for the annelids as our kidneys do for us, are called the

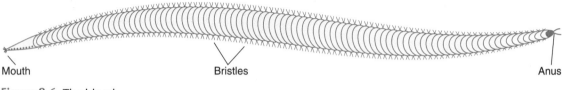

Mouth   Bristles   Anus

Figure 8-6 The blood-worm.

nephridia. The sandworm reproduces sexually; the sexes are separate (unlike the earthworm, which is a hermaphrodite).

The Bloodworm and Other Polychaetes

Another segmented worm found in marine sediments is the **bloodworm** *(Glycera)*. (See Figure 8-6.) Like the sandworm, the bloodworm is a polychaete. However, unlike the sandworm, the bloodworm has an open circulatory system in which the blood circulates through the tissue spaces rather than through blood vessels. You can see a bloodworm's blood through its skin, hence its common name. Polychaete worms also include a variety of other "bristle worms," such as the plumeworm and the paddle worm.

Sandworms and bloodworms usually burrow in sand and hide in seaweed to avoid detection by predators. Some polychaete worms have developed a different strategy to avoid being eaten—they live inside a tube, which they make themselves. Most of these tube-dwelling worms are small and threadlike in appearance. The fan worm *(Sabella)* constructs its hard tube by mixing sand grains with mucus, which it secretes from special sacs. (See Figure 8-7.) The parchment worm *(Chaetopterus variopedatus)* secretes a tube formed from a tough fibrous material, in which it lies buried in sediments below low tide. Both of these worms thrust their feathery gills out of the tube into the water to obtain oxygen, give off carbon dioxide, and capture plankton. The tube of the trumpet worm *(Pectinaria gouldii)* looks like an ice-cream cone fashioned from sand grains and bits of shell cemented together. This worm extends its ciliated tentacles into the sediment to obtain food. Another polychaete tube worm, called the Atlantic tube worm *(Hydroides)*, secretes a hard tube composed of calcium carbonate, which is cemented to the surfaces of mollusks, rocks, and corals.

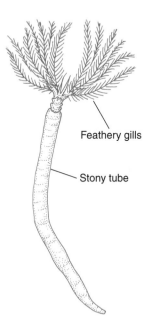

Feathery gills

Stony tube

Figure 8-7 The fan worm.

The Leech

An example of a segmented worm without bristles is the **leech**, which is placed in class Hirudinea. (See Figure 8-8.) Some leeches are free-living, while others are parasites that can be found attached to the gills and mouths of fish. The leech attaches to its host by means of two suckers, located at the anterior (front) and posterior (back) ends of its body. Sharp teeth in the sucker at the anterior end pierce the host's skin. The leech then draws blood from its host. At the same time, the leech secretes a chemical anticoagulant, called **hirudin**, into the wound. This prevents the host's blood from clotting, thus enabling the blood to flow freely into the leech. While feeding, leeches can increase their body weight threefold. After feeding, the leech drops off the fish and fasts while it digests its blood meal.

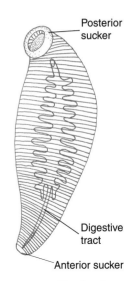

Figure 8-8 The leech.

8.2 SECTION REVIEW

1. In what part of the marine environment are roundworms usually found? What do they feed on there?

2. Describe two adaptations polychaetes have to avoid predators.

3. How do polychaete tube worms feed and protect themselves?

8.3 GIANT TUBE WORMS AND ARROW WORMS

Giant Tube Worms

You have just read about the segmented worms that live in tubes. There are other types of worms that live in protective tubes, and some of these grow quite large. An interesting group of deep-sea gutless worms belongs to the phylum Pogonophora. In 1977, a new species of worm belonging to this phylum was discovered living near hot-water vents on the deep seafloor, thousands of meters below the ocean's surface. First photographed by the submersible *Alvin*, these **giant tube worms** *(Riftia pachyptila)* measure up to 1 meter long and live clustered in water that is rich in hydrogen

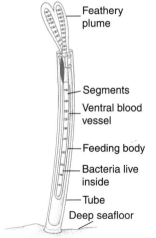

Feathery plume

Segments

Ventral blood vessel

Feeding body

Bacteria live inside

Tube

Deep seafloor

Figure 8-9 The giant tube worm.

sulfide. The tubes in which they live (which are made of proteins and minerals) may measure up to 2 meters in length. (See Figure 8-9.)

Marine biologists have discovered unique bacteria that live inside the giant tube worms. These bacteria are able to use the hydrogen from the hydrogen sulfide in the water and combine it with carbon dioxide from seawater to produce sugars. The energy-rich sugars are also used by the worms. This way of obtaining energy makes the giant tube worms different from most other animals that live on Earth. Remember, most animal life depends on the energy of the sun, energy that is captured by plants and algae during the process of photosynthesis. Yet sunlight cannot reach the great depths at which the giant tube worms live. So the worms live by utilizing carbohydrates made by the bacteria, which get their energy from the oxidation of hydrogen sulfide. (The worms excrete the sulfur from the hydrogen sulfide compound.) The process by which the bacteria produce energy-rich compounds from inorganic chemicals is called *chemosynthesis*.

Arrow Worms

At the other end of the water column, near the surface of the ocean, lives the tiny, transparent **arrow worm** *(Sagitta)*, shown in Figure 8-10. The arrow worm is classified in the phylum Chaetognatha (meaning "bristlejaw"). Arrow worms are just a few centimeters long. They have tiny fins that enable swimming, but they mostly drift as part of the plankton community. An active hunter, the arrow worm uses its mouth bristles, which are modified as hooks, to prey on other animal plankton such as copepods, fish eggs, and fish larvae.

Arrow worms have a one-way digestive tract. Food is digested in the worm's narrow intestine, and undigested wastes are eliminated through the anus. Like the flatworms and roundworms, the arrow worm has no circulatory or respiratory systems. Digested nutrients diffuse into its cells. Respiratory gases and cellular wastes are exchanged between the arrow worm and the outer environment through the worm's thin skin. Arrow worms have a simple nervous system that lets them respond to stimuli. Two eyes in the head region are sensitive to light. Sensory projections (called papillae)

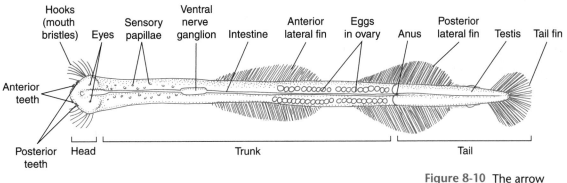

Figure 8-10 The arrow worm.

located along the surface of the worm's body are sensitive to touch. Ganglia are located in the head and trunk regions.

Arrow worms are hermaphrodites, and reproduction is sexual. Each worm produces both sperm and eggs, but self-fertilization does not occur. The sperm and eggs are shed directly into the water, where the sperm from one arrow worm fertilizes the eggs of another.

8.3 SECTION REVIEW

1. How are giant tube worms able to feed at such great depths?

2. In what part of the water column are arrow worms found? What do they eat?

3. How are the arrow worms' means of locomotion and feeding adaptive in their particular habitat?

Laboratory Investigation **8**

Adaptations of the Sandworm

PROBLEM: How is the sandworm adapted for carrying out life functions?

SKILL: Observing features of the sandworm by means of dissection.

MATERIALS: Preserved sandworm, dissecting pan, water, hand lens, forceps, scissors, pins.

PROCEDURE

1. Place a sandworm in a dissecting pan and cover it with water so that you can see the tissues more clearly. Notice that the body is divided into segments. Count the number of segments and record the number.

2. Which end of the sandworm is the front, or anterior, end? Use your hand lens to locate the mouth, which is located in the first two segments at the anterior end. With the forceps, gently squeeze the area. Two sharp hooks (jaws) should stick out from the mouth, showing that this is the worm's front.

3. Which side of the sandworm is the upper surface and which is the lower? Notice that the lower surface is lighter in color than the upper surface. The lower surface is the ventral side, and the upper surface is the dorsal side.

4. Notice the many tiny fiberlike appendages projecting from each segment. Examine one of the appendages with your hand lens. The appendage, called a parapodium, has a fleshy paddle shape. The parapodia contain bristles called setae. These appendages are used for swimming and for burrowing.

5. Sandworms breathe through the thin, moist skin of the parapodia. Use your hand lens to examine the parapodia.

6. Now place the sandworm in a dry dissecting pan with its dorsal surface up. Pin the sandworm to the tray at the worm's anterior and posterior ends. Beginning at the posterior end, cut through the skin with the scissors. Make the cut just slightly to the right of the midline (center of worm, lengthwise). Carefully cut the skin without cutting the underlying tissues. As you cut, pin the skin on both sides to the tray. Cut all the way up to the anterior end.

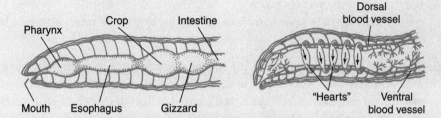

Figure 8-11 The internal anatomy of a sandworm.

7. To see the internal structures more clearly, pour just enough water over the sandworm to cover it. (See Figure 8-11.) Look at the inside body wall. The wall is composed of longitudinal and circular muscles for movement.

8. Use your hand lens to locate the "brain," or cerebral ganglia, at the anterior end near the dorsal surface. The brain is a white structure with two lobes. Locate the connection of the brain to a double ventral nerve cord. Trace the ventral cord and notice that it extends the entire length of the sandworm. Locate the tiny lateral nerve branches going to the muscles in each segment. The ventral nerve cord relays messages to and from the brain.

9. Food enters the worm's mouth; then it is digested in a food tube, the digestive tract. The mouth is connected to a wider part of the digestive tract called the pharynx.

10. The pharynx connects with a narrow esophagus, located in segments 6 through 14. Food moves from the esophagus into the crop. The crop, in segments 15 and 16, temporarily stores the food. Then the food passes into the gizzard, where it is ground up with sand that is ingested with it.

11. Next, the food passes into the intestine, where it is further digested and then absorbed into the blood. Finally, solid wastes are eliminated through the anus, located in the last two segments. Note that the digestive tract is separated from the skin by the fluid-filled space called the coelom.

12. Nutrients and oxygen are transported in the blood. Use your hand lens to locate the aortic arches ("hearts") and the dorsal blood vessel (on top of the food tube), which connects with the arteries, veins, and capillaries.

13. Locate the ventral blood vessel under the digestive tract. The smaller blood vessels connect the dorsal and ventral blood vessels. The blood vessels are the circulatory system of the sandworm. Blood flows only inside this network of blood vessels; thus the sandworm has a closed circulatory system.

14. Wastes are removed by paired, coiled tubes called nephridia, located in most segments. Use your hand lens to look for tiny white tubes attached to the inside body wall. (You may have to push the digestive tract aside.)

15. Two sandworms are required for sexual reproduction to occur; the sexes are separate. With your hand lens, locate a pair of testes in segments 10 and 11, or the ovaries in segments 12 and 13. Both fertilization and development are external; the larvae are planktonic.

OBSERVATIONS AND ANALYSES

1. What adaptations does the sandworm have for breathing?

2. Describe the structures of the sandworm that enable locomotion.

3. How is the sandworm adapted for carrying out ingestion and digestion?

Chapter 8 Review

Answer the following questions on a separate sheet of paper.

Vocabulary

The following list contains all the boldface terms in this chapter.

annelids, arrow worm, bilateral symmetry, bloodworm, coelom, flatworms, ganglia, giant tube worms, hirudin, host, leech, nematodes, nephridia, parapodia, parasite, planarian, proboscis, ribbon worm, roundworms, sandworm, segmented worms, setae, tapeworm, trematode, water column

Fill In

Use one of the vocabulary terms listed above to complete each sentence.

1. A worm's nerve cell clusters that act like a brain are the _____.

2. Sandworms wriggle through the sand by means of _____.

3. The segmented worms are also referred to as the _____.

4. A sharp organ used by ribbon worms to spear prey is the _____.

5. The chemical secreted by a leech to make its victim's blood flow is _____.

Think and Write

Use the information in this chapter to respond to these items.

6. Diagram and discuss the life cycle of the parasitic trematode.

7. Describe the structures that enable sandworms to move.

8. Why is chemosynthesis important for giant tube worms?

Inquiry

Base your answers to questions 9 through 12 on the diagram below, which shows the anatomy of an arrow worm, and on your knowledge of marine biology.

9. Does the arrow worm have a one-way or a two-way digestive system? Explain your answer.

10. Which is an accurate statement regarding reproduction in the arrow worm?   *a.* Arrow worms are separate males and females that reproduce sexually.   *b.* Arrow worms are separate males and females that reproduce asexually.   *c.* Arrow worms are hermaphrodites that reproduce sexually.   *d.* Arrow worms are hermaphrodites that reproduce asexually.

11. Based on the diagram, you could conclude that arrow worms *a.* have no adaptations for food-getting   *b.* have a closed circulatory system   *c.* have a simple nervous system   *d.* are normally tube-dwelling worms.

12. What kind of locomotion is the arrow worm capable of?   *a.* ameboid   *b.* free-swimming   *c.* deep-diving *d.* crawling

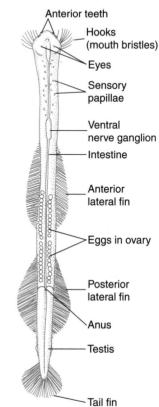

Anterior teeth
Hooks (mouth bristles)
Eyes
Sensory papillae
Ventral nerve ganglion
Intestine
Anterior lateral fin
Eggs in ovary
Posterior lateral fin
Anus
Testis
Tail fin

Multiple Choice

Choose the response that best completes the sentence or answers the question.

13. Two examples of parasitic flatworms are the   *a.* tapeworm and leech   *b.* tapeworm and fluke   *c.* sandworm and tapeworm   *d.* sandworm and leech.

14. The worms are not all classified into a single phylum because
    a. there are too many worms   *b.* they live in such varied
    habitats   *c.* they are too difficult to classify   *d.* they are too
    varied in structure to be in one group.

15. Members of phylum Annelida are all characterized by
    a. having a thick skin   *b.* the absence of appendages
    c. living in the same habitat   *d.* having a segmented body.

16. What part of a worm's anatomy is shown in the diagram
    below?   *a.* the anterior end   *b.* the posterior end
    c. the dorsal side   *d.* the ventral side

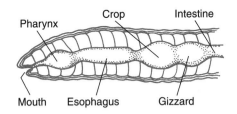

17. Segmented worms such as the sandworm have a fluid-filled
    space called the   *a.* water column   *b.* proboscis
    c. coelom   *d.* nephridia.

18. Which of the following worms always lives as a parasite?
    a. trematode   *b.* bloodworm   *c.* sandworm
    d. ribbon worm

19. A worm that does *not* live inside a substrate is the   *a.* arrow
    worm   *b.* sandworm   *c.* tube worm   *d.* roundworm.

20. The largest free-living marine worm is the   *a.* arrow worm
    b. giant tube worm   *c.* ribbon worm   *d.* sandworm.

21. The most numerous worm species in the sea are the   *a.* tube
    worms   *b.* roundworms   *c.* bloodworms   *d.* sandworms.

22. The pairs of coiled tubes that excrete wastes in annelids are
    the   *a.* setae   *b.* parapodia   *c.* ganglia   *d.* nephridia.

23. Another term used to refer to roundworms is
    a. platyhelminthes   *b.* nematodes   *c.* annelids   *d.* tube
    worms.

24. What worms would you most likely see under the microscope in a sample of aquarium gravel?   *a.* ribbon worms *b.* roundworms   *c.* arrow worms   *d.* sandworms

25. Which of the following marine worms depend on chemosynthesis by bacteria to obtain their food? *a.* roundworms   *b.* flatworms   *c.* sandworms *d.* giant tube worms

Research/Activity

Roundworms are commonly found in wet beach sand from the intertidal zone. If available, obtain a sample and prepare a wet mount slide. Examine the slide under a microscope's low power. Look for the typical whipping action by which the roundworm propels itself. Design an investigation to determine if changes in temperature, pH, or salinity affect its motion.

9 Mollusks

When you have completed this chapter, you should be able to:

DESCRIBE the basic structures and functions of the bivalves.

DISTINGUISH among the variety of gastropod forms and functions.

DISCUSS the unique shapes and adaptations of the cephalopods.

The recent improvement in water quality in coastal cities such as New York produced an unexpected result. Wood pilings that supported the piers began to collapse. The culprit was a marine invertebrate called the shipworm. Like a termite, the shipworm lives in wood and tunnels through it. Not surprisingly, shipworms prefer wood that is not soaked with pollutants. The return of the shipworm to cleaner harbors has prompted officials to use pilings made of recycled plastic.

The shipworm is really not a worm; it is a *mollusk,* a soft-bodied animal like other clams, snails, and mussels. These mollusks have an outer shell, which serves to protect the soft-bodied animal within. The whelk—a marine snail shown laying its eggs in the photo above—is a mollusk with an outer shell.

However, not all mollusks live in shells. The squid and the octopus are classified as mollusks because they are soft-bodied animals with structural and developmental characteristics similar to those of the shelled mollusks. There are more than 100,000 species of mollusks, all classified into several distinct groups within the phylum Mollusca. In this chapter, you will discover the diversity among the mollusks.

9.1 BIVALVES

The next time you go to the beach, observe the shells that are strewn along the shore. Many of them are the shells of clams, mussels, scallops, and oysters. What do the two mollusks shown in Figure 9-1 have in common? Notice that both mollusks have two shells. Mollusks with two shells are called **bivalves** (meaning "two shells"). All bivalves (also called pelecypods) are grouped together in the class Bivalvia.

Structures of a Typical Bivalve

All **mollusks** have soft, bilaterally symmetrical bodies composed of a head, foot, and coiled visceral mass (internal organs), and most have either an external or internal shell. They also have "advanced" features, first seen in the annelids, such as a coelom and a brain. The two shells of a bivalve are hinged together at one end and are kept closed by short, tough **adductor muscles**. There are two scars on the inside of bivalve shells where the adductor muscles are attached. The clam, oyster, and mussel have two adductor muscles, and the scallop has one. In fact, when you have scallops for dinner, you are eating the adductor muscle.

The clam (*Mercenaria*) is a common bivalve. There are more than 15,000 species of clams. Most clams live buried in sand in the intertidal and subtidal zones. Sea stars and predatory snails feed on them. After their soft bodies have been consumed, the clams' empty shells are washed up on the beach by the incoming tide. (See Figure 9-2.)

Interestingly, you can often tell the age of a bivalve by looking at its shell. You may have noticed the lines on clamshells found on the beach. Each line is a new layer of shell that the clam produces as it grows. The lines form bands. Each band represents about a year's growth, much like the growth rings in a tree. For example, a clam that has four bands would be about four years old. Some bands are wider than others. Can you explain why? Wider bands indicate more growth. In some years, conditions for growth are more favorable, so the clam grows more. (*Note:* You can examine a clamshell when you do the lab investigation at the end of this chapter.) One

Mussel

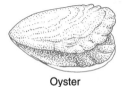

Oyster

Figure 9-1 Two representative bivalves, mollusks with two shells.

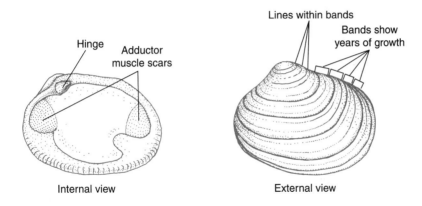

Hinge

Adductor muscle scars

Internal view

Lines within bands

Bands show years of growth

External view

Figure 9-2 The internal and external features of a clam's shell.

of the oldest living clam specimens known is a giant clam *(Tridacna gigas)* from the tropical waters of the South Pacific; it is over 60 years old and weighs more than half a ton.

Mollusk shells are hard due to the presence of the compound calcium carbonate ($CaCO_3$). How does a bivalve build its shell? A thin membrane called the **mantle** lines the insides of both shells and protects the internal organs. The mantle contains shell glands that secrete calcium carbonate, thereby producing the shell.

Life Activities of Bivalves

Normally, the shells of a bivalve are shut tight, with only a small gap between them. The bivalve cannot feed if its shells are completely closed. However, there are water flow passageways in a **siphon** tube, which protrudes through a gap between the clam's shells. This tube can be extended for feeding and breathing, while the rest of the bivalve lies protected under the sand. The siphon has two openings, an **incurrent siphon** and an **excurrent siphon**. Water that contains food and oxygen enters through the incurrent siphon. Waste products of digestion and respiration are eliminated through the excurrent siphon. Since bivalves filter their food from the water, they are examples of filter feeders. As filter feeders, bivalves are responsible for filtering, and thereby cleansing, great quantities of seawater.

Bivalves are adapted to breathe underwater by using gill membranes. In function, gills are like your lungs. They are membranes that take in oxygen and give off carbon dioxide. Water brought in through the incurrent siphon flows to the gills. The surface of the gills contains specialized cells with microscopic cilia. These ciliated cells beat back and forth, creating the currents of water that enter and exit the clam. Oxygen that is dissolved in the water flows over the gills' surface and diffuses through the gill membranes. Carbon dioxide diffuses from the gills back into the water. Thus, gas exchange at the gill surface is how a bivalve breathes. You can observe the ciliated cells in a bivalve by doing the laboratory investigation at the end of this chapter.

How does a bivalve filter feed? During filter feeding, currents of water that contain plankton and organic debris pass into the clam through its siphon, propelled by the ciliary action of the gill surfaces. And how does the clam actually ingest its food? The food particles in the water get stuck in mucus that coats the surface of the gills and mantle. The ciliated cells move the food along to the clam's mouth, which is located opposite its siphon. Food is digested in a one-way digestive tract. Wastes exit through the excurrent siphon. Bivalves have an open circulatory system. Nutrients and oxygen are transported through their body by a colorless blood.

Bivalves have a variety of adaptations that, by securing them to a substrate, enable them to filter feed more effectively. The mussel *(Mytilus)*, shown in Figure 9-1, lives in the turbulent intertidal zone, where there is constant wave action. Tough **byssal threads**, which are made of a fibrous protein, attach the mussel firmly to rocks and other hard substrates. A gland inside the mussel's foot secretes the byssal threads, which extend out from the mussel. Sticky pads at the end of the threads enable the mussels to cling to surfaces.

The oyster *(Crassostrea)*, shown in Figure 9-1, also lives attached to a substrate. The shells of the oyster are rough and uneven. The flat upper shell fits like a lid on top of the more curved lower shell. The lower shell secretes a cement that adheres to rocks and other hard substrates. Dental scientists are interested in the chemical properties of oyster cement, because it might be useful in developing a new type of filling for teeth.

Besides secreting their shells and cement, under the proper conditions oysters and other bivalves can produce natural pearls. A pearl is a small, round, shiny bead that has the same composition as the smooth interior lining of a shell. A natural pearl starts to develop when a sand grain gets into an oyster and lodges between the mantle and the shell. The mantle tissue reacts to the sand grain as a foreign body and secretes layers of shell around the grain, forming a pearl.

Movement in Bivalves

Not all bivalves adhere to substrates; some bivalves move. In fact, bivalves display a variety of adaptations for locomotion. The scallop is the fastest of the bivalves. This fan-shaped mollusk can scoot across the seafloor in sudden spurts. The quick movements are caused when a scallop repeatedly contracts and relaxes its large adductor muscle. The scallop's shells then open and close, forcing water out from between them, which pushes the bivalve in the opposite direction. (See Figure 9-3.)

Bivalves can also move rapidly through substrates. The razor clam *(Ensis directus)* and the soft-shell clam *(Mya arenaria)* move quickly through the sand by using their muscular foot as a digging tool. (See Figure 9-4.) Burrowing quickly enables the clam to escape from its enemies, including clam diggers—people who dig for soft-shell clams in the intertidal zones at low tide. Clam diggers locate the buried clams by looking for holes in the sand made by the

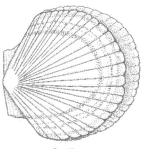

Scallop

Figure 9-3 The scallop: a quick-moving bivalve.

Figure 9-4 Two examples of burrowing clams.

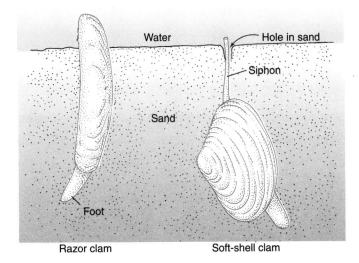

Razor clam   Soft-shell clam

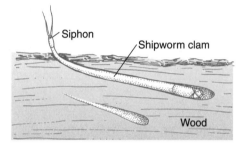

Figure 9-5 The teredo, or shipworm, is a clam that burrows through wood.

Siphon

Shipworm clam

Wood

extended siphon tubes. Some clams, such as the shipworm *(Teredo)*, can even burrow through a solid substrate, such as the wood of a ship's hull or wharf piling. (See Figure 9-5.)

Reproduction in Bivalves

The bivalves have separate sexes. Both fertilization and development are external. Females release their eggs into the water, where they are fertilized by sperm cells released by the males. During the early stages of development, the bivalves live as part of the plankton population. When they form their shells, the tiny bivalves sink to the seafloor, where they settle and mature into adults.

9.1 SECTION REVIEW

1. How does a bivalve breathe?
2. Explain how a bivalve feeds.
3. Compare locomotion in the clam and the scallop.

9.2 GASTROPODS

Look at the mollusks illustrated in Figure 9-6. What do these two animals have in common? They both have a single shell and are marine snails, a type of gastropod. The **gastropods** are a diverse group, comprising about two thirds of all mollusk species. These mollusks are also referred to as univalves, meaning "one shell," which is one of their distinguishing characteristics. All gastropods

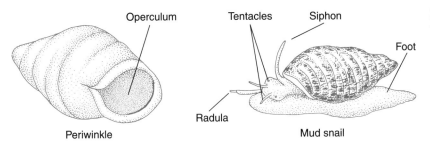

Figure 9-6 Two representative gastropods (snails), mollusks with one shell.

Operculum

Tentacles Siphon

Foot

Radula

Periwinkle

Mud snail

are placed in the class Gastropoda (meaning "stomach-foot"). The snail is described below, as representative of a typical gastropod.

Structures of a Typical Gastropod

Snails, which are the most common gastropods, have a single coiled shell. (See Figure 9-7.) How does a gastropod carry out its life functions? Look at Figure 9-8, which shows the internal structures of a typical snail. The snail glides along the surface of a substrate on its large muscular foot. Food is ingested through the mouth and digested in a one-way digestive tract. Wastes are eliminated through the anus. Nutrients are transported through the body in an open circulatory system, consisting of a one-chambered heart and tiny blood vessels. Its colorless blood is pushed through tissue spaces by contractions of heart and body muscles. Kidneys excrete cellular (metabolic) wastes. When not feeding or moving about, a snail

Snail

Figure 9-7 The snail, a common marine gastropod, has a single coiled shell.

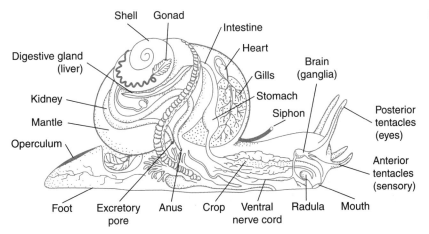

Figure 9-8 Internal structures of a typical snail.

Shell Gonad
Intestine
Heart
Digestive gland (liver)
Brain (ganglia)
Gills
Kidney
Stomach
Mantle
Siphon
Posterior tentacles (eyes)
Operculum
Anterior tentacles (sensory)
Foot Excretory pore Anus Crop Ventral nerve cord Radula Mouth

retracts its soft body inside its shell. The shell's opening is then covered by the snail's **operculum**, a thick pad of tissue that closes like a trapdoor over its foot. The operculum is usually composed of a type of protein; in some snail species, it is calcified.

Life Activities of Snails

The snail breathes by taking in oxygenated water through its siphon tube. Gills inside the snail take up the oxygen and give off carbon dioxide. The anterior tentacles in the head region are the receptors that the snail uses for sensations of touch. Two posterior tentacles, or eyestalks, are used for vision. Impulses from these receptors travel along sensory nerves to a small brain, made up of several pairs of ganglia. Movement is carried out by the nervous and muscular systems working together—impulses from the brain reach the muscles in the foot by way of the motor nerves.

Snails are adapted to crawl and climb in search of food. Most snails move along the seafloor, where they graze on tiny organisms. The periwinkle *(Littorina)*, shown (shell only) in Figure 9-6, grazes on algae that grow on the surface of substrates such as rocks and marsh plants. Typical of all gastropods, the periwinkle has a ribbonlike, toothed structure called a **radula**, which it uses to scrape off and ingest the algae.

Other snails, such as the mud snail *(Ilyanassa)*, shown in Figure 9-6, are scavengers—they feed on dead, or dying, organisms. If you put an open (dead or dying) clam or mussel into an aquarium with mud snails, they will quickly swarm all over the food. Mud snails feed by using their radula to tear and shred the dead matter into small pieces.

Some snails are predators; they actively hunt and kill their food. Two examples are shown in Figure 9-9. The moon snail *(Neverita)* feeds on live clams. Like many other predatory snails, it secretes chemicals from a gland in its foot to soften the shell of a clam. It then uses its radula to drill a small hole in the hinge area of the clamshell. The moon snail inserts its mouth through the hole and eats the soft-bodied animal inside. You can find clamshells that have these telltale holes washed up on the beach. (If you collect them, you can make a necklace.)

Another type of predatory snail, the cone snail *(Conus)*, has

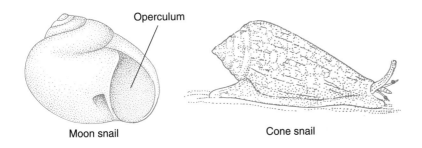

Operculum

Moon snail

Cone snail

Figure 9-9 Two types of predatory snails.

generated great interest for medical researchers around the world. This group of snails, numbering about 500 species, uses toxins to kill its prey. A harpoonlike radula at the end of the snail's proboscis is used to spear prey, such as small fish, and deliver the poison. The potent toxins that cone snails produce are being studied because they show a variety of effects, such as pain relief, on vertebrate nervous systems.

Reproduction in Snails

Some snail species have separate sexes, while many others are hermaphrodites. Fertilization is internal, and development is external. Snails produce large numbers of offspring. In some species, the females deposit their fertilized eggs directly into the water. Others enclose the developing eggs in a protective covering. The whelk *(Busycon)*, another type of predatory snail, produces an egg case composed of several capsules strung together, like beads on a string. (See Figure 9-10.) Each capsule contains one to two dozen embryos developing within it, making a total of several hundred embryos in the whole egg case. You may find the egg case of a whelk on the beach. You can carefully open up one of the capsules with sharp scissors and observe the contents with a hand lens or dissecting microscope. Young egg cases contain the embryos, whereas mature egg cases have "baby" whelks within their tiny shells.

In the moon snail, the eggs develop into larvae within a thin, leathery membrane called a sand collar, which is attached to the shell. The sand collar consists of sand grains cemented together by mucus that is secreted by the moon snail. Within three to five weeks, each larva produces a shell and then settles to the sea

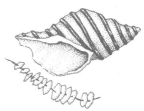

Whelk and egg case

Figure 9-10 The shell and egg case of a whelk, another predatory snail.

bottom, where it grows into an adult moon snail. Sexual maturity is reached in about three years.

The mud snail deposits flattened, transparent jelly capsules on substrates in the intertidal zones. Inside the capsules are 50 to 250 fertilized eggs. Within 6 to 7 days, each mud snail egg hatches into a swimming, ciliated larva called a **veliger**. The veligers join the plankton community, where they feed on tiny plankton and are, in turn, eaten by other animals. The production of so great a number of eggs by snails is important, since so many larvae are eaten while in the plankton community.

Gastropod Diversity

Look at the three gastropods shown in Figure 9-11. Compare them with the gastropods shown in Figures 9-6 through 9-10. As you can see, the shells in the previous figures are spiral in shape, whereas those in Figure 9-11 are more flat. The biggest of the flat-shelled gastropods is the abalone *(Haliotis),* which inhabits the rocky Pacific Coast, where it grazes on algae that grow on rocks. The inside of an abalone shell has a shiny rainbow pattern of colors referred to as iridescence. Just as the mantle tissue in an oyster can produce pearls, the abalone's mantle secretes this pearly lustrous material, which is called *mother-of-pearl.* (This material is often used to make jewelry.)

A common gastropod of the intertidal zone is the slipper shell *(Crepidula)*. The underside of the slipper shell resembles a slipper or shoe, hence its name. The slipper shell is a filter feeder that strains microorganisms and organic debris from the water. This mollusk lives attached to almost any hard substrate, including the backs of other snails, horseshoe crabs, even old tires and bottles. As the slipper shell grows, it changes its sex from male to female. The slipper shells are often found stacked, one on top of the other, with the

Figure 9-11 Three types of flat-shelled gastropods.

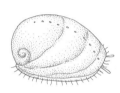

Abalone

Slipper shells
(stacked up)

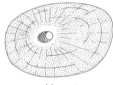

Limpet

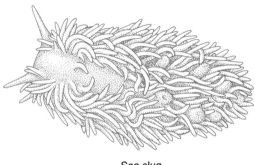

Figure 9-12 The sea slug, or nudibranch, is a gastropod that has no shell.

Sea slug

uppermost one being a male. Each level represents a different generation, with the oldest ones (females) on the bottom.

Another flat-shelled gastropod that inhabits the rocky coast in the intertidal zone is the limpet. Like snails, the limpets graze on algae that grow on the rocks. The limpet shell is shaped like a slightly flattened cone. Some species, such as the keyhole limpet *(Diodora),* have a hole at the top of the shell, which allows wastes to exit.

An unusual type of gastropod is shown in Figure 9-12, the sea slug *(Dendronotus),* also called a **nudibranch.** The sea slug and the sea hare *(Aplysia)* are gastropods that either lack shells or have reduced shells. These shy, membranous, beautiful creatures glide along the seafloor in the intertidal and subtidal zones. They breathe through their skin and through decorative tufts of gills on their backs. The sea slug feeds on hydroids and anemones; it uses their stinging cells as part of its own body's defense by storing them in special projections on its skin. The slug's dramatic coloration serves as a warning to would-be predators that it would be painful to eat! Sea hares graze on algae; some even turn green from the chlorophyll they consume. Like the cephalopods (see Section 9.3), some sea hares can release dark ink to confuse, and escape from, predators. The California sea hare, which can grow up to 75 cm in length and weigh up to 16 kg, is the biggest gastropod in the ocean.

9.2 Section Review

1. Describe two different feeding methods of snails.

2. Compare the feeding methods of slipper shells and limpets.

3. By what means do sea slugs and sea hares breathe?

9.3 CEPHALOPODS

Look at the two mollusks shown in Figure 9-13. What do these unusual-looking animals have in common with one another? Unlike the gastropods and bivalves, these mollusks are excellent swimmers. The swimming mollusks belong to the class Cephalopoda (meaning "head-foot"), with their prominent features being the head and the tentacles (the "foot"). Referred to as the **cephalopods**, they swim by a kind of jet propulsion. How are the cephalopods adapted for swimming? Most cephalopods have a streamlined body shape and lack an external shell. Water is drawn into the mantle cavity. When the mantle contracts, water is expelled through the siphon. The cephalopod is propelled in the opposite direction (much as a balloon flies around a room when its air escapes). By moving its siphon tube into different positions, the cephalopod can change directions suddenly. This mobility is coordinated by the cephalopod's highly developed nervous system.

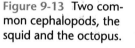

Figure 9-13 Two common cephalopods, the squid and the octopus.

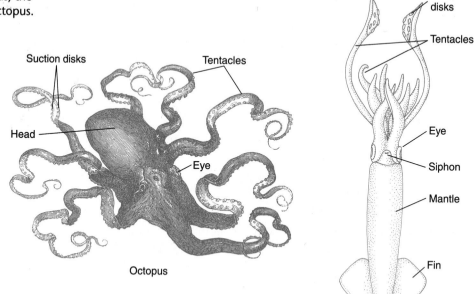

Octopus

Squid

Suction disks

Tentacles

Head

Eye

Suction disks

Tentacles

Eye

Siphon

Mantle

Fin

RESEARCH
The Smartest Invertebrate

The octopus is an animal that leads a largely solitary existence. Scientists in Naples, Italy, wanted to find out if octopuses can learn behaviors from one another. The scientists were interested because, even though octopuses are not social animals, they are known to be the most intelligent of the invertebrates.

The researchers set up an aquarium that contained one red ball and one white ball submerged in it. Several octopuses were placed in the tank. They were conditioned to attack the red ball by rewarding them with a fish if they did so. They were taught to avoid the white ball by giving them a mild electric shock if they approached it. After the octopuses were trained to attack the red ball and avoid the white ball, an untrained octopus was put into an adjacent tank to watch them. When the untrained octopus was later tested without being rewarded or shocked, it picked the red ball 129 times out of 150 trials. That's about an 85 percent success rate!

Scientists were not too surprised by these findings. For its size, the octopus has the largest brain of any invertebrate and has shown signs of intelligence in other situations before. We now know that some of their brainpower is used for learning adaptive behaviors from others in new and challenging situations.

QUESTIONS

1. Calculate the failure rate of the untrained octopus. What can explain this outcome?

2. Why is the octopus considered to be so smart?

3. Why was electric shock used in this experiment? Why were fish used?

Structures of a Typical Cephalopod

The cephalopod captures prey, such as fish or crabs, with its tentacles—killing the animal with a bite from its parrotlike beak. In the case of the octopus (*Octopus*), paralyzing venom is often injected into the prey along with the bite. (The blue-ringed octopus is so venomous that it can cause death in humans.) Food is digested in a one-way digestive tract; wastes are eliminated through the anus. Nutrients are distributed through a closed circulatory system.

Compare the number of tentacles in the octopus with that of the squid (*Loligo*), shown in Figure 9-13. The octopus (meaning "eight feet") has eight tentacles, whereas the squid has ten tentacles (two of which are longer than the others). Notice the numerous suction disks on the tentacles. The suction disks are used for grasping and holding onto prey. In the octopus, the suction disks are also used for climbing and crawling along the seafloor. One species of octopus even crawls onto land to search for food along the shore!

Life Activities of Cephalopods

How do the squid, octopus, and other cephalopods survive without the protection of an external shell? What they lack in external protection, these cephalopods have evolved to make up for in speed and other adaptations. The squid (which retains a long, thin internal shell called a "pen") is the fastest of all cephalopods. It swims in large schools, which may also give it some protection. The octopus, cuttlefish (*Sepia*), and nautilus *(Nautilus)* are more solitary animals than the squid.

Besides being able to move quickly, these cephalopods use camouflage to avoid being detected. By using camouflage, an animal such as an octopus can change its appearance so that it blends in with the natural surroundings. Special pigmented cells in the skin, called **chromatophores**, expand and contract, causing changes in skin pattern and coloration that make the animal match its background. Another defense used by these cephalopods is the discharge of a thick cloud of ink into the water. This dark ink cloud can surprise and confuse would-be predators, enabling the cephalopod to escape. Recently, scientists have identified numerous species of drab-colored pygmy octopuses, which rely on their small size, rather than camouflage, to escape detection within coral reefs and kelp forests.

In addition, the octopus's brain and eye are both highly developed. The cephalopod's intelligence and acute vision enhance its ability to seek prey, avoid predators, and communicate.

Elusive Cephalopods

Two other types of cephalopods are illustrated in Figure 9-14. Notice that one of them, the nautilus, has an external shell. The **chambered nautilus**, which inhabits the deep waters of the South Pacific Ocean, has a spiral-shaped shell that is divided into compartments. The innermost compartments are gas-filled, which helps the animal regulate its buoyancy. The nautilus lives in the outermost compartment, slowly moving up and down in the water, with its sticky tentacles (approximately 40 to 90 of them) extended to capture crabs and other prey. However, it is not a very active hunter, since it swims more slowly than other cephalopods.

The **cuttlefish** is a bottom-dwelling cephalopod that feeds on invertebrates in the sand. An internal shell composed of calcium carbonate, known as the cuttlebone (which is either coiled or flat, depending on the species), adds support to the cuttlefish's soft, streamlined body. Like the squid, the cuttlefish has ten tentacles, two of which are long and slender with suckers on the end. (See Figure 9-14.)

The largest of the swimming mollusks, and also the largest invertebrate, is the giant squid *Architeuthis* (meaning "chief squid"). Over the years, fishermen have seen a few live giant squid, and dead ones have been netted and found washed up on the beach. However, an adult giant squid has never been captured alive. The giant

Figure 9-14 Two "elusive" cephalopods, the nautilus and the cuttlefish.

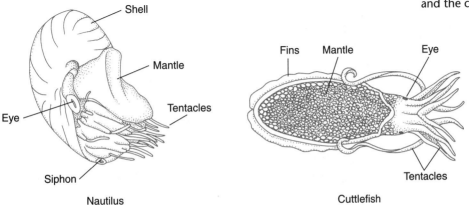

Nautilus

Cuttlefish

squid, which resembles the common squid in structure, but which can grow to a length of about 20 meters (including tentacles), inhabits the deep parts of the ocean, between 300 and 600 meters. There, deep in the sea, the sperm whale *(Physeter)* hunts and eats it. What evidence shows that the sperm whale feeds on giant squid? Body parts, such as beaks, from giant squid have been found in sperm whale stomachs. Also, large circular marks made by giant squid suction disks have been seen on sperm whales, proving that these two giants do fight. Marine biologists are now using submersibles and robots to try to locate live giant squid in the deep. In addition, squid specialists are trying to capture and raise juvenile giant squid from the species' breeding grounds off the coast of New Zealand. Unfortunately, the tiny squid are hard to find and even harder to keep alive in captivity.

Reproduction in Cephalopods

With the exception of the nautilus, cephalopods breed in shallow water. Males and females are separate. Fertilization is internal and development is external. The male cephalopod delivers a packet of sperm to the female, often using a tentacle to place it within her mantle cavity. The female squid deposits clusters of fertilized eggs that look like rice grains on rocks and shells. Most of the squid die after mating, leaving the eggs to develop on their own. Hatching occurs after two weeks, and the young emerge looking like miniature adult squid. The female octopus deposits clusters of thousands of fertilized eggs that look like grapes, attaching them to rocks and seaweed. The octopus protects and cleans her eggs, staying with them until they hatch, which may take as long as four months. After the eggs hatch, the mother usually dies of starvation, having not eaten during the incubation period.

9.3 SECTION REVIEW

1. In what ways does the soft-bodied squid protect itself from predators?

2. Describe the unusual method of cephalopod locomotion. Why is it adaptive?

3. How does the octopus catch and subdue its prey?

9.4 OTHER MOLLUSKS

Chitons

So far you have studied three classes of mollusks: bivalves, gastropods, and cephalopods. How would you classify another mollusk, the **chiton**, shown in Figure 9-15? Unlike many mollusks, chitons have no eyes or tentacles on their heads. Also, notice the overlapping shells, a characteristic not found in any of the other mollusks. The chiton is assigned to its own class, Polyplacophora (meaning "many plates"). Chitons inhabit the rocky intertidal zones around the world. They vary in size from 1 to 40 cm in length. Chitons possess eight overlapping shells, which give the animal some flexibility. The shells cover a muscular foot that is used to grasp and glide over the surfaces of the rocky substrate. As the chiton moves along, it feeds (like a snail) by scraping algae off the rocks with its radula.

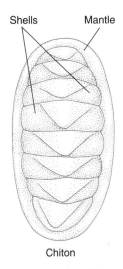

Figure 9-15 The chiton has eight overlapping shells.

Scaphopods

Another small class of mollusks, called Scaphopoda, consists of the **tusk shells**, named for their tapering shell shape. (See Figure 9-16.) Most species of tusk shells, also called **scaphopods**, burrow in the sand of deep waters, while some live in the sediments of shallow tropical waters. Their long foot helps anchor them in the sand. Tusk shells have numerous long, thin tentacles with sticky ends, which they use to capture tiny worms and plankton found in the sand. The tentacles then push the food to the tusk shell's mouth. Empty tusk shells are often found washed up on the beach; they are sometimes made into decorative ornaments. Native Americans used the tusk shells to make necklaces and valued them as a form of currency known as wampum.

9.4 SECTION REVIEW

1. How is the chiton adapted to live in the rocky intertidal zone?
2. Describe the feeding method of the tusk shell.
3. In what kinds of environments do tusk shells live?

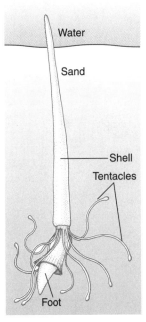

Figure 9-16 The scaphopod, or tusk shell, burrows in the sand.

Laboratory Investigation 9

Feeding in a Bivalve

PROBLEM: How does a bivalve filter feed?

SKILL: Observing the action of ciliated cells under a microscope.

MATERIALS: Live clams or blue mussels, shallow bowls, seawater, medicine dropper, food coloring, hand lens, dissecting trays, newspapers, small rock, forceps, carmine powder, slides, coverslips, dissecting needles, microscope.

PROCEDURE

1. Put a live clam or mussel in a bowl and cover it with seawater. Using a medicine dropper, squeeze out a few drops of food coloring near the edge of its shell. Notice what happens. Enter your observations in your notebook. If the bivalve is alive (and it should be!), the dye should enter through the siphon. Observe the siphon with your hand lens. Make a sketch of the bivalve; draw and label its siphon. (See Figure 9-17.)

2. Remove the bivalve from the bowl and place it on a tray lined with newspapers or hold it in the palm of your hand. Open the bivalve by gently tapping one shell with a small rock until it cracks. Remove the pieces of shell with forceps. Try not to pull away the underlying tissues. (*Note:* It may work best to hold the bivalve in your hand, so that both shells do not crack.)

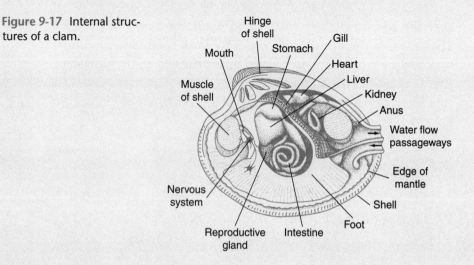

Figure 9-17 Internal structures of a clam.

Hinge of shell
Mouth
Stomach
Gill
Heart
Liver
Kidney
Anus
Water flow passageways
Edge of mantle
Shell
Foot
Intestine
Reproductive gland
Nervous system
Muscle of shell

3. Put the bivalve back in a bowl and cover it with seawater. The top layer of tissue is a thin membrane called the mantle. Parts of the mantle may be stuck to the shell fragments. Glands in the mantle secrete the shell.

4. Under the mantle lie several overlapping membranes coated with a thick fluid or mucus. These membranes are the gills. Sprinkle a few particles of carmine powder on the gills. Wait and see what happens. The particles are moved along by the action of special cells on the surface of the gills.

5. With your forceps, pull out a tiny piece of gill membrane and put it on a glass slide. Use dissecting needles to tease the piece of gill into even smaller pieces. Add a few drops of seawater to the teased pieces and put a coverslip over the slide.

6. Observe this wet mount under the low power of the microscope. Look for currents of moving water. Focus on the cells. Notice the tiny hairs beating back and forth. The hairs that are attached to the surface of these specialized cells are the cilia. The beating of the cilia causes the currents of water.

7. Turn to high power. Notice that each cell has a single hair, or cilium. Draw and label a row of ciliated cells.

OBSERVATIONS AND ANALYSES

1. How does a bivalve filter feed? What is the siphon's role?

2. Describe the action of the cilia. Why are they so important?

3. Why was food coloring used in this experiment?

Chapter 9 Review

Answer the following questions on a separate sheet of paper.

Vocabulary

The following list contains all the boldface terms in this chapter.

adductor muscles, bivalves, byssal threads, cephalopods, chambered nautilus, chiton, chromatophores, cuttlefish, excurrent siphon, gastropods, incurrent siphon, mantle, mollusks, nudibranch, operculum, radula, scaphopods, siphon, tusk shells, veliger

Fill In

Use one of the vocabulary terms listed above to complete each sentence.

1. A membrane called the _____ lines the inside of bivalve shells.

2. Oxygen and food are taken in through a clam's _____.

3. Gastropods use a toothed structure called a _____ to feed.

4. A mussel secretes _____ to attach its shell to a substrate.

5. Respiratory and digestive wastes exit from a clam's _____.

Think and Write

Use the information in this chapter to respond to these items.

6. Describe the special adaptation that mussels have for surviving strong wave action along the shore.

7. Why do you think the squid and the octopus have camouflage, whereas the bivalves and gastropods do not?

8. Compare the feeding methods of bivalves and gastropods.

Inquiry

Base your answers to questions 9 through 12 on the following data and on your knowledge of marine science.

A marine science student learned in her biology class that the normal movement of cilia inside the human respiratory system helps rid the lungs of harmful airborne substances. Upon doing further research, she also learned that tobacco smoke can slow down or even stop the beating of cilia (possibly leading to lung cancer). The student decided to test the effect of cigarette tobacco on the movement of cilia in the ribbed mussel, since its gills are lined with numerous cilia. She placed the gill tissue from a ribbed mussel in two Petri dishes: one containing seawater only (the control) and the other containing seawater and a tobacco extract (the experimental group). The gill tissue in the experimental group was exposed to the chemical extract for a period of 10 days. Observations were made on ciliary activity in both groups, recorded four times a day on each of the 10 days. Results are shown in the table below.

EFFECT OF TOBACCO EXTRACT ON CILIARY ACTIVITY IN RIBBED MUSSELS (*MODIOLUS DEMISSUS*)*

Day	Activity Levels of Experimental Group (exposed to tobacco extract)				Activity Levels of Control Group (not exposed to tobacco extract)			
	9:00 A.M.	12 NOON	3:00 P.M.	6:00 P.M.	9:00 A.M.	12 NOON	3:00 P.M.	6:00 P.M.
1	6	6	6	6	6	6	6	6
2	3	3	3	3	6	6	6	6
3	3	3	3	3	6	6	6	6
4	1	1	0	0	6	6	6	6
5	0	0	0	0	6	6	6	6
6	0	0	0	0	6	6	6	6
7	0	0	0	0	6	6	6	6
8	0	0	0	0	6	6	6	6
9	0	0	0	0	6	6	6	6
10	0	0	0	0	6	6	6	6

*Key to activity levels: 6 = very active; 3 = moderately active; 1 = least active; 0 = no activity.

9. The results of this experiment show that ciliary activity
   a. stops in the control group after the fourth day   *b.* stops in the experimental group after the fourth day   *c.* increases in the control group after the fourth day   *d.* increases in the experimental group after the fourth day.

10. Which is an accurate statement regarding this experiment? *a.* The experiment was done without prior research. *b.* It was not carried out as a controlled experiment. *c.* The student did not quantify her results. *d.* The biology of the ribbed mussel was used as a model for understanding human biology.

11. Which statement represents a valid conclusion that can be drawn from this experiment? *a.* Tobacco extract has no effect on ciliary activity. *b.* Ciliary activity in gills cannot be studied outside the living animal. *c.* Tobacco extract does have an effect on the level of ciliary activity in ribbed mussels. *d.* There is not enough information to draw any tentative conclusions.

12. In a complete sentence or two, describe what effect, if any, the tobacco extract has on ciliary activity in gill tissue, and provide a reasonable scientific explanation for the results that were observed and recorded.

Multiple Choice

Choose the response that best completes the sentence or answers the question.

13. A snail closes its shell opening with a pad of tissue called the *a.* shell *b.* foot *c.* operculum *d.* mantle.

14. Which of the following is a bivalve mollusk? *a.* oyster *b.* snail *c.* whelk *d.* nautilus

15. An example of a univalve mollusk is the *a.* clam *b.* mussel *c.* scallop *d.* snail.

16. The structure in a clam that secretes the shell is the *a.* siphon *b.* foot *c.* gills *d.* mantle.

17. A filter-feeding mollusk is the *a.* moon snail *b.* mussel *c.* mud snail *d.* whelk.

18. Of the following mollusks, which one is *least* related to the others? *a.* snail *b.* scallop *c.* mussel *d.* clam

19. All the following are cephalopods *except* the *a.* octopus *b.* chiton *c.* squid *d.* nautilus.

20. The gastropod that lacks a shell is the   *a.* cuttlefish
b. nudibranch   *c.* nautilus   *d.* squid.

21. All the following live mostly in or on a substrate *except* the
a. mussel   *b.* scaphopod   *c.* cephalopod   *d.* cone snail.

Refer to the following figure to answer questions 22 and 23.

22. The mollusk shown here would be
classified as a   *a.* univalve   *b.* bivalve
c. pelecypod   *d.* cephalopod.

23. What is the function of the structures
on top of its head?   *a.* vision
b. respiration   *c.* defense   *d.* food-getting

24. All of the following are gastropods *except* the   *a.* slipper shell
b. limpet   *c.* scallop   *d.* moon snail.

25. The class of mollusks that contains many species without
shells is   *a.* bivalves   *b.* scaphopods   *c.* chitons
d. cephalopods.

26. What characteristics do the two mollusks shown here have in
common? *a.* Both are filter feeders.   *b.* Both are soft-bodied
animals.   *c.* Both have stinging tentacles.   *d.* Both have
external shells.

Research/Activity

If you live near the coast, go to the beach and pick up a variety of
shells. (Make sure they are no longer inhabited!) Try to include
both bivalves and univalves in your collection. Identify the shells
with their common and scientific names. Make a display of the
shells and present your project to the class.

10 Crustaceans

When you have finished this chapter, you should be able to:

LIST the basic characteristics of the crustaceans.

DESCRIBE the structures and functions of lobsters and crabs.

IDENTIFY important features of the smaller crustaceans and other marine arthropods.

Hiding in the recesses of a rocky shoreline is a northern lobster, ready to grab its next meal with its strong claws. The lobster's claws, and the rest of its body covering, are composed of a tough fibrous material. This outer skeleton is like a suit of armor that protects the lobster.

Other animals with tough outer skeletons include such diverse forms as crabs, shrimps, barnacles, spiders, and insects. These animals also have numerous segmented legs, or appendages, which they use for locomotion and food-getting. Numbering more than a million species, and found both on land and in water, this largest group of animals is classified as the phylum Arthropoda.

With their tough body covering and movable appendages, lobsters and other members of this phylum can move about with some degree of security. In this chapter, you will learn how the marine arthropods have successfully adapted to life in the sea.

10.1 INTRODUCTION TO CRUSTACEANS: THE LOBSTER

Members of phylum Arthropoda are commonly called the **arthropods** (meaning "jointed feet"). The characteristic movable limbs, which give the phylum its name, are referred to as jointed appendages. The tough body covering, or outer skeleton, is the animal's **exoskeleton.** It is made of **chitin,** a type of carbohydrate. Chitin varies from flexible to hard in different arthropod species. The exoskeleton functions not only as a protective cover for arthropods but also as a place of attachment for their muscles. The arthropods are so diverse in appearance that scientists differ on whether to divide the group into several classes, subphyla, or completely separate phyla. For our purposes, we refer to the major subdivisions as classes, the most important in marine habitats being the class Crustacea (meaning they have a "crust" or "shell").

Crustacean Features

Look at the lobster and crab illustrated in Figure 10-1. What do the two animals have in common? These sea creatures are **crustaceans,** animals that have a hard outer covering. The bodies of these crustaceans have bilateral symmetry and are divided into two main segments—the **cephalothorax** (which comprises the head and chest regions) and the abdomen (including a tail, if present). The part of the exoskeleton that covers the head and chest regions is called the **carapace.** Crustaceans such as lobsters, crabs, and shrimps have five pairs of legs located under their carapace. Therefore, they are referred to as the decapods (meaning "ten legs"). The claws, which are used in food-getting, are the first pair of legs; the four other pairs are the walking legs. The head contains two eyes, two pairs of antennae, and special mouthparts used for feeding. The thorax contains the food-getting appendages and the walking legs. Some crustaceans, such as the lobster, can glide along the sea bottom by using their small paddlelike appendages, called **swimmerets,** which are located under the abdomen.

How does a crustacean grow, if it is encased in a rigid exoskeleton? Crabs, lobsters, and shrimps shed their outer covering once or more

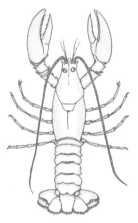

Northern lobster

Fiddler crab

Figure 10-1 Typical crustaceans—a lobster and a crab.

each year in a process called **molting**. To molt, the crustacean secretes a new exoskeleton inside the old one, which it splits. The animal then pushes its body out through a gap between the thorax and the abdomen. After molting, the crustacean has a soft new exoskeleton that gradually hardens; but until that time, it is vulnerable to predators. While they are still soft, crabs such as the blue claw crab (*Callinectes sapidus*) are harvested and sold commercially as soft-shell crabs.

The Lobster

Two common lobster species are the northern lobster (*Homarus americanus*) and the spiny lobster (*Panulirus argus*). The northern lobster, also called the Maine lobster, has two large claws, which are absent in the spiny lobster. The northern lobster lives in the rocky subtidal zone, from Labrador to Virginia. The spiny, or rock, lobster is found in the waters of Florida, the Gulf of Mexico, and California. (There is also a European species of spiny lobster.)

Lobsters are aggressive and often fight amongst themselves. If one lobster grabs the claw of another lobster, the latter can escape by releasing its arm from its socket. Lobster trappers can also be left holding just a lobster arm if they make the mistake of grabbing a lobster by its claws. The ability of a lobster to sacrifice a body part to escape from its enemy is an adaptation for survival. The arm grows back, because arthropods can regenerate appendages. This ability is of interest to medical scientists, who are investigating the possibility of limb regeneration in humans.

Structures and Life Activities of Lobsters

Lobsters are predators, able to feed on other invertebrates such as mussels and sea urchins, which they grab with their claws. They also scavenge on the remains of dead animals. Food is digested in a one-way digestive tract consisting of a mouth, esophagus, stomach, and intestines. Wastes are eliminated through the anus. (See Figure 10-2, which shows the internal and external anatomy of a typical northern lobster.)

How does a lobster breathe? Like the mollusks, lobsters use gills for breathing. The gills are featherlike structures located in a water-

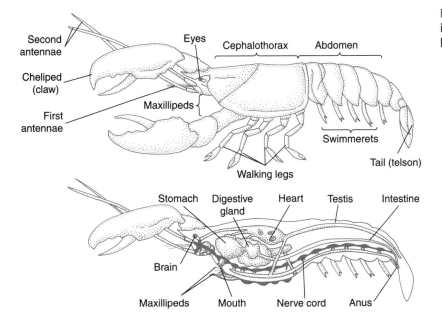

Figure 10-2 External and internal anatomy of a lobster.

External anatomy labels: Second antennae, Eyes, Cephalothorax, Abdomen, Cheliped (claw), Maxillipeds, First antennae, Swimmerets, Tail (telson), Walking legs

Internal anatomy labels: Stomach, Digestive gland, Heart, Testis, Intestine, Brain, Maxillipeds, Mouth, Nerve cord, Anus

filled chamber under the carapace. Each gill is attached to the upper end of a walking leg. The rapid beating of the mouthparts sends currents of water over the gills. Oxygen in the water diffuses into the gills, and carbon dioxide passes from the gills out to the water.

Oxygen and nutrients are transported around the lobster's body in its blood. The blood is blue in color due to the presence of a pigment called hemocyanin. Hemocyanin contains copper, which binds oxygen in much the same way that the iron in our hemoglobin binds oxygen (and imparts a red color to our blood).

Blood is pumped through the body by a one-chambered heart. Veins and arteries carry blood to and from the heart, aided by the body's muscular contractions. However, there are no capillaries connecting the arteries and veins, so the blood just passes through the tissue spaces. Thus, lobsters (and all other arthropods) have an open circulatory system.

The lobster's nervous system enables it to carry out a variety of responses. Its eyes are mounted on movable stalks. Two pairs of antennae actively feel out the environment. The impulses from these receptors are carried by sensory nerves to the brain, or cerebral ganglia. Responses are carried out when the brain sends impulses back via the ventral nerve cord to muscles in the legs and in the abdomen.

Lobsters reproduce sexually. Fertilization is internal, and development is external. The male deposits sperm cells into the female's abdomen, where they are stored in a chamber called the seminal receptacle. As the eggs are released, they are fertilized by the sperm. The female carries the large mass of fertilized eggs on its abdomen, attached to its swimmerets, for nearly a year before they hatch. The embryos go through a larval phase typical of most other crustaceans, floating as part of the plankton population, and molting as they grow and develop into the adult form.

10.1 SECTION REVIEW

1. How can an arthropod grow if it is enclosed in an exoskeleton?

2. How does a lobster carry out the life activity of breathing?

3. Explain why some lobsters may have one claw that is larger than the other claw.

10.2 ANOTHER IMPORTANT CRUSTACEAN: THE CRAB

There are many species of crabs alive in the world today. Look at the three crabs shown in Figure 10-3. They look very different from one another. What do you think is the reason for this? Crabs live everywhere—on land and in the sea, and from surface waters down to the great ocean depths. The great diversity that exists among them is due to the fact that crabs have successfully adapted to these many different habitats. In this section, you will learn about several of these crab species and their unique adaptations.

The Crab

The fiddler crab (*Uca*) digs tunnels in the sand along the shores of bays and inlets. When the tide comes in, fiddler crabs retreat to the tunnels and plug up the entrances with sand. At low tide, the fiddlers leave their tunnels to search for food. If you approach them, they will scurry back into the nearest hole. The fiddler crab is named

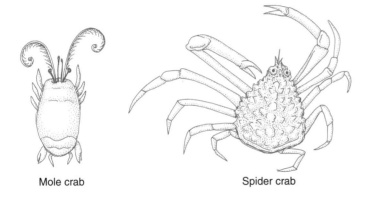

Mole crab   Spider crab

Hermit crab

Figure 10-3 Three representative crabs.

for its large claw, which resembles the arm position of a person playing the fiddle. (Refer to Figure 10-1.) The males have the one large claw plus a smaller one, and the females have two small claws. The male uses the large claw to threaten or fight with other males over territory. If a male loses his large claw in a fight, it regenerates a new small claw, while the other claw then grows into a large claw.

The mole crab (*Emertia*) lives in the turbulent surf zone along sandy beaches. How is this animal adapted to live in such an unstable environment? The mole crab has a smooth, streamlined body, which diminishes the impact of waves and lets it burrow and move (by use of its swimmerets) efficiently through swirling mixtures of sand and water. Mole crabs also have featherlike antennae, which they use to capture the microscopic organisms that live between sand grains and in the surf.

The hermit crab (*Pagurus longicarpus*) is born with a soft abdomen that lacks an exoskeleton, making it vulnerable to attack. For protection, the hermit crab finds an empty snail shell to live in. Hermit crabs inhabit shallow coastal waters where they scavenge on particles of food. As the hermit crab increases in size, it outgrows its old snail shell and has to find a larger one. So hermit crabs are always trying on a new shell for a better fit.

Not all crabs are active and aggressive. The spider crab (*Libinia emarginata*) is slow in its movements. Notice that it has no paddle-like appendages for quick swimming or digging in the sand. The pointed legs are used for crawling very slowly along the ocean bottom. As a consequence, organisms such as algae and barnacles have time to attach and grow on the back of the spider crab. The spider crab inhabits Atlantic and Pacific waters, where it scavenges on food particles. The biggest crab in the ocean is the giant spider crab (*Macrocheira kaempferi*), found in deep waters off the coast of Japan. Some specimens have measured, leg tip to leg tip, up to 4 meters in width.

Structures and Life Activities of Crabs

What structures does the crab possess that enable it to carry out its life functions? As mentioned above, the crab's body is divided into segments (the cephalothorax and the abdomen), which are covered by the carapace. (There is no tail.) The abdomen in the crab is small and flat and is folded between the crab's walking legs on its ventral side. You can tell the sex of a crab by the shape of its abdomen; the female has a U-shaped abdomen, and the male has a V-shaped abdomen. (See Figure 10-4.) When you do the lab investigation at the end of this chapter, you will observe the structures of the crab more closely.

Crabs eat mainly dead plant and animal matter, although some graze on algae and others are predatory. They use their two sharp claws to tear and shred food. The food is passed to the mouth, where it is cut into even smaller pieces by the mouthparts. As in the lobster, food is digested in a one-way digestive tract, consisting of the mouth, esophagus, stomach, and intestines. Digestive wastes are eliminated through the anus.

Like the lobster, crabs breathe by means of their gills and they

Figure 10-4 A male and a female crab (ventral view)—they can be distinguished by the shape of the abdomen.

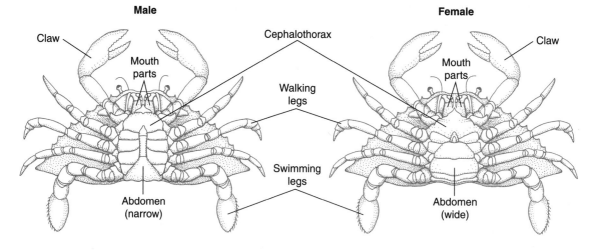

transport nutrients and oxygen through an open circulatory system. Their head region has two eyes (on stalks) and antennae for perceiving touch and temperature stimuli. Their well-developed nervous system enables them to respond to stimuli and control muscular activities, such as locomotion, via the ventral nerve cord.

Crabs produce large numbers of offspring. The female carries a mass of eggs between her abdomen and thorax. Sperm from the male fertilizes the eggs internally. The embryos develop externally, passing through the larval phase that is typical of decapod crustaceans. It is most adaptive for the crab to produce an abundance of fertilized eggs, since so many are eaten during their larval phase in the plankton population.

10.2 SECTION REVIEW

1. How is the mole crab adapted to live in the turbulent surf zone?

2. Why does the hermit crab live inside an empty snail shell?

3. Why might it be particularly beneficial for the giant spider crab to be so large?

10.3 DIVERSITY AMONG CRUSTACEANS: THE SHRIMP AND OTHERS

Crustaceans range in size from nearly microscopic to absolutely huge. You are more aware of the larger ones, such as the lobster and crab, because they are popular seafood items. In this section, you will learn about the shrimp and some of the smaller crustaceans that may be less familiar to you.

The Shrimp

The shrimp looks somewhat like a small version of the lobster. The pink Gulf shrimp (*Penaeus duorarum*), which grows up to 17 cm long, is caught for the seafood industry by fishing trawlers off the

Figure 10-5 Two species of shrimps.

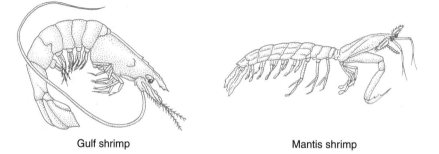

Gulf shrimp   Mantis shrimp

coast from the Carolinas to the Gulf of Mexico. The smaller common shore shrimp (*Palaemonetes vulgaris*) lives among the grasses and seaweeds in salt marshes, where it scavenges on dead plant matter and other organic debris. One of the smallest shrimps is the cleaning shrimp (*Periclimenes*), several species of which inhabit coral reefs. The cleaning shrimp, which looks like a tiny, colorful version of the shore shrimp, survives by eating parasites that are found on the skin of reef fish—a symbiotic relationship that benefits both fish and shrimp. The mantis shrimp (*Squilla empusa*) is the largest of all shrimp, growing up to 25 cm in length. From its burrow in the sand or mud, the mantis shrimp spears such prey as worms and small fish with its spiny front appendages, which unfold like a jackknife. It is found from Cape Cod down to Brazil. (See Figure 10-5.)

Copepods and Krill

Do you recognize either of the crustaceans in Figure 10-6? The most abundant crustacean in the ocean is actually the tiny **copepod** (*Calanus*)—there are more than 1000 species of copepods in the sea. The copepod, which is less than half a centimeter long, eats mainly diatoms. You may recall from an earlier chapter that the copepod is a very important part of the zooplankton community, because it forms the vast bulk of the base of the oceanic food chain for so many other species. Copepods reproduce sexually; the developing larvae undergo numerous molts before reaching maturity. During the spring and summer, when copepods are abundant, you can catch many of them in a plankton net. For a closer look, you can observe them under a dissecting microscope.

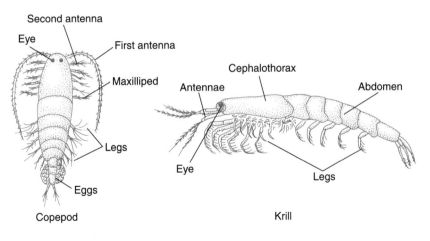

Figure 10-6 Copepods and krill are important members of the zooplankton community.

Second antenna
Eye
First antenna
Maxilliped
Legs
Eggs
Copepod

Cephalothorax
Antennae
Abdomen
Eye
Legs
Krill

A cold-water relative of the copepod is a shrimplike animal called **krill** (*Euphausia*). (Although it resembles a shrimp, the krill has more than 10 legs, so it is not classified with the decapods.) Krill grow to about 5 cm in length. Most krill live in Antarctic waters, although some are found in other oceans. Like the copepod, the krill is a planktonic animal that eats diatoms and floats in large masses near the ocean's surface. Krill are the principal food source for the filter-feeding (baleen) whales and, along with the copepod, provide food for countless fish, birds, and seals.

Amphipods and Isopods

Whereas copepods and krill float in the open ocean, other shrimp-like crustaceans are found near the edge of the sea, living in or on a substrate. (See Figure 10-7.) If you turn over a rock or some debris in the intertidal zone, anywhere from the Arctic to the Chesapeake Bay, you are likely to find a bottom-dwelling crustacean called the

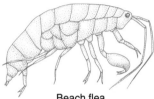

Beach flea

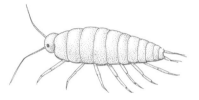

Sea roach

Figure 10-7 Amphipods and isopods live in the intertidal and subtidal zones, respectively.

scud (*Gammarus*). This little crustacean feeds on small invertebrates that live in the wet sand. And on sandy beaches along the East and West coasts, under the moist seaweeds along the strandline, a tiny crustacean called the beach flea (*Talorchestia*) can be found. If you turned over the seaweed, you would see hundreds of these creatures hopping and darting about. The beach fleas use the seaweed both for cover and for food. Crustaceans such as the scud and the beach flea, which look like tiny shrimps that have flattened sides, are called **amphipods**.

Some crustaceans, such as the sea roach (*Ligia*), resemble species of arthropods that live on land. (The pill bug, often found under rotting logs, is related to these crustaceans. Over time, land animals may evolve from sea animals—and vice versa—by first adapting to intertidal conditions.) Sea roaches swim and crawl in coastal waters among the seaweeds; they feed on algae. Like the cockroach, the sea roach is active at night and hides during the day. The sea roach inhabits the shallow coastal waters from California to Central America and from Cape Cod to Canada. Notice that compared with the beach flea, the sea roach's body is flattened from top to bottom (rather than from side to side). These tiny crustaceans, with flattened bodies and seven pairs of legs, are called **isopods**. Other species of isopods are parasitic; they live by attaching to the gills and skin of fish such as cod and halibut.

The Barnacle

Figure 10-8 Barnacles are sessile, encrusting organisms; here they have used an empty glass bottle as their substrate.

An unusual crustacean that is often mistaken for a mollusk is the **barnacle**. The acorn, or rock, barnacle (*Balanus*) lives in the upper intertidal zone, attached to rocks and other hard surfaces. Its overlapping, sharp calcium carbonate plates, which resemble a mollusk's shell, protect the animal inside.

Barnacles attach to almost any substrate, from a ship's hull to a whale's skin. Like the yellow boring sponge described in Chapter 6, the barnacle is a type of encrusting organism. (See Figure 10-8.) Ships must be dry-docked periodically to be scraped clean of barnacles, because they add weight, increase friction, and thus hinder a ship's ability to move smoothly and quickly. In effect, a barnacle-encrusted ship uses more fuel to move through the water.

Structures and Life Activities of Barnacles

How does the barnacle feed? The barnacle's body is actually folded up within its shell, so that its appendages (legs) can protrude from the opening. When the tide comes in and covers it, the barnacle opens its shell plates and extends its six pairs of feathery appendages, called **cirri**. The cirri whip about in rhythmic fashion to catch phytoplankton and other food particles, which are then brought into the mouth. Barnacles are filter feeders because they filter plankton and organic debris from the water. Food is digested in a one-way digestive tract, consisting of the mouth, stomach, and intestine. Wastes are eliminated through the anus. (See Figure 10-9.)

Water that contains dissolved oxygen is swept into a barnacle's shell by the movement of its cirri. The oxygen and carbon dioxide gases are exchanged across the barnacle's skin membrane. At low tide, when the barnacle is exposed to air, its overlapping plates shut tight, thus retaining moisture to keep the animal from drying out.

Notice that the barnacle contains both ovaries and testes; thus it is a hermaphrodite. Although each animal contains both sets of reproductive organs, self-fertilization does not occur. Mating occurs when a penis that is carrying sperm from one barnacle is inserted into a neighboring barnacle. Fertilization occurs inside the barnacle, and the fertilized eggs develop into swimming larvae, which are shed into the water. The barnacle larvae, like those of other crustaceans, are part of the plankton population. When the larvae come

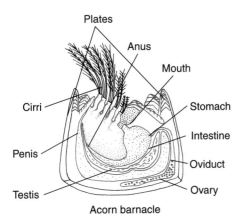

Acorn barnacle

Figure 10-9 The barnacle keeps its body folded up within its shell of overlapping plates; only its cirri extend out of the shell for filter feeding.

RESEARCH
Sticking with the Barnacle

The barnacle is a source of fascination for marine biologists. It can be found attached to a variety of substrates, such as rocks, wood pilings, ships' hulls, and even the bodies of live turtles and whales. Storms and powerful waves cannot dislodge the barnacle from its attachment to these surfaces. How can this tiny arthropod stick so firmly to a substrate? Researchers have found that the barnacle produces natural cement that is more than twice as strong as any factory-made adhesive. This sticky glue, which is produced underwater when the barnacle is still in its larval stage, does not dissolve in strong acids, alkaline substances, or organic solvents.

How does this strong attachment develop? Once the barnacle larva settles down on a suitable substrate, it secretes a clear liquid adhesive from glands in its body. As the barnacle grows, the glands enlarge to produce more cement, which hardens into an opaque, rubbery solid. Biochemists have determined that the cement is composed of protein molecules that provide a strong framework to anchor the barnacle.

Scientists are conducting further research to try and unravel various mysteries of the barnacle. Some biologists are focusing on its larval stage in order to understand how it selects a suitable location on which to settle. Medical researchers are interested in the barnacle's adhesive—if scientists can figure out how to manufacture this powerful substance, they might be able to use it to mend broken bones and fill teeth.

QUESTIONS

1. How does a barnacle attach to the shell of a sea turtle?

2. Why are scientists fascinated with the barnacle?

3. Describe two areas of further research on the barnacle.

into contact with a suitable substrate—preferably near other barnacles—they attach to it by secreting glue from special glands, and then develop into adult barnacles.

Barnacles also live in clusters among the mussel beds in the lower intertidal zone along the West Coast. Called gooseneck, or goose, barnacles (*Pollicipes*) for the long stalks by which they are attached to the substrate, they are able to bend with the currents to capture large plankton that drift by. Their six pairs of cirri extend from the shell at the top of the stalk, enabling them to compete with the mussels for living space and food. (See Figure 10-10.)

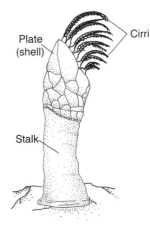

Gooseneck barnacle

Figure 10-10 The gooseneck barnacle can bend with the ocean currents, extending its cirri to capture drifting plankton.

10.3 SECTION REVIEW

1. Describe three methods of food-getting in shrimps.

2. Why are krill and copepods important members of the plankton population?

3. How do barnacles become encrusted on substrates, such as the hull of a ship?

10.4 DIVERSITY AMONG THE ARTHROPODS

As you have learned, the most important arthropods in the ocean are the crustaceans. However, the arthropods are a very diverse group. In addition to crustaceans, there are other types of arthropods that are adapted to the marine environment. Two unusual groups of arthropods found in or near the sea are the horseshoe crab and the marine insects. In this section, you will learn how these arthropods are uniquely adapted to their marine habitats.

The Horseshoe Crab

An arthropod that is often mistaken for a crab is the **horseshoe crab** (*Limulus polyphemus*). The horseshoe crab is not a true crab; it lacks antennae and mouthparts, and it has six pairs of legs, or appendages. (A true crab has five pairs of legs.) The horseshoe crab is actually more closely related to spiders and scorpions than to crustaceans. As such, it is placed in it own class, Merostomata.

The horseshoe crab inhabits the waters along America's Atlantic and Gulf coasts and along the Asian Pacific coast, where it searches for food in the mud, preying on small mollusks and crustaceans, and scavenging on dead matter. The horseshoe crab has four eyes—two simple eyes and two compound eyes—located on the top of its carapace. The compound eyes contain many visual units, which are grouped together for better vision.

Structures and Life Activities of Horseshoe Crabs

On the underside of the horseshoe crab are its legs. The first pair of appendages is a set of pinching claws (the cheliceras); the five other pairs are the walking legs. In males, the first claw is shaped like a boxing glove. The body and legs are covered by a domed carapace, which is followed by a long spiked tail. (See Figure 10-11.)

Behind the legs are numerous overlapping membranes that

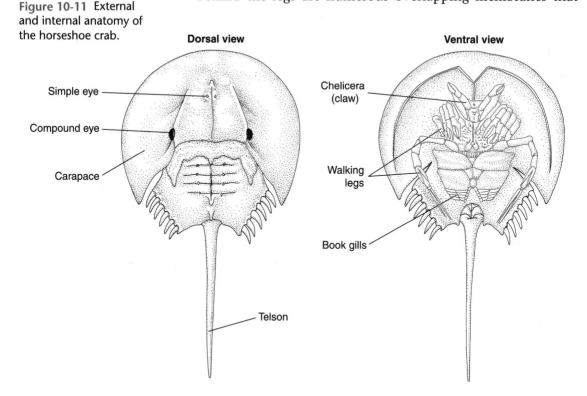

Figure 10-11 External and internal anatomy of the horseshoe crab.

Dorsal view

Simple eye

Compound eye

Carapace

Telson

Ventral view

Chelicera (claw)

Walking legs

Book gills

resemble pages in a book. These are the **book gills**, which are used for breathing and locomotion. The movement of the book gills enables larval horseshoe crabs to swim upside down, so that they can feed more easily on suspended food particles and plankton. Like crustaceans, the horseshoe crab has copper-based hemocyanin in its blood for transporting oxygen through its body.

Contrary to what most people think, the horseshoe crab's spiked tail, or **telson**, is not used as a weapon. It is used in locomotion. Also, when tossed on its back by a wave, the horseshoe crab sticks its telson in the sand to use as a lever to flip itself over.

Life Cycle of the Horseshoe Crab

In late spring, when the tide is high, thousands of horseshoe crabs invade the sandy beaches and marshes. The females, carrying the smaller males clutched to their backs, congregate at the high-tide mark where they dig holes in the sand and lay hundreds of tiny pale green eggs. The males, still attached to the females, externally fertilize the eggs, which are then covered with sand. (This abundance of horseshoe crab eggs in the sand provides a feast for shorebirds migrating north in the spring. In fact, the harvest of horseshoe crabs by people is limited, since their eggs are such an important food source for the migrating birds.) About two weeks after fertilization, the eggs hatch into tiny, swimming juvenile horseshoe crabs (unlike crustacean larvae) that are carried out to sea by the tide. Within a month, they develop the spiked tails that are characteristic of the adults.

As the horseshoe crab grows, its outer skin hardens to form the carapace. The young animal settles to the bottom, where it feeds on invertebrates in the sand. The horseshoe crab reaches sexual maturity in about eight years. During this period of growth, the horseshoe crab undergoes many molts, casting off its outer shell each time it grows. These cast-off shells are often found on beaches and mistaken for dead crabs, because they look so complete. Horseshoe crabs can live for as long as 20 years. Fossils of the horseshoe crab show that this animal has not changed very much throughout its more than 400-million-year history. For this reason, the horseshoe crab is often described as a "living fossil."

Marine Insects

Arthropods also include the marine insects. While insects tend not to inhabit water with a high salinity (such as the open ocean), some are found in habitats such as salt marshes. The ones that you are most likely to come into contact with are the biting insects that live in inland bays and marshes. They all possess a chitinous exoskeleton and jointed appendages, the characteristic features of all arthropods. Unlike the crustaceans, insects have only three pairs of legs, and a body made up of three segments (head, thorax, and abdomen). Insects also differ from some other arthropods in that they have just one set of antennae and one pair of eyes. They are placed in their own class, Insecta, which comprises nearly a million known species.

The most familiar of marine insects is the marsh mosquito. Mosquitoes draw blood from their hosts by using a specialized mouthpart, the proboscis, like a hypodermic needle. Another common marine insect is the greenhead fly, which is seen in salt marshes and above sand dunes. One biting marine insect, the sand fly (*Culicoides*), commonly called the "no-see-um," gets its name from the fact that the insect is so small that you can be bitten by it without even seeing it. This fly can transmit a fever to the people that it bites.

Mosquitoes, and some species of flies, inhabit inland marine habitats (such as estuaries) where the wave impact is slight. These calm waters are less saline than the ocean; and they provide a flat surface on which the insects can lay their eggs. The eggs develop into larvae, which then develop and hatch into the adult insects. The salt-marsh mosquito is a pest to humans, but it is food to the fish and marine invertebrates that feed on its larvae.

10.4 Section Review

1. Why is the horseshoe crab not classified as a true crab?

2. Explain why the horseshoe crab is considered a living fossil.

3. Why are mosquitoes classified as arthropods, along with such animals as lobsters?

Laboratory Investigation 10

Adaptations of Crabs

PROBLEM: How is the crab adapted for carrying out its life functions?

SKILL: Observing adaptive features of a crab's external anatomy.

MATERIALS: frozen crabs (thawed), trays, hand lens, probe.

PROCEDURE

1. Put a crab on a tray with the dorsal side facing up. Tap the shell with your pen. Notice the hardness of the shell. The shell is the crab's exoskeleton. Because the exoskeleton is rigid, the crab has to shed it, or molt, several times during its lifetime as its body size increases. (See Figure 10-12.)

2. The body of a crab is divided into segments: the cephalothorax and the abdomen. The cephalothorax is composed of two parts, the head and the chest. The shell that covers the cephalothorax is the carapace. The abdomen is located on the ventral side. Turn the crab over and look at the flat abdomen, located between the legs. In the male, the abdomen is narrow and V-shaped. In the female, it is wide and U-shaped.

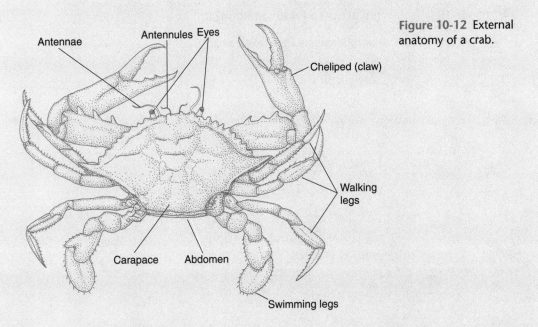

Figure 10-12 External anatomy of a crab.

Antennae Antennules Eyes Cheliped (claw) Walking legs Carapace Abdomen Swimming legs

3. How does the crab move? Crabs use their legs, or appendages, for crawling and swimming. Count the number of legs. There are five pairs (ten legs); hence the name of the order to which crabs and lobsters belong: Decapoda (*deca* meaning "ten"; *pod* meaning "foot"). Why are some of the legs pointed and others flat? The pointed ones are used for crawling, and the flat ones are used like paddles for swimming. Examine the first pair of legs, which are modified as claws, called chelipeds. The chelipeds catch and hold food and bring it to the mouth. Sketch the appendages in your notebook. Identify which ones are used for swimming, crawling, and feeding.

4. How does the crab ingest food? Food is brought to the mouth by the claws. Locate the mouth using your probe and hand lens. The mouth is surrounded by several pairs of mouthparts, which are used for tasting, moving, and shredding the food into smaller pieces.

5. How does the crab sense its environment? Use the hand lens to observe its eyes and its antennae in the head region. The two eyes are mounted on stalks. The two pairs of antennae are used to sense the environment. They function as receptors for touch, temperature, sound, and smell.

OBSERVATIONS AND ANALYSES

1. What are the advantages and disadvantages of an exoskeleton?

2. What body parts does the crab use to ingest food?

3. How is the crab adapted for locomotion?

4. How does the crab sense its environment?

Answer the following questions on a separate sheet of paper.

Vocabulary

The following list contains all the boldface terms in this chapter.

amphipods, arthropods, barnacle, book gills, carapace, cephalothorax, chitin, cirri, copepod, crustaceans, exoskeleton, horseshoe crab, isopods, krill, molting, swimmerets, telson

Fill In

Use one of the vocabulary terms listed above to complete each sentence.

1. A characteristic of all arthropods is their tough _____.
2. A crustacean's head and chest regions make up the _____.
3. Paddlelike _____ are used for movement by crustaceans.
4. Crustaceans shed their shell in a process called _____.
5. A tiny crustacean at the base of ocean food chains is the
   _____.

Think and Write

Use the information in this chapter to respond to these items.

6. Compare and contrast feeding in the barnacle and the crab.
7. Identify the important features that are characteristic of all arthropods.
8. Describe a main difference in isopod and amphipod body shapes.

Inquiry

Base your answers to questions 9 to 12 on the experiment described below and on your knowledge of marine science.

A student observed the movement of cirri in three groups of rock barnacles kept at different water temperatures. He recorded the number of cirri beats per minute for one (random) barnacle from each group in six separate trials. All other conditions in the three groups' environments, such as levels of salinity and sunlight, were kept constant for each group throughout the six trials. The results are shown in the table below.

Trial	Group A Cirri beats per minute at 5°C	Group B Cirri beats per minute at 15°C	Group C Cirri beats per minute at 28°C
1	12	25	25
2	11	26	30
3	14	23	35
4	9	28	26
5	16	19	31
6	10	23	33
Total	72	144	180
Average	12	24	30

9. Which statement is most correct based on the data in the table? *a.* Barnacles in group A showed the most cirri activity. *b.* Barnacles in group B were, on average, twice as active as those in group A. *c.* Barnacles in group C were the least active of all the barnacles. *d.* All three groups of barnacles showed little or no cirri activity.

10. The results of this experiment show that *a.* as temperature increases, cirri activity decreases *b.* as temperature decreases, cirri activity increases *c.* as temperature increases, cirri activity increases *d.* there is no relationship between temperature and cirri activity.

11. A tentative conclusion that can be drawn from the data is that *a.* temperature has an effect on the life activities of barnacles *b.* food is the only factor that affects cirri activity *c.* barnacles are warm-blooded animals *d.* barnacles show no response to changes in their environment.

12. When the student who performed this experiment wrote up his results, he called it "The Effect of Temperature Change on

Cirri Movement in Barnacles." In one or more complete sentences, describe and provide a reasonable scientific explanation for the results obtained.

Multiple Choice

Choose the response that best completes the sentence or answers the question.

13. In lobsters, the part of the shell that covers the cephalothorax is the   *a.* swimmeret   *b.* carapace   *c.* book gills *d.* telson.

14. An important food source for filter-feeding whales is the crustacean known as the   *a.* scud   *b.* krill   *c.* barnacle *d.* isopod.

15. The process of molting is related to the life function of *a.* growth   *b.* reproduction   *c.* sensitivity   *d.* digestion.

16. The horseshoe crab's telson is used in   *a.* reproduction *b.* defense   *c.* locomotion   *d.* food-getting.

17. To escape predators, fiddler crabs will   *a.* cover themselves with seaweed   *b.* hide in their sand tunnels   *c.* change their shell color   *d.* lie motionless in the water.

18. Barnacles obtain food by means of appendages called *a.* tentacles   *b.* cirri   *c.* swimmerets   *d.* cilia.

19. Which arthropod is considered to be a living fossil? *a.* hermit crab   *b.* copepod   *c.* horseshoe crab *d.* sea roach

Refer to the following drawing of an arthropod to answer questions 20 and 21.

20. This organism is classified as a type of *a.* true crab   *b.* horseshoe crab *c.* amphipod   *d.* copepod.

21. The structure labeled "A" is used for *a.* swimming   *b.* digging into sand *c.* mating   *d.* food-getting.

22. Which statement about crustaceans is true? *a.* All have the same number of appendages. *b.* All have an internal skeleton. *c.* All have an exoskeleton. *d.* Only some have jointed appendages.

23. All of the following arthropods are crustaceans *except* the *a.* lobster *b.* barnacle *c.* horseshoe crab *d.* beach flea.

24. The arthropod that lives as an encrusting organism is the *a.* barnacle *b.* lobster *c.* hermit crab *d.* copepod.

25. The horseshoe crab breathes by means of its *a.* carapace *b.* telson *c.* chelicera *d.* book gills.

Research/Activity

■ To find out if the presence of food particles increases feeding response in barnacles, sprinkle some fish food into a container of live barnacles. Count the number of times the cirri move per minute (count beats for 15 seconds; then multiply by 4). Compare with a control group that has not been fed. Record your data in a table and draw a graph.

■ Barnacles cause problems when they coat the hulls of ships. Unfortunately, the "antifouling" chemicals that have been used to deter them are toxic to other forms of marine life. Report on what is being done to try to find safer alternatives to these chemicals.

11 Echinoderms

When you have completed this chapter, you should be able to:

IDENTIFY the important characteristics of all echinoderms.

DISTINGUISH among the five main groups of echinoderms.

DISCUSS some important life functions of the echinoderms.

While slowly moving across the surface of a coral reef, the crown-of-thorns sea star *(Acanthaster planci)* devours the coral animals in its path. A voracious predator, the crown-of-thorns is responsible for the destruction of coral reefs around Hawaii and other tropical islands in the South Pacific.

Sea stars, or starfish, are invertebrates that have a spiny skin covering, among other unique features. Such spiny-skinned animals are classified in the phylum Echinodermata (meaning "spiny skinned"). This group also includes such animals as the sea urchin, brittle star, and sea cucumber. In this chapter, you will learn how these exclusively marine animals are adapted to the ocean environment.

11.1 STARS IN THE SEA

Stroll along a beach and you might see a "starfish" clinging to rocks at the water's edge. These bottom-dwelling invertebrates are not fish at all; they have neither scales nor a backbone. In fact, starfish, or **sea stars**, as they are now more appropriately called, are types of **echinoderms**—spiny-skinned animals that lack body segmentation but have radial symmetry (usually five-part) and an internal skeleton. In radial symmetry, all similar body parts are regularly arranged around the central point of an animal's body.

There are more than 5000 species of echinoderms, which are placed in five main classes: sea stars; sea urchins and sand dollars; brittle stars; sea lilies and feather stars; and sea cucumbers. This first section describes the familiar sea stars, members of the class Asteroidea, as representative of this phylum.

Types of Sea Stars

Sea stars are found from the subtidal zone to the deepest parts of the ocean. These echinoderms usually have five (or multiples of five) appendages, or arms, radiating out from a central body—hence the "star" in their name. However, there is great variety among the sea stars.

The common Atlantic sea star *(Asterias)*, which looks typical, is found in mussel and clam beds along the East Coast. (See Figure 11-1.) Likewise, the West Coast sea star *(Pisaster)* is found in beds of California mussels. The seafood industry regards sea stars as pests, because they can eat large numbers of commercially important bivalves. The bat star *(Patiria)*, whose five arms are connected in a weblike structure like the wings of a bat, is commonly found in kelp beds along the West Coast, from Alaska to California. Another echinoderm from the Pacific, the sun star *(Solaster)*, has 10 to 15 arms. The sun star lives on a variety of ocean bottoms, from low tide to depths of more than 400 meters. Sun stars are atypical in that they prey on other sea stars and even eat members of their own species. (See Figure 11-2.)

Figure 11-1 The common Atlantic sea star *Asterias.*

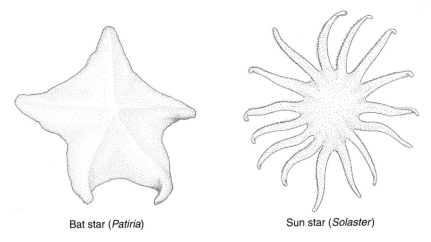

Figure 11-2 Two Pacific sea stars—the bat star *Patiria*, which has a web-like structure, and the sun star *Solaster*, which has up to 15 arms.

Bat star (*Patiria*)   Sun star (*Solaster*)

11.1 Section Review

1. Why is it more accurate to say "sea star" than "starfish"?

2. List some important characteristics of the sea stars.

3. Why do some people consider sea stars to be pests?

11.2 ADAPTATIONS IN THE SEA STAR

Sea stars often lose an appendage in struggles with other marine animals. When a sea star loses an arm, it can grow another one back, or regenerate it, as evidenced by the fact that one arm will be noticeably shorter than the others.

The spines that give sea stars their characteristic rough skin are composed of calcium carbonate ($CaCO_3$). The spines are connected to an internal skeleton, or **endoskeleton**, within the skin, also composed of $CaCO_3$. The spiny covering helps support and protect the echinoderm.

Sea stars breathe through their skin and through their tube feet. On the dorsal surface of the skin are small, ciliated fingerlike projections called **skin gills**. Oxygen from the water diffuses through the thin membrane of the tube feet and skin gills into a fluid-filled space under the skin called a coelom. The coelom is lined with ciliated cells that beat back and forth to circulate oxygenated fluid around the body. Cell wastes and carbon dioxide diffuse from the coelom through the skin gills and tube feet to the outside. In effect, the sea star has an open circulatory system.

Feeding and Locomotion in the Sea Star

Sea stars use their arms for locomotion and for food-getting. The underside, or ventral surface, of each arm contains numerous little **tube feet** located in a groove. At the end of each tube foot is a suction disk. When the suction disk comes into contact with a hard surface, it clings to that surface. Muscles in the tube feet control the clinging and pulling actions that enable the sea star to move. This "walking" motion helps the sea star find its food. In addition, as discussed above, for most echinoderms the thin walls of the tube feet serve as an important respiratory surface for the exchange of gases. (See Figure 11-3.)

Bivalve mollusks are a favorite food of the sea star. How does a sea star open up a clam? The sea star uses its hundreds of tube feet to grasp the clam and cling onto each of its shells. The tube feet exert a force that pulls the two shells in opposite directions. When that force is applied for several hours, the adductor muscles inside the clam become tired, and the clam opens.

Figure 11-3 External anatomy of a sea star (ventral view).

Ventral surface

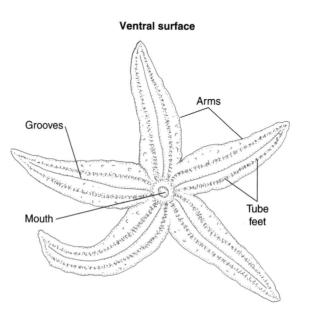

Grooves

Arms

Mouth

Tube feet

How does the sea star consume the clam? Since clams are usually too big to fit through a sea star's mouth (located in the center of its underside), the sea star pushes its thin, membranous stomach out through its mouth to engulf the food. (In some cases, the sea star's stomach can be pushed into a shell that is not tightly shut, without the tube feet first prying the shell open.) Digestive enzymes secreted by the sea star's stomach digest the food externally. The sea star then pulls back its stomach, which contains the digested food particles. Nutrients are absorbed and transported to its body cells in the fluid-filled coelom. Wastes are eliminated through the anus. (Undigested wastes, such as shell fragments, are eliminated through the mouth.)

Locomotion is necessary for food-getting by sea stars. How is movement accomplished? A network of water-filled canals and tubes, called the **water vascular system**, enables movement in sea stars. Tracing the pathway of water through this system will help you to understand how it works. (See Figure 11-4.) Water enters the sea star (when there is a loss of internal liquid) through a small filter called the **sieve plate**, also called the *madreporite*. The sieve plate is found on the topside, or dorsal surface, of the sea star near its cen-

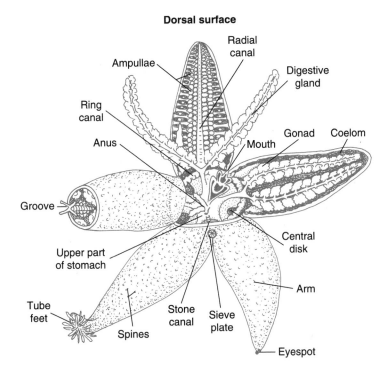

Dorsal surface

Radial canal

Ampullae

Digestive gland

Ring canal

Gonad   Coelom

Anus

Mouth

Groove

Central disk

Upper part of stomach

Arm

Tube feet

Stone canal   Sieve plate

Spines

Eyespot

Figure 11-4 External and internal (cut-away) anatomy of a sea star (dorsal view).

ter, an area referred to as the central disk. After entering, the water passes down through a short stone canal, then into a circular ring canal within the central disk. From the ring canal, the water flows through the radial canals. There is one radial canal in each arm. Many tube feet are connected to each of the radial canals.

Movement occurs when water enters the tube feet. At the top of each tube foot is an **ampulla**, a structure that resembles the rubber bulb on a medicine dropper. After the ampulla fills with water from the radial canal, it contracts. This contraction of the ampullae (by ampullar muscles) forces water into the tube feet, causing them to extend. Then, when the tube feet make contact with a substrate, circular and longitudinal muscle fibers within them contract, forcing water back into the ampullae. This exit of water from the tube feet creates the suction that holds the sea star to a substrate or clamshell. The sea star uses this suction force to push and pull itself along or to open a bivalve shell.

Sea Star Response, Reproduction, and Regeneration

Sea stars are sluggish creatures and slow to respond to stimuli because they have a simple nervous system. However, they can respond to stimuli such as changes in the amount of light. Tiny light receptors, called **eyespots**, are located at the end of each arm. The eyespots convert light into electrical impulses, which are carried by nerves to a central nerve ring that encircles the mouth. The nerve ring coordinates the movements of the arms by sending messages to and from radial nerves located in the arms.

Sea stars have separate sexes, but the sexes look identical so you cannot know the sex by looking at them. You have to examine the sea star internally. Look at the cross section of part of a sea star's arm. (Refer to Figure 11-4). Gonads are located inside each arm, near the central disk. Ovaries and testes shed the eggs and sperm, respectively, into the water through openings found between the appendages. Both fertilization and development occur externally.

Sea stars can also increase their numbers through regeneration. If an arm is torn off during a struggle (for example, with a predator), a new arm can be regenerated; and a whole new sea star can grow from the severed appendage, provided part of the central disk is present.

The lab investigation at the end of this chapter will give you a better understanding of the sea star's external anatomy.

11.2 SECTION REVIEW

1. How does a sea star open a bivalve such as a clam?

2. Explain how a sea star uses its tube feet to move.

3. Describe ingestion and digestion in a sea star.

11.3 SEA URCHINS AND SAND DOLLARS

The echinoderm with the most impressive spines is definitely the **sea urchin**, a member of the class Echinoidea. The sea urchin's movable spines are attached to its internal skeleton, which is formed by bony plates that are fused. (As in the sea star, both the spines and endoskeleton are made of $CaCO_3$.) This endoskeleton, which remains when a sea urchin dies, is sometimes found washed up on a beach. It has an attractive pattern of raised bumps, evidence of the former attachment points for the spines.

The animals in this class, which also includes sand dollars and sea biscuits, are characterized by oval or round bodies that lack arms. They are the only echinoderms that use both their spines and tube feet to move. Sea urchins inhabit the intertidal and subtidal zones along rocky coasts. They move very slowly along the rock surfaces, scraping off algae with their unique five-toothed mouth structure, called an **Aristotle's lantern** (because of its resemblance to an ancient Greek lantern). Along the rocky coasts of Maine, California, the Pacific Northwest, and elsewhere in the world, sea urchins do such a good job of grazing that they often scrape the rocks bare of seaweeds.

Predation and Protection Among Sea Urchins

In shallow tropical waters, be careful where you walk—you could step on the long-spined sea urchin *(Diadema)*. The sharp spines can inflict a very painful puncture wound. In some species, the spines may be hollow and contain toxins as well. Other species of sea urchin, such as the purple sea urchin *(Arbacia)* and the green sea

Green sea urchin

Hatpin sea urchin

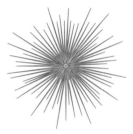

Long-spined
sea urchin

urchin *(Strongylocentrotus),* which graze on seaweeds along the Pacific Coast, have shorter, thicker spines. For protection from predators and strong wave action, sea urchins often use their spines to wedge themselves in the spaces between rocks. (See Figure 11-5.) The rock-boring urchin *(Echinometra)* that inhabits the Caribbean takes this a step further—it uses its teeth to bore into the rock, forming a cup to hide in.

The spines of the sea urchin are a natural protection against most predators, except the California sea otter. The sea otter (see Chapter 14) is a marine mammal that dives to the ocean floor to hunt for sea urchins. After picking up a sea urchin, the sea otter swims to the surface, rolls over on its back, then places the sea urchin on its chest. Using a rock that it also picks up from the seafloor, the sea otter cracks open the sea urchin and eats the contents. Humans also eat sea urchins. In many countries, sea urchins are considered a delicacy because of the eggs they contain.

Life Cycle of the Sea Urchin

There are male and female sea urchins but, as with the sea star, you cannot tell the animal's sex just by looking at it. During the breeding season, the female sea urchin releases a great number of large eggs into the water. Sperm released from the male sea urchin fertilizes the eggs externally.

The processes of fertilization and development in the sea urchin can be easily observed under the microscope. For this reason, biologists use the sea urchin in embryological studies. A useful feature of their development is that up to the blastula stage, all the cells of the embryo are identical—if separated from the embryo, each cell can

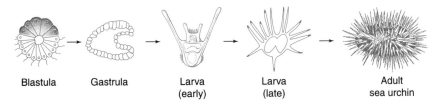

Figure 11-6 Development of the sea urchin, from zygote to adult stage.

| Blastula | Gastrula | Larva (early) | Larva (late) | Adult sea urchin |

develop into a separate, identical animal. Much of what we know today about embryology has come from studies done on the sea urchin. Like those of the other echinoderms, embryos of the sea urchin go through a free-swimming larval phase. The larvae, which are bilaterally symmetrical, live as part of the plankton community until they settle on the seafloor and develop into adult sea urchins. (See Figure 11-6.)

Sand Dollars and Sea Biscuits

The **sand dollar** *(Echinarachnius)* looks like a large coin (hence its name), and has short spines covering its skin. Sand dollars use their spines to burrow in the sand, where they feed by catching plankton and organic debris in sticky strings beneath their spines. The food is then pushed toward the mouth. Members of this class have a well-developed intestine and anus, through which the food is digested and eliminated, respectively. When a sand dollar dies and its soft parts decay, the flat internal skeleton of calcium carbonate remains. People often collect these attractive "shells," which have a distinctive star-shaped pattern on them. (See Figure 11-7.)

Closely related to the sand dollar is the sea biscuit *(Plagiobrissus)*. However, this echinoderm is more rounded (like a biscuit), has longer spines, and inhabits the sandy seafloor around coral reefs. Sea biscuits feed on organic debris and algae.

11.3 SECTION REVIEW

1. Compare food-getting in sea stars and sea urchins.

2. By what method do sand dollars feed? What do they eat?

3. Why is the sea urchin considered a good organism for embryological studies?

Sand dollar

Figure 11-7 The sand dollar uses its short spines for burrowing in the sand; its "shell," or internal skeleton, has a unique pattern.

11.4 ECCENTRIC ECHINODERMS

The sea urchin and the sea star are probably the most commonly encountered echinoderms. Species of echinoderms that may be less familiar to you are described below.

More "Stars" in the Sea

One of the most curious of the echinoderms is the brittle star, which is placed in its own class, Ophiuroidea. Although they are actually the most abundant of the echinoderms (in terms of both numbers of species and individuals), **brittle stars** (such as *Ophiopholis*, *Ophiocoma*, and *Ophioderma*) are not very obvious because they are nocturnal, bottom-dwelling animals that hide under rocks during the day. Brittle stars live in the intertidal zone, from the arctic to the tropics. A subgroup of brittle stars, called basket stars *(Gorgonocephalus)*, have coiled, branching arms and live on the deep ocean floor, thousands of meters below the surface.

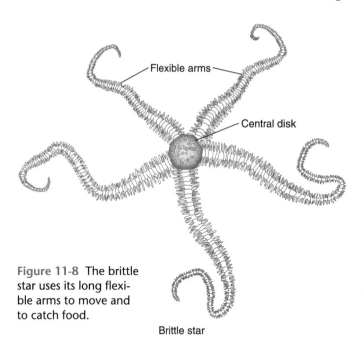

Flexible arms

Central disk

Figure 11-8 The brittle star uses its long flexible arms to move and to catch food.

Brittle star

Unlike the sea stars, brittle stars have a distinct, flattened central disk; and they do not use their tube feet for movement. Rather, they have muscles in their long, narrow flexible arms that enable them to scurry rapidly about on the seafloor, looking for morsels of food. (See Figure 11-8.) The brittle star is so named because of its delicate appearance and its ability to detach its arms when attacked, thus evading predators. Like the sea stars, brittle stars can regenerate their missing arms.

Brittle stars have more than one feeding method. They can use their arms to gather organic debris from the seafloor, to capture live invertebrates, to filter-feed by trapping bits of food in sticky strands,

CONSERVATION

In a "Pickle" over the Sea Cucumber

It doesn't look very appetizing, this spiny-skinned, oblong-shaped animal. Yet in Southeast Asia, a single cooked and dried sea cucumber is considered a delicacy and sells for $80. With interest in, and profits from, the sea cucumber so high, the demand for these echinoderms has far outstripped their numbers in local South East Asian waters. So, the sea cucumber fishing federation turned to the Galápagos Islands, located over 900 km off the coast of Ecuador, as a potential source of this item. Although much of the area has been declared a national park by Ecuador, the islands are home to 15,000 people, most of whom make their living from the sea.

By 1992, about 30 million sea cucumbers had been collected from the waters around the Galápagos Islands. Scientists were concerned that the echinoderm was in danger of being over-harvested. So, the government of Ecuador imposed a one-year ban on the harvest, followed by a partial ban. Then, in the mid-1990s, Ecuador established a fishing season and quotas to reduce over-harvesting. Unfortunately, these conservation measures were not successful. By the late 1990s, more than 6 million sea cucumbers were being harvested each year. In 1999, a complete ban on commercial fishing of sea cucumbers was enacted. This led to strikes and protests by the local fishermen, and to an increase in the illegal harvest of sea cucumbers.

The ban was lifted again in 2002, based on the outcome of a scientific study of the sea cucumber population and on a meeting that included local fishermen, government officials, and the scientific community. Stricter guidelines and new quotas for the harvest were established. Now, all fishermen will be licensed; the harvest will be permitted in designated areas only; and monitors will be hired to check for compliance.

Hopefully, a compromise has been reached that will allow the development of a sustainable harvest of Galápagos sea cucumbers (one that does not threaten their survival). Then, all parties concerned will no longer be in a "pickle" over these unlikely objects of desire.

QUESTIONS

1. Why are some people concerned about the harvest of Galápagos sea cucumbers?

2. What groups of people are involved in this controversy? Defend the position of one group.

3. Describe a possible compromise (solution) that might satisfy all the parties involved.

4. How is sustainability of the harvest related to survival of the sea cucumber?

or to capture suspended food bits with their tube feet—all of which is brought into their jawed mouth.

"Lilies" and "Feathers" in the Sea

The sea lilies and feather stars—members of the class Crinoidea—look much more like flowers than like animals. Known as **crinoids**, they are the most ancient group of echinoderms, having originated hundreds of millions of years ago. The body of a crinoid is composed of dozens of feathery arms, usually perched atop a jointed stalk. Crinoids generally have just a limited ability to move. The **sea lilies** are sessile; they live attached by a stalk to the ocean bottom. (See Figure 11-9.) The **feather stars** mostly crawl along coral reefs, but some swim by flapping their arms. Using a type of feeding similar to that of the brittle stars, crinoids filter feed by waving their arms, thereby capturing bits of zooplankton in their tube feet (which then pass the food to the mouth). Like the brittle stars, crinoids do not use their tube feet for locomotion.

Sea lily

Figure 11-9 The sea lily is a sessile crinoid with feathery arms, used for filter feeding.

"Cucumbers" on the Seafloor

At first glance, members of this last group of echinoderms do not look much like echinoderms; in fact, they do not even look like animals! However, on closer examination you can see that the sea cucumber—whose soft, oblong body lacks arms—has tube feet that are arranged in five rows, similar to the five-part radial pattern seen in the sea star. The **sea cucumbers**, which are placed in the class Holothuroidea, have lost the endoskeleton and spines typical of their phylum, retaining only small bony pieces in the skin. They live on sandy and rocky seafloors in intertidal and subtidal zones and are most abundant at great depths. (See Figure 11-10.)

Sea cucumbers such as *Holothuria* use their sticky, branching tentacles—which are actually enlarged tube feet—to trap microscopic organisms. The tentacles, which are located around the mouth, are extended during feeding and retracted when the animal is disturbed. Members of the genus *Cucumaria* that live on the East and West coasts have five rows of tube feet along their bodies, which are used for slowly moving along the substrate and for trapping food

Sea cucumber

Figure 11-10 The sea cucumber has five rows of tube feet, used for feeding and movement.

particles in the sand. Sea cucumbers have a one-way digestive tract; wastes are excreted through the anus. Whereas most echinoderms exchange gases through their tube feet and skin gills, sea cucumbers take in and release water through their anus. Gas exchange then occurs inside the coelom across the membranes of a structure called the "respiratory tree." Another unusual feature of the sea cucumber is that it can release its digestive organs when disturbed by a predator, thus leaving a meal for the predator while it escapes. It later regenerates the lost organs.

11.4 SECTION REVIEW

1. Describe some feeding methods of the brittle stars.

2. What is the basic structure of a crinoid? How does it feed?

3. What features of the sea cucumber show it is an echinoderm?

Laboratory Investigation 11

Adaptations of Sea Stars

PROBLEM: How is the sea star adapted for carrying out its life functions?

SKILL: Identifying relationships between body structures and life functions.

MATERIALS: Living sea star, pan of seawater, hand lens, fresh clam or mussel.

PROCEDURE

1. Put a sea star, dorsal side up, in a shallow pan and cover it with seawater. Use the sea star diagrams in Figures 11-3 and 11-4 as a guide. How many arms or appendages does the sea star have? Make a sketch of your sea star. Label one of the arms in your drawing.

2. Feel the skin of the sea star. Then examine the skin with a hand lens. Notice the short spines, which you were able to feel. The spines are connected to an endoskeleton, which is composed of calcium carbonate (like the shells of mollusks). Label the spines in your drawing.

3. How does the sea star breathe? Examine the skin with your hand lens. Look for tiny fingerlike projections, called skin gills. Oxygen diffuses from the water through the thin membrane of the skin gills and into the coelom.

4. Locate the sieve plate, or madreporite, which is a white or orange spot on the dorsal surface. Water enters through the sieve plate, then passes through a network of canals that ends in the tube feet.

5. Locate the tube feet by turning the sea star over. The many tube feet are in grooves that run down the center of each arm. Touch the tube feet; you will notice that they cling to your finger. Each tube foot looks like a tiny plunger. Put the sea star back in the pan of water, with the tube feet facing down. Notice the clinging and pulling action of the tube feet used in locomotion. Make a sketch of a tube foot and describe its function.

6. Now place the sea star ventral side up in the pan of seawater. Make a sketch of the sea star that shows its ventral side. Describe the motion of the tube feet. Can the sea star turn itself over? Which arms does it use to turn over? Record your observations in a copy of Table 11-1 in your notebook.

TABLE 11-1 SEA STAR STRUCTURES AND FUNCTIONS

Sea Star Observations	Structure	Function	Behavior
Dorsal Side			
Ventral Side			

7. How does the sea star feed? Look for the mouth in the center of the sea star on its ventral side. The mouth is too small to ingest a whole clam. Instead, the sea star pushes its thin, membranous stomach out through its mouth and into the clam's shell, where it digests the food externally. Open up a mussel or clam shell and put it in a pan of seawater. Place a sea star that has not been fed for a few days next to the clam. Record your observations.

8. How does a sea star open up a clam? Put your hand underwater and place a sea star on top of it. Gently try to pull the sea star off your hand. Notice how it clings to your skin. The tube feet, with their suction disks, generate a pulling force. When the arms of a sea star are draped over the two shells of a clam, hundreds of tube feet pull the shells in opposite directions. The adductor muscles in the clam become fatigued, causing the shells to open.

OBSERVATIONS AND ANALYSES

1. How does a sea star move?

2. How does the sea star ingest and digest food?

3. Compare the "skeleton" of a mollusk with that of an echinoderm.

Answer the following questions on a separate sheet of paper.

Vocabulary

The following list contains all the boldface terms in this chapter.

ampulla, Aristotle's lantern, brittle stars, crinoids, echinoderms, endoskeleton, eyespots, feather stars, sand dollar, sea cucumbers, sea lilies, sea stars, sea urchin, sieve plate, skin gills, tube feet, water vascular system

Fill In

Use one of the vocabulary terms listed above to complete each sentence.

1. Delicate echinoderms found on the seafloor are the _____.

2. The _____ uses its short spines to burrow in the sand.

3. Water enters a sea star through its madreporite, or _____.

4. The _____ are the most ancient group of sessile echinoderms.

5. In sea stars, the clinging and pulling of muscles in _____ allows movement.

Think and Write

Use the information in this chapter to respond to these items.

6. Describe what happens if a sea star loses one of its arms.

7. What functions do spines serve in the sea urchins and sand dollars?

8. Compare and contrast the lifestyles of sea lilies and feather stars.

Inquiry

Base your answers to questions 9 through 12 on the results of the experiment described below and on your knowledge of marine science.

A marine biology student hypothesized that a brittle star would have a slower turnover response than an Atlantic sea star. To test this idea, he placed the two species of echinoderms upside down in separate containers of seawater under the same experimental conditions. The time it took for each animal to turn over (in each of six trials) is shown in the table below.

Brittle Star		Atlantic Sea Star	
Trial	Turnover Response Time (minutes)	Trial	Turnover Response Time (minutes)
1	0.15	1	6.0
2	0.17	2	10.0
3	0.33	3	2.0
4	0.25	4	1.75
5	0.23	5	2.50
6	0.15	6	2.0
Average	0.21	Average	4.04

9. Which part of the scientific method is represented by the data in the table? *a.* hypothesis *b.* materials *c.* results *d.* conclusion

10. Which is an accurate statement regarding the data in the table? *a.* The data support the hypothesis. *b.* The hypothesis is not supported by the data. *c.* The average turnover response for the brittle star is 21 seconds. *d.* The turnover response was recorded in seconds, not minutes.

11. A tentative conclusion that can be drawn from the data in the table is that *a.* the brittle star moves more quickly than the Atlantic sea star *b.* the Atlantic sea star moves more quickly than the brittle star *c.* turnover response in echinoderms cannot be measured in minutes *d.* there is no significant difference in turnover response time between the sea star and the brittle star.

12. Which of the following suggests the best way to verify the results of this experiment? *a.* Perform the experiment again, but with fewer trials. *b.* Perform the experiment again, but with more trials. *c.* Add food to give each animal an incentive for movement. *d.* Use brittle stars only in both containers of seawater.

Multiple Choice

Choose the response that best completes the sentence or answers the question.

13. The small ciliated projections that enable breathing in this animal are called
    a. spines
    b. skin gills
    c. ampullae
    d. eyespots.

14. You notice that a sea star in an aquarium has one very short arm. The best explanation for this is that   *a.* its growth hormones have been suppressed   *b.* the appendage was lost and is regenerating   *c.* its tube feet are not functioning   *d.* the arm is not really needed.

15. The side of a sea star on which its sieve plate is found is the   *a.* dorsal   *b.* ventral   *c.* anterior   *d.* posterior.

16. What prevents a sea star from falling off the side of an aquarium tank?   *a.* clinging action of its tube feet   *b.* suction by its mouth   *c.* adhesive properties of its spines   *d.* water pressure

17. The symmetry of echinoderms is referred to as   *a.* bilateral   *b.* radial   *c.* spiral   *d.* unilateral.

18. A sea star can open up a clam because of the functioning of its   *a.* tube feet   *b.* spines   *c.* stomach   *d.* madreporite.

19. The crown-of-thorns sea star is considered a pest because it   *a.* destroys coral reefs   *b.* consumes bivalve mollusks   *c.* is harmful to humans   *d.* is harmful to fish.

20. The function of the water vascular system in the sea star is to enable   *a.* locomotion   *b.* digestion   *c.* sensitivity   *d.* respiration.

21. Which of these echinoderms moves most rapidly on the seafloor?   *a.* sea star   *b.* brittle star   *c.* sea lily   *d.* sea urchin

22. The echinoderm that uses both its spines and its tube feet to move is the *a.* sea urchin *b.* brittle star *c.* feather star *d.* sea star.

23. Sea urchins scrape the algae from rock surfaces with a specialized mouthpart called the *a.* sieve plate *b.* madreporite *c.* Aristotle's lantern *d.* skin gill.

24. The echinoderm that differs from all others in that it lacks an endoskeleton, and only retains small bony pieces in its skin, is the *a.* sand dollar *b.* sea lily *c.* brittle star *d.* sea cucumber.

Research/Activity

■ Observe sea stars moving in an aquarium. Examine the underside (ventral surface) of the sea stars as they move along the sides of the tank. Describe the motion of their tube feet and explain how sea stars use their arms to grip surfaces and turn over.

■ Use the Internet to get an update on the sea cucumber harvest in the Galápagos Islands or to research the latest findings on the damage done to coral reefs by the crown-of-thorns sea star. Write a report and present your findings to the class.

UNIT 4

MARINE VERTEBRATES

Along the equatorial shores of the Galápagos Islands, marine iguanas graze on the algae growing on underwater rocks. Sea lions swim offshore while Galápagos sharks cruise just beyond. Galápagos penguins "fly" undersea in pursuit of fish. And gannets cry overhead as frigate birds dive-bomb them, forcing them to give up their hard-won catch of fish.

The unique and much-studied ecosystem of the Galápagos Islands includes animals from several major classes. All interact within, and depend on, the marine environment for their survival. In this unit, you will study marine vertebrates, a diverse group that includes the fishes, reptiles, birds, and mammals—all of which are represented in the Galápagos, as well as in other marine ecosystems around the world.

12 Marine Fishes

When you have completed this chapter, you should be able to:

LIST the distinguishing features of the three classes of fishes.

IDENTIFY some important adaptations of fishes to ocean life.

DISCUSS various unusual adaptations that fishes have evolved.

12.1
Protochordates and Jawless Fishes

12.2
Cartilaginous Fishes

12.3
Bony Fishes

12.4
Unusual Adaptations in Fish

The barracuda waits for its next meal in the shadows of a coral reef. Suddenly, the barracuda strikes an unlucky fish that happens to swim by. Well-adapted for hunting other fish, the barracuda has a stream-lined body that enables it to swim quickly, and needlelike teeth for grasping slippery prey.

Probably no other animal more closely defines our thoughts of life in the sea than a fish. In fact, fish are the most common and diverse group of animals with backbones in the ocean—and in the world. Fish are an ancient group of animals whose origins date back more than 500 million years. There are about 20,000 different species of fish found worldwide (both freshwater and marine), and they show an astonishing variety of shapes, sizes, and colors—from the large protective schools of silver-gray fish typical of the cold northern waters to the exquisitely colored reef fish found in tropical waters. In this chapter, you will learn about the similarities and differences among the fishes and how these aquatic creatures are adapted to their environment.

12.1 PROTOCHORDATES AND JAWLESS FISHES

Look at the two marine animals shown on page 282. They look like invertebrates, but during some stage of their development they possess a hollow dorsal nerve cord, a notochord, and pharyngeal gill slits. (The notochord is a flexible, rodlike structure that supports the spinal cord.)

The presence of these structures in both animals means that they are **chordates**, related to such animals as fish, and classified within the phylum Chordata. However, unlike fish, they lack certain advanced structures and so are referred to as primitive (invertebrate) chordates, or **protochordates** (meaning "first chordates"), a subgroup of the chordate phylum. Fish, which are higher chordates, are also the first **vertebrates**, the group of animals characterized by having a skeleton, backbone, skull, and brain. The higher chordate subgroup includes all the vertebrates: fish, amphibians, reptiles, birds, and mammals. In vertebrate embryos, the notochord develops into a backbone. The protochordates are of great scientific interest because their ancestors served as an evolutionary link between the invertebrates and the vertebrates. Let's take a closer look at the protochordates—the closest we can get to animals living today that resemble the ancestors of modern fish from long ago.

Protochordates

If you were to scrape off some encrusting organisms from the underside of a floating dock or wharf piling, you might notice the grape-like clusters of animals known as sea squirts (*Molgula*). Sea squirts are found worldwide in a variety of habitats; they get their name from the fact that they squirt water when disturbed.

Sea squirts are also referred to as **tunicates** because they are covered by a clear, tough membrane that resembles a tunic. (See Figure 12-1.) Like mollusks, the sea squirts have an incurrent and an excurrent siphon. Water that contains food particles enters the incurrent siphon, and wastes exit through the excurrent siphon. Each sea squirt possesses both male and female reproductive organs, but

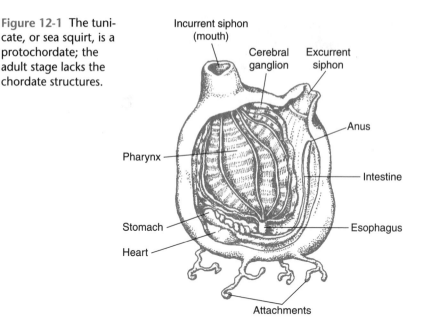

Figure 12-1 The tunicate, or sea squirt, is a protochordate; the adult stage lacks the chordate structures.

self-fertilization generally does not occur. After the sea squirts shed their gametes into the water, fertilization occurs externally. The fertilized egg develops into a tiny fishlike larva. The larva has a dorsal nerve cord, a notochord, and gill slits, which identify this animal as a primitive chordate. These structures disappear after the larva attaches to a substrate and grows into an adult sea squirt.

Another primitive chordate is the **lancelet** (*Amphioxus*). (See Figure 12-2.) This tiny, transparent fishlike animal lives half-buried in the sand, with its head sticking out to filter plankton from the water. Notice that the adult lancelet retains the dorsal nerve cord, notochord, and gill slits. There are separate sexes in the lancelet. Fertilization and development are both external.

A third primitive chordate is the **acorn worm** (*Saccoglossus*). Superficially, this animal looks like any other invertebrate, since it resembles a worm. Yet the acorn worm has a dorsal nerve cord and

Figure 12-2 The lancelet is a marine protochordate; the adult stage retains the chordate structures.

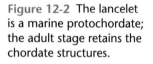
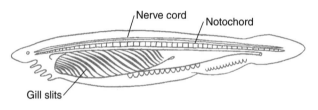

gill slits (but no notochord). Acorn worms burrow in the sand in the intertidal and subtidal zones, where they feed on organic debris that also includes sand. The indigestible sand is expelled in small lumps, or "casts," on the seafloor.

The study of the acorn worm, lancelet, and sea squirt is important because these protochordates represent an evolutionary link between the invertebrates and the vertebrates. Scientists think that the primitive chordates and fishes probably share an invertebrate ancestor.

Jawless Fishes

When you look at a typical fish in a school or home aquarium, you can observe its mouth opening and closing as it breathes. Most fish possess movable mouthparts called jaws, but a few species of fish do not. Hundreds of millions of years ago, before the appearance of the modern groups of fishes, the oceans were inhabited by **jawless fishes**. In fact, the first fish to evolve were the jawless fish. Many of these early fish had bodies covered with armor made up of bony plates. Among this group were the ancestors of the jawless fishes that survive today. Jawless fishes are the most primitive of the vertebrates, because they are jawless and they do not have a true backbone. The adults retain the larval notochord for support of their long, flexible bodies.

There are only two types of jawless fishes alive today—the sea lamprey and the hagfish. These fishes, which are placed in the class Agnatha (meaning "without jaws"), live as parasites. Some jawless fishes can grow up to one meter in length. The sea lamprey (*Petromyzon marinus*), which inhabits estuaries from Maine to Florida, uses the sucking disk on its mouth to attach to living trout and other host fish in rivers. (See Figure 12-3.) It feeds by using its teeth and rasping tongue to make a hole in the body of another fish; and then it

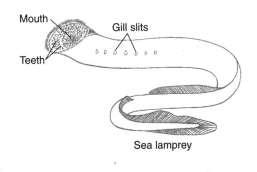

Figure 12-3 The sea lamprey is a jawless fish; note the gill slits, and the sharp teeth lining its circular mouth.

sucks out the blood and tissues of that host. Unlike most other fish, the lamprey does not have any scales. The Pacific hagfish (*Eptatretus stouti*) and the Atlantic hagfish (*Myxine glutinosa*) actually use the sharp teeth in their round mouths to burrow into the bodies of dead or dying fish.

12.1 SECTION REVIEW

1. How can you tell tunicates and lancelets are protochordates?
2. Why are all fishes classified as vertebrates in phylum Chordata?
3. Why are lampreys and hagfish referred to as primitive fishes?

12.2 CARTILAGINOUS FISHES

Feel the tip of your nose. The end feels soft because it is made of **cartilage**, a flexible connective tissue composed of cells and protein. The human skeleton is composed mostly of bone, but some parts are made of cartilage, such as the outer ear and the ends of bones where movable joints are found.

Fish whose entire skeleton is composed of cartilage are called **cartilaginous fishes**. They are placed in the class Chondrichthyes (meaning "cartilage fishes"). This ancient class includes such interesting fishes as the sharks, skates, and rays.

Characteristics of Cartilaginous Fishes

There are fewer than 700 species of cartilaginous fishes, as compared with more than 25,000 species of bony fishes. The cartilaginous fishes, which were the first jawed fish, have several unique characteristics. Unlike a bony fish, which has mucus-covered scales that come off easily, a cartilaginous fish (such as the shark) has **placoid scales**, which are actually tiny teeth that are deeply embedded in the skin. All of these scales point backward. If you rubbed the skin of a shark in a tail-to-head direction, it would feel like you were touching rough sandpaper.

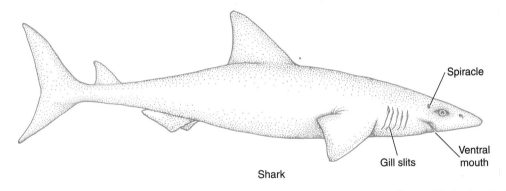

Shark

Spiracle

Gill slits

Ventral mouth

Figure 12-4 The shark, a cartilaginous fish, has the spiracles, gill slits, and ventral mouth typical of these fish.

Cartilaginous fish have visible **gill slits** for breathing, whereas bony fish have gill slits that are concealed by a flap. The position of the gills in rays, skates, and some bottom-dwelling sharks is ventral, on the underside of the body. How can these fish breathe when their gills are buried in the sand? The cartilaginous fish have a pair of breathing holes called **spiracles**, located on their dorsal side behind each eye. Water passes through the spiracles and flows to the gill chamber. Some bottom-dwelling sharks use their spiracles and mouth to actively pump water over their gills, since their limited swimming does not pass enough water through. The mouth in cartilaginous fish also is located on the ventral side. (See Figure 12-4.) A ventral mouth is usually an adaptation for bottom feeding, so the development of dorsal spiracles is a successful adaptation for this method of feeding. (*Note:* Most sharks are not bottom feeders, but all sharks have a ventral mouth as a shared characteristic.)

The fins of cartilaginous fishes are more rigid than those of bony fishes. A cartilaginous fish such as the shark relies on the lift provided by its winglike pectoral fins to help prevent sinking and to glide in the water. In rays and skates, the pectoral fins are even more highly developed. The up-and-down movements of the huge pectoral fins of a manta ray (*Manta*) resemble the wings of a bird in flight. While "flying" through the water, the giant manta—which can have a "wingspan" up to 7 meters across—opens its enormous mouth to filter feed on plankton. (See Figure 12-5 on page 286.)

Many species of rays and skates are bottom-dwellers. The stingray (*Dasyatis*) is found—often very well concealed—in the sand of the Gulf of Mexico and along the Atlantic coast from the Carolinas to Brazil. A sharp spine located near the base of its tail can inflict

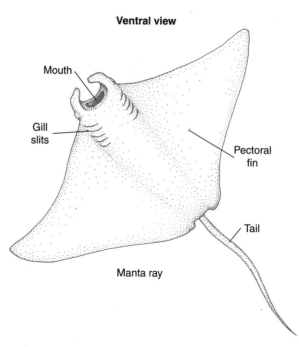

Figure 12-5 The manta ray, a cartilaginous fish, has ventral gill slits and a large mouth for filter feeding.

Ventral view

Mouth

Gill slits

Pectoral fin

Tail

Manta ray

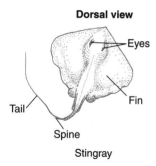

Dorsal view

Eyes

Tail

Fin

Spine

Stingray

Figure 12-6 The stingray is a bottom-dwelling cartilaginous fish that has a sharp spine at the base of its tail.

a very painful stab wound. (See Figure 12-6.) The skate (*Raja*), which does not have a spine on its tail, is found in temperate waters along the Atlantic and Pacific coasts. Rays and skates eat crustaceans and mollusks that live in the subtidal zone.

One of the more unusual cartilaginous fishes is the sawfish (*Pristis*). The sawfish inhabits coastal waters from Virginia to Brazil and in the Gulf of Mexico. Its long, bladelike snout contains 24 or more teeth that stick out on each side. The sawfish uses its snout like a weapon, swinging it back and forth as it swims through a school of fish to stun and kill them, after which it can easily eat them.

The Sharks

There are about 350 known species of sharks. Sharks vary greatly in size. The smallest shark is the pigmy shark (*Squaliolus laticaudus*), which is about 25 cm long. The biggest shark—in fact, the largest of all fish—is the whale shark (*Rhincodon typus*), a warm-water inhabitant that can grow to more than 15 meters in length. (See Figure 12-7.) The whale shark is actually harmless to humans and other

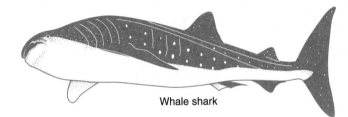

Figure 12-7 The whale shark, a plankton eater, is the largest shark—and also the largest fish—in the world.

Whale shark

vertebrates. Whale sharks are "strainers" (filter feeders) that consume enormous quantities of plankton as they swim near the surface with their mouths wide open. (The large basking shark feeds on plankton in the same manner in cooler waters.) Some sharks, such as the nurse shark and the leopard shark, are bottom-dwellers that have crushing teeth for feeding on shelled organisms such as mollusks. One of the most dangerous sharks to humans is the great white shark (*Carcharodon carcharias*). Though it preys mostly on marine mammals, such as seals and sea lions, this predatory shark has attacked and killed humans. The biggest great whites ever caught weighed over 1200 kg and measured from 5 to nearly 6.5 meters long. Other dangerous sharks that have killed humans include the tiger shark (which preys on sea turtles, seals, and smaller sharks), the bull shark (an aggressive species that is found in some freshwater habitats as well), and various hammerhead sharks (*Sphyrna*), which have two lateral projections on their heads with an eye at each end. (See Figure 12-8.) The more predatory sharks have sharp, often serrated, teeth for catching and cutting up their large, fast-moving prey.

Figure 12-8 The hammerhead is a predatory shark that has two lateral projections on its head, with an eye at each end.

Hammerhead shark

CONSERVATION
Shark Attack!

No, it's not what you think. This is not another report on a hapless victim falling prey to a shark attack. In fact, there were only five human fatalities from shark attacks reported worldwide in 2001. The problem is, actually, just the opposite. It is the sharks that are being preyed on by humans in unprecedented numbers. With bony fish stocks declining worldwide, commercial fisheries in many countries are turning to sharks to put food on the table. In the West, mako shark steaks have become a trendy item on restaurant menus. In the Pacific, sharks are slaughtered in large numbers just for their fins, which are used to make shark-fin soup. Ground-up shark cartilage is sold in pill form by the health food industry, because it is believed that the cartilage has anti-cancer properties—a myth based on the fact that sharks rarely get cancer.

NOAA estimates that, every year, millions of sharks are harvested worldwide to meet the demand for shark meat and shark products. Many thousands more are accidentally caught in fishing nets and then discarded. The depletion of shark populations worries fisheries experts because sharks take a long time to reach sexual maturity. Some sharks take as long as 15 years to mature, after which they reproduce only once every two years and produce few offspring. A declining shark population may take decades to recover, if it ever does.

The United States government is on the attack to protect sharks. NOAA officials now manage the shark fisheries in Atlantic and Gulf waters. Scientists are tagging sharks in order to monitor their population sizes. Finally, catch limits and quotas for commercial and recreational fishing have been put into effect to protect various shark species.

The shark is called an "apex predator," meaning that it is at the top of ocean food chains and, as such, has an important role to play in maintaining a balanced ecosystem. We humans too are apex predators who, as stewards of this planet, have to carefully manage and protect the marine species we depend on—including the fearsome, yet vulnerable, shark.

QUESTIONS

1. Why are some people more concerned about shark survival than about shark attacks?

2. Why are shark populations in decline worldwide?

3. Describe two conservation measures enacted to protect sharks.

4. Explain what is meant by the term "apex predator."

Structures and Behavior of Sharks

As a group, sharks have survived for more than 300 million years. Sharks are often called "living fossils" because they closely resemble some early ancestral forms. What features do sharks possess that have enabled them to survive for so long? Look at the structures of a typical shark, shown in Figure 12-9. Sharks have very sensitive receptors for the detection of stimuli. The **lateral line organ**, shown as a faint line along each side of a shark's body, can pick up sound vibrations over great distances. The shark's sense of smell is so acute that it can detect a small amount of blood nearly half a kilometer away. In fact, nearly two-thirds of the shark's brain is devoted to its sense of smell. Sharks can also detect weak electric fields given off by other organisms. Tiny pores in the shark's snout contain nerve receptors called **ampullae of Lorenzini**, which sense the electric fields generated by the muscles of fish and other potential prey.

After detecting prey, sharks are well equipped to catch them. The streamlined body shape enables the shark to move quickly through the water. The contraction of powerful muscles along the sides of the body, in combination with the thrust from the caudal (tail) fin, produces speedy acceleration. Predatory sharks have several rows of razor-sharp teeth, which constantly move forward to replace teeth that are lost. At any given time, a shark may have hundreds of teeth in its mouth.

Many sharks have to be in constant motion because if they do not swim, they have a tendency to sink to the bottom. (Swimming also provides a continuous stream of oxygenated water for the gills.)

Figure 12-9 External anatomy of a typical shark; note the lateral line organ, used to detect sound vibrations.

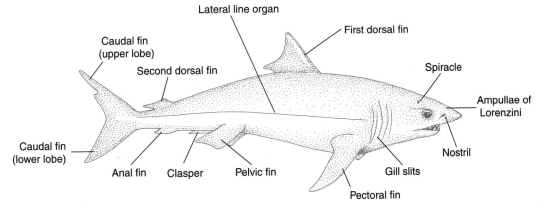

Lateral line organ

Caudal fin (upper lobe)

Second dorsal fin

First dorsal fin

Spiracle

Ampullae of Lorenzini

Caudal fin (lower lobe)

Nostril

Anal fin   Clasper   Pelvic fin

Gill slits

Pectoral fin

As mentioned before, the shark's stiff winglike pectoral fins provide some lift. More important, the shark's very large, oily liver increases its body's buoyancy.

Reproduction in Sharks

Mermaid's purse (egg case)

Figure 12-10 Egg case of a shark; empty "purses" are sometimes found washed up on beaches.

Sharks are different from most other fish in their mode of reproduction. Fertilization is external in most bony fishes, but the cartilaginous fishes have internal fertilization. Male sharks have two organs called **claspers**, located between their pelvic fins. The claspers transfer sperm into the female's reproductive tract. Some sharks have internal development and live-bear their young. For example, the young of the dogfish shark develop for two years within the mother shark before being born. Interestingly, some types of sharks begin their predatory behavior while still developing—they devour some of their siblings before they hatch! Other sharks (and skates) have external development in which one or more embryos develop within a black, leathery egg casing called a "mermaid's purse." These embryos take more than a year to develop. Their empty egg casings are often seen washed up on the beach. (See Figure 12-10.)

12.2 SECTION REVIEW

1. Why are the sharks, skates, and rays placed in their own class?

2. In what kind of a habitat are skates and rays found? What do they usually eat?

3. What special receptors does the shark have for detecting prey?

12.3 BONY FISHES

More than 95 percent of all fish on Earth belong to the class Osteichthyes (meaning "bony fishes"). As the name implies, the **bony fishes** have a skeleton that is made up of bone. Since they are vertebrates, bony fishes have a backbone that is made up of a chain of

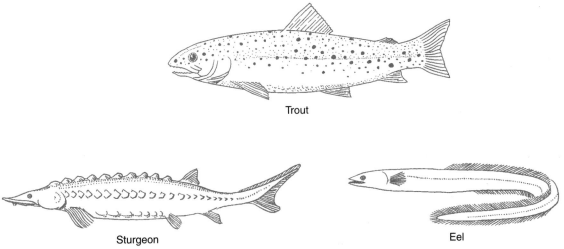

Trout

Sturgeon

Eel

Figure 12-11 Three examples of bony fishes (freshwater and marine).

individual bones called *vertebrae*. The vertebrae surround and protect the spinal cord.

Bony fishes are found in every type of aquatic environment, from lakes and rivers to tropical reefs and polar oceans. (See Figure 12-11.) How have bony fishes adapted so well to life in the sea? Look at the structures of a bony fish, in Figure 12-12 on page 292. As you can see, bony fish have a protective covering of scales, which are loosely attached to the skin and can even rub off in your hands. Bony fish feel slimy to the touch because their skin, which is living, secretes a protective mucus coating over their scales. The mucus serves two functions: it acts as a barrier against infection and it reduces friction so the fish can move easily through the water.

Scales can indicate the approximate age of a fish. On the scales are growth rings called circuli. As the fish grows older, new circuli are produced. When the circuli are close together, they form bands. Although there is variation in the bands of each fish, a single band may represent about one year's growth, much like the bands on a clamshell.

Breathing in Bony Fish

Like so many other marine animals, fish use gills for breathing. The gills are located in the head region, on either side of the body. In bony fishes, the gills are covered by a flap of tissue called the

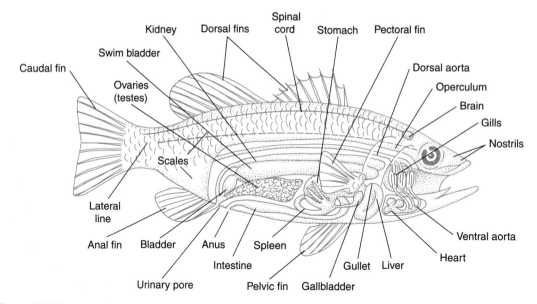

Figure 12-12 External and internal anatomy of a typical bony fish.

operculum. Under the operculum are red overlapping membranes, which are the gills. The operculum opens and closes every time a bony fish breathes.

How exactly does a fish use its gills for breathing? The next time you see a fish in an aquarium, observe it closely. When a fish breathes, its mouth and operculum open and close in unison. Water that contains dissolved oxygen enters the mouth and passes over the blood-filled gills, which are composed of arches that have gill filaments attached to them. Oxygen diffuses from the water, passes through the membranes of the gill filaments, and flows into the blood. Carbon dioxide diffuses in the opposite direction, from the gills into the water, exiting through the openings called gill slits. The gill arches also have gill rakers, which have bony spines that channel incoming food particles into the esophagus, thereby preventing them from passing through the gill slits and damaging the gill filaments.

Locomotion in Fish

One of the defining characteristics of fish is that they have fins, which are mainly used in swimming. Marine animals, such as fish, that have the ability to swim are referred to as **nekton** (as opposed to plankton, which mainly float or drift near the surface). Look at the fins shown in Figure 12-12. Notice that some are single and others are paired. The pectoral fins and pelvic fins are paired. (*Note:* The pectoral fins correspond to the forelimbs of other animals, while the pelvic fins correspond to the hind limbs.) These fins are attached to body muscles. When the muscles contract, the fins move. The actions of the pectoral and pelvic fins enable the fish to move in all directions—up, down, forward, and backward. Both the single dorsal fin and the single anal fin work to stabilize the fish, thereby preventing its rolling from side to side. Sometimes there is a second dorsal fin behind the first one. For protection, some fish, such as the beautiful tropical lionfish, have venomous spines in their dorsal fins.

Which is the fastest fish in the ocean? According to some scientists, the swordfish is the fastest, followed by the tuna, the dolphinfish, and the barracuda. The open water, or **pelagic**, species of fish tend to be much faster moving and wider ranging than the bottom-dwelling species such as the sole and flounder. What makes one fish able to swim faster than another? The speed of a fish depends, to a great extent, on its body shape. Look at the body shapes of two of the fastest fish, shown in Figure 12-13. The tuna and the swordfish both have a fusiform shape, meaning they are tapered at both ends, whereas the sole and flounder have a flattened shape. The fusiform shape is very streamlined and produces little resistance to movement through water. For this reason, submarines and torpedoes are built with a fusiform shape. (*Note:* Sharks also have a fusiform shape.)

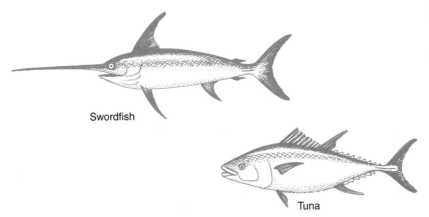

Figure 12-13 Fast-moving pelagic fish such as these have a streamlined, or fusiform, shape.

Swordfish

Tuna

The fastest speeds are attained when a fish uses its tail, or caudal fin, to propel it through the water. Both the shape and height of a caudal fin affect speed. Faster fish, such as the tuna and shark, have a greater fin height than do slower fish, such as the sole. Fish also have powerful muscles located under the skin along both sides of the body. These muscles contract in sequence on alternate sides, from the anterior to the posterior end, causing the caudal fin to move from side to side.

Another factor that affects speed is temperature. Nearly all fish are **ectothermic** (formerly called cold-blooded) animals; that is, their body temperature is determined by the temperature of the water around them. Cooler water temperatures decrease a fish's metabolic rate, which decreases muscular activity. However, predatory fish such as the tuna and the great white shark have evolved the ability to retain the heat that their muscles generate. As a result, they have an increased metabolic rate, which enables them to keep moving fast, in spite of the lower temperatures of the surrounding water. Such fish have to ingest more calories to maintain the higher metabolic rate.

How does the side-to-side motion of a fish cause it to move forward? Scientists have researched this question and found that when a fish's tail pushes against the water, swirling currents called vortices are produced. The scientists have hypothesized that these vortices produce thrust, which propels the fish forward. To test this hypothesis, researchers at the Massachusetts Institute of Technology (MIT), in Boston, have built a life-sized robotic tuna, called *RoboTuna*, in order to analyze the mechanics of motion in fish and better understand the relationship between structure and function in organisms.

Buoyancy in Fish

Swimming can be tiring, so fish need to rest periodically by floating in place. The ability to float or rise in a liquid is called buoyancy. (See Chapter 1.) Bony fish are buoyant because they have an internal gas-filled organ called a **swim bladder**, as shown in Figure 12-12. When the body muscles around the swim bladder contract, the swim bladder gets smaller and the fish sinks. When the body muscles relax, the swim bladder enlarges and the fish rises. By regulating the volume of gas in its swim bladder, a fish can rise, sink, or maintain a steady position in the water. The ability to maintain a steady position at any water depth is called neutral buoyancy. (Cartilaginous fishes such as sharks do not have a swim bladder; instead, as mentioned above, their huge oily liver helps to increase their buoyancy.)

Food-Getting in Fish

Fish exhibit a variety of feeding methods. You have already read about the habits of parasitic fishes (such as the lampreys and hagfish) that feed by sucking out the blood and tissues of their host's body. A method seen in a few species of both cartilaginous and bony fishes is a type of filter feeding called straining, whereby the fish swims with its mouth wide open and filters out great quantities of plankton on its gill rakers; the water exits the gills and the plankton is swallowed. (The whale shark and basking shark, mentioned above, feed in this manner.) Some bottom-dwelling fish are adapted for sucking their food—much like a vacuum cleaner—into their round mouths, which are equipped with small teeth. Other fish have beaklike mouths that are adapted for nibbling on chunks of algae-covered coral, or strong flat teeth suitable for crushing the shells of invertebrates. And, of course, there are the predatory fish, such as sharks and barracuda (see page 280), which have large sharp teeth for catching fish and other fast-moving prey. In addition, filter feeders that have to take in large quantities of plankton-rich water, and predatory fish that attack large prey, benefit by having large, wide-opening mouths.

Digestion and Transport in Fish

The fish has a one-way digestive system, as shown in Figure 12-12. Food enters the mouth and passes through a food tube, or alimentary canal (consisting of the pharynx, esophagus, stomach, and intestines), where it is digested. Solid wastes from the digestive tract are eliminated through the anus. Most metabolic wastes from the cells are brought to the kidneys by the blood and are excreted through the urinary pore; some wastes exit the gills by diffusion.

How do the nutrients get to all of the fish's cells? As in your own body, nutrients are brought to the body's cells in the blood. Fish, like humans, have a closed circulatory system, which means that the blood travels within blood vessels. Fish have a two-chambered heart. (In contrast, the human heart has four chambers.) The heart pumps the blood into blood vessels called arteries. The blood moves through the arteries into smaller blood vessels called capillaries, and then is returned to the heart by the veins. Nutrients in the blood diffuse through the thin capillary walls into the body's cells. You can observe blood circulation in a fish's tail by doing the lab investigation at the end of this chapter.

Sensitivity in Fish

Fish are very sensitive to changes in their environment. They have a well-developed nervous system to carry out their responses. Fish have a good sense of hearing. Although they have no external ears, fish have an inner ear located behind their brain, and sound receptors located in tiny openings in the skin along each side of the body. This line of sound receptors, known as the lateral line organ, has sensory cells that contain cilia, which are attached to a nerve that connects with the dorsal spinal cord. Sound waves cause the cilia to vibrate, which produces electrical impulses that are transmitted to the spinal cord and then to the brain. These sound receptors help the fish respond to vibrations and sounds in the water. The swim bladder is also sensitive to vibrations in the water, and aids in the hearing process.

Fish have an excellent sense of smell. Notice the two openings in front of the eyes called the nostrils, or nares. Sensitive nerve endings inside the nares can detect tiny quantities of chemicals in the water. Vision is also good in fish; in most species, the eyes are located one on each side of the head, giving the fish a wide field of vision. Fish that live in clear waters near the surface have the best color vision, whereas fish that live in deep, darker waters have larger eyes that can take in more light.

Reproduction in Bony Fish

In fish, the sexes are usually separate. (Interestingly, in various species of fish, the individuals change sex at some point in their life cycle.) Most species of fish have external fertilization and development. In the vastness of the ocean, it is necessary for egg-laying fish to produce large numbers of eggs to ensure survival of at least some of the offspring. The male gonads, called testes, produce the sperm; and the female gonads, called ovaries, produce the eggs. The gametes are released (during external fertilization) in a process called **spawning**. Fertilized eggs of bony fish usually develop and hatch within a few days, either while drifting or attached to a substrate. In the species of fish that have internal fertilization, the females either lay the fertilized eggs within a casing (seen in some cartilaginous fishes) or live-bear the young fish (also seen in some cartilaginous fishes and freshwater bony fishes).

The life cycle of a typical bony fish includes five major stages: the egg stage, larval stage, postlarval (prejuvenile) stage, juvenile stage, and adulthood. (See Figure 12-14.) The egg stage encompasses spawning (egg-laying), fertilization, embryological development, and hatching from the egg case. During the larval stage, which lasts a few weeks, the young hatchling is about 2 mm in length. The hatchling lives as part of the plankton population, drifting near the ocean surface while its body develops the muscles and fins necessary for more independent locomotion. (The very young larva still obtains nourishment from its yolk sac.) During the postlarval (prejuvenile) stage, muscle and fin development accelerate as the young fish actively swims about. During the juvenile stage, the young fish

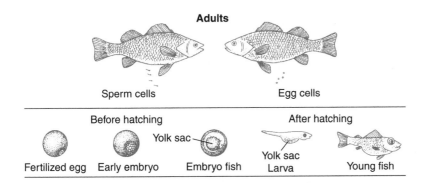

Adults

Sperm cells   Egg cells

Before hatching   After hatching

Yolk sac

Yolk sac

Fertilized egg   Early embryo   Embryo fish   Larva   Young fish

Figure 12-14 Life cycle (reproduction and development) of a typical bony fish.

resembles an adult in form but is still small and immature. At adulthood, the fish is capable of reproduction.

12.3 Section Review

1. Explain how the different fins of a fish enable it to move.

2. How does a fish use its gills to breathe?

3. Describe several methods of food-getting in fish.

12.4 UNUSUAL ADAPTATIONS IN FISH

In the struggle for survival, fish have evolved diverse adaptations in a variety of marine environments. For example, the porcupine fish (*Diodon*) has two features that work together to deter predators. Like the puffer fish (*Chilomycterus*), the porcupine fish can inflate its stomach with water when threatened, resulting in a balloon-shaped body. In addition, while the porcupine fish is inflated, sharp spines project outward from its skin to further discourage predators from trying to swallow it.

Camouflage in Fish

Figure 12-15 The bottom-dwelling flounder camouflages itself by matching the patterns on the seafloor.

Fish can also protect themselves by changing color. Flatfish such as the flounder (*Platichthys*) can use camouflage to match their surroundings. Like the octopus, these bottom-dwelling fish have special chemicals called *pigments* in their cells. When the natural background is dark, the pigments inside the skin expand, darkening the fish. When the surroundings are light, the skin pigments contract, making the fish lighter in color. Sometimes they can even change color to match mottled patterns in their environment. (See Figure 12-15.) Another adaptation of the flounder for its benthic lifestyle is that its two eyes are on the side of its body that faces up. The flounder is born with an eye on each side of its head, because as a young fish it swims through the water. However, before the young flounder settles to the bottom, one eye migrates to join the other eye on what then becomes the upper side of the fish.

Another interesting type of camouflage is that of the sargassum fish, which has evolved to resemble the shape, color, and texture of the sargassum seaweed in which it lives. This camouflage helps the sargassum fish blend in with its environment so that it can avoid predators and catch unsuspecting prey.

Strange Shapes and Behaviors of Fish

One of the most curious looking fish in the ocean is the sea horse (*Hippocampus*). Sea horses live in the shallow waters along the Atlantic, Pacific, and Gulf coasts. To prevent itself from being tossed up on the beach by waves, this 10- to 15-cm-long bony fish wraps its flexible tail around a piece of seaweed, marsh grass, or coral. Using its horselike snout like a straw, the sea horse sucks up plankton and other food particles. The sea horse also has an unusual mode of reproduction. After the male and female sea horses wrap their tails around each other, the female releases her eggs. The eggs are fertilized by the male and then transferred to a brood pouch on his abdomen. After about two weeks of development, the baby fish hatch from their "pregnant" father's pouch. (See Figure 12-16.)

Have you ever seen or heard of flying fish? In his book about the *Kon-Tiki* expedition, explorer Thor Heyerdahl described how some flying fish flew onto his raft during the night to escape schools of bonitos and dolphinfish. How is the flying fish (*Cypsilurus*) adapted for flight? The flying fish vibrates its caudal fin at more than 40 beats per second to propel itself out of the water. Once airborne, its pectoral fins expand like wings to give the fish lift. The flying fish can glide through the air for a distance of about 30 meters, and "fly" as high as 12 meters above the water. At the end of its flight, the flying fish splashes down, ready to take off again whenever it needs to escape other predatory fish.

Creatures that look more like something that people would expect from a science fiction movie than from Earth's oceans are the deep-sea fishes. Found at depths of about 500 to 2000 meters, the various dragonfish, viperfish, and anglerfish (*Melanocetus*) have

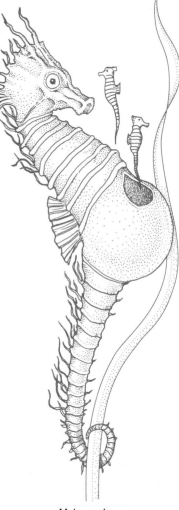

Male sea horse
and offspring

Figure 12-16 Baby sea horses hatch from a brood pouch on the adult male sea horse's abdomen.

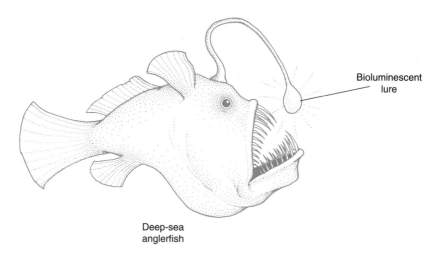

Figure 12-17 The angler-fish (*Melanocetus*), typical of various deep-sea fishes, has special adaptations for feeding in the depths, such as a large mouth, huge spiky teeth, and a light-producing lure to attract prey.

Bioluminescent lure

Deep-sea anglerfish

wide-opening mouths equipped with huge spiky teeth, to catch and swallow whatever rare prey they encounter at those depths. In addition, they often have the ability to produce light in their bodies, either to lure prey or to attract mates. (*Note:* The production of light by a living organism, called *bioluminescence,* was discussed in Chapter 4.) Many deep-sea fishes have a bioluminescent lure that dangles over their mouth; unsuspecting prey are attracted to the lure, and then seized and swallowed whole. (See Figure 12-17.)

Another strange-looking fish is the mola (*Mola mola*), or ocean sunfish. Attaining a body length of 3 meters and a weight of up to 2275 kg, the mola is the biggest bony fish in the ocean. Molas often lie on their side near the water's surface off the Atlantic and Pacific coasts of North America, where they feed on jellyfish, small fish, and plankton. On occasion, they pursue their prey to greater depths. Molas can swim by using their elongated dorsal and anal fins for movement. The caudal fin and posterior region of the body are much reduced in size, making the mola look like a big "swimming head." (See Figure 12-18.) Though huge, these docile fish are slow moving and thus are vulnerable to predation by sea lions, killer whales, and even humans. Fortunately, a female mola can produce as many as 300 million eggs, which helps to ensure that this sunfish can still be seen floating at the ocean surface.

An unusual and scientifically important fish was "rediscovered" in the early 1900s, when it was caught in the deep waters off the Comoros Islands in the Indian Ocean. In the 1990s, a similar fish was caught in the waters off the island of Sulawesi, in Indonesia.

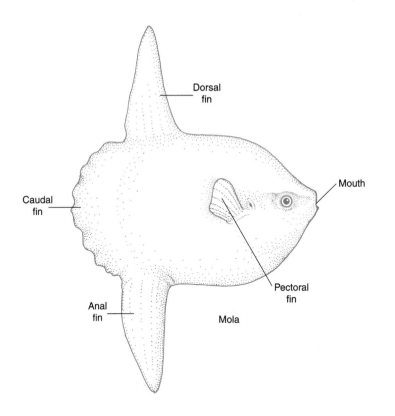

Figure 12-18 The huge mola, or ocean sunfish, often lies on its side near the ocean surface, feeding on small fish and plankton.

Dorsal fin

Mouth

Caudal fin

Pectoral fin

Anal fin

Mola

This unusual fish, the coelacanth (*Latimeria chalumnae*), which grows to nearly 2 meters long, was thought to have been extinct for over 60 million years. The coelacanth is a significant find because it has paddlelike pectoral and pelvic fins that resemble those seen in fossils of the ancient lobefin fish—the most probable ancestor of the earliest amphibians. (See Figure 12-19.) Although several specimens have been found, the coelacanth is considered rare and is protected by law. (Another very unusual fish, the California grunion, is discussed in Chapter 20; it is unique because it comes ashore to breed!)

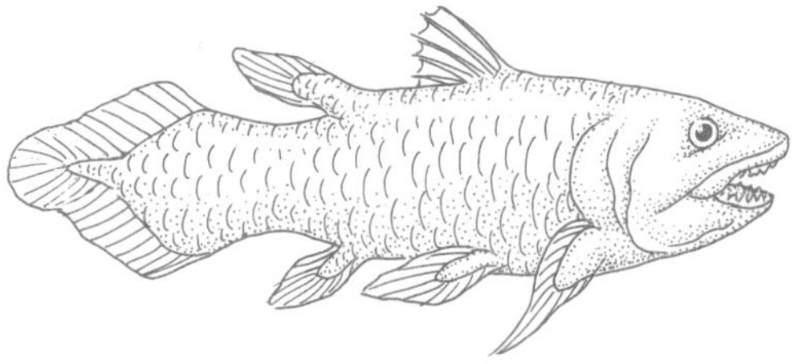

Coelacanth

Figure 12-19 Once thought to be extinct, the coelacanth is unique because it has paddlelike fins that resemble those of the ancient lobefin fish.

12.4 SECTION REVIEW

1. Describe two adaptations of the flounder to life as a benthic organism.

2. What are some unusual characteristics of the sea horse?

3. What features do deep-sea fishes have for capturing prey?

Laboratory Investigation 12

Breathing and Transport in a Fish

PROBLEM: How does a fish carry out respiration and circulation?

SKILLS: Calculating a fish's breathing rate; observing circulation in a fish.

MATERIALS: Killifish (*Fundulus*), seawater, bowl, petri dish, cotton (or gauze), watch, microscope, medicine dropper, dip net.

PROCEDURE

1. Place a killifish in a small bowl filled with seawater. Notice the fish's operculum, which covers the gills on either side of the head. You can see that the fish is breathing because its operculum continually opens and closes.

2. The fish breathes by taking in water through its mouth. Notice that the fish's mouth opens and closes at regular intervals. The water passes over the gill membranes, where an exchange of gases occurs.

3. What is the breathing rate of the fish? You can measure the breathing rate by counting how many times the operculum moves per a unit of time. Have your lab partner count movements of the operculum while you time it for 15 seconds. Multiply the number by 4 to get the movements (rate) per minute. Do six trials; calculate the total and divide by 6 to get an average rate.

4. Now observe circulation in a fish. Thoroughly moisten two pieces of cotton or gauze by dipping them in seawater. Flatten out the two pieces and put them in a petri dish.

5. Use a dip net to remove a killifish from the aquarium. (See Figure 12-20.) Carefully place the fish on one piece of cotton, with its tail hanging over the edge of the cotton and lying flat against the petri dish. (*Note:* To protect its scales, wet your hands thoroughly before touching the fish.)

6. Cover the fish's head and gills with the other piece of wet cotton. Leave the mouth and tail exposed. Put a few drops of seawater on the fish's tail.

7. Place the petri dish on the microscope stage under the low-power lens. Position the fish so that its tail is directly under the low-power lens. (You may want to place a slide carefully over the fish's tail to help keep it in position.)

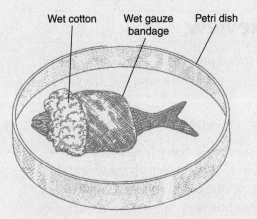

Wet cotton   Wet gauze   Petri dish
              bandage

Figure 12-20 Observing circulation in a fish's tail.

8. Focus under low power. The fish may move its tail. Check the position of the tail and then refocus. Look for moving streams of blood. Notice the small red blood cells, which carry oxygen, moving in the bloodstream.

9. Observe the blood vessels. The vessels in which the blood spurts are the arteries. The spurting of blood is a result of the pumping action of the heart. The vessels in which the blood moves more slowly are the veins. Look for very narrow blood vessels, the capillaries, which connect arteries to veins. Notice that red blood cells in the capillaries move very slowly in single file.

10. Observe your fish for only 5 minutes or so; then return it to the aquarium. If the fish is sluggish, gently prod it with your finger to help get it moving.

OBSERVATIONS AND ANALYSES

1. How does a fish breathe? What is the role of the gills in breathing?

2. What are the three types of blood vessels in the fish's circulatory system?

3. Why was it important to cover the fish's head and gills with wet cotton?

Chapter 12 Review

Answer the following questions on a separate sheet of paper.

Vocabulary

The following list contains all the boldface terms in this chapter.

acorn worm, ampullae of Lorenzini, bony fishes, cartilage, cartilaginous fishes, chordates, claspers, ectothermic, gill slits, jawless fishes, lancelet, lateral line organ, nekton, operculum, pelagic, placoid scales, protochordates, spawning, spiracles, swim bladder, tunicates, vertebrates

Fill In

Use one of the vocabulary terms listed above to complete each sentence.

1. The flexible connective tissue found in sharks is _____.

2. The flap of tissue that covers a bony fish's gills is the

   _____.

3. Openings on a shark's head that aid in breathing are

   _____.

4. The _____ along a fish's body detects sound vibrations.

5. Bony fishes maintain neutral buoyancy with their_____.

Think and Write

Use the information in this chapter to respond to these items.

6. Discuss some differences in reproduction between the bony and the cartilaginous fishes.

7. Describe the basic life cycle of a typical bony fish.

8. Compare the adaptive features of a bottom-dwelling fish with those of a pelagic fish.

Inquiry

Base your answers to questions 9 through 11 on the experiment described below and on your knowledge of marine science.

A student observed the breathing rates of two small intertidal fish, *Fundulus*, each kept at a different temperature. She hypothesized that the breathing rate, which is measured by counting the number of movements of the fish's operculum per minute, would be higher in warmer water than in colder water. All other variables, such as salinity, light, and tank size, were kept constant for both fish. The results of six trials (for each fish) are shown in the table below.

Experimental Group (Temperature 23°C)		Control Group (Temperature 7°C)	
Trial	Breathing rate (movements/min)	Trial	Breathing rate (movements/min)
1	108	1	36
2	116	2	60
3	112	3	48
4	104	4	64
5	113	5	60
6	112	6	45
Average	110.83	Average	52.16

9. The results of the experiment show that   *a.* water temperature has no effect on the breathing rate in *Fundulus*   *b.* an increase in water temperature causes an increase in breathing rate   *c.* an increase in water temperature causes a decrease in breathing rate   *d.* the difference in breathing rates could have been caused by a difference in salinity.

10. Which statement represents a valid conclusion that can be drawn from this experiment?   *a.* The student's hypothesis is supported by the data.   *b.* The student's hypothesis is not supported by the data.   *c.* The student's hypothesis could not be tested because the other variables were not controlled.   *d.* There were too few trials to draw any possible conclusions.

11. What recommendation would you make to check and verify the results of this experiment?   *a.* Repeat the experiment, but with fewer trials.   *b.* Repeat the experiment, but include more trials.   *c.* Change the salinity in the experimental and in the control groups.   *d.* Use two different species of fish (one in each tank).

Multiple Choice

Choose the response that best completes the sentence or answers the question.

12. The general term that describes marine animals that can swim is   *a.* pelagic   *b.* placoid   *c.* nekton   *d.* tunicate.

13. Fast-moving fish species that live in open water are referred to as   *a.* cartilaginous   *b.* bony   *c.* spawning   *d.* pelagic.

Base your answers to questions 14 through 16 on the numbered drawing and on your knowledge of marine biology.

14. The structure that enables a fish to detect sound vibrations is   *a.* 1   *b.* 2   *c.* 3   *d.* 4.

15. The body part that is used for quick acceleration is the   *a.* 1   *b.* 2   *c.* 3   *d.* 4.

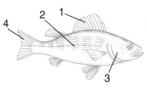

16. Water exits this structure during breathing.   *a.* 1   *b.* 2   *c.* 3   *d.* 4

17. The chordates are animals that contain a notochord, gill slits, and a   *a.* ventral nerve cord   *b.* dorsal nerve cord   *c.* skull   *d.* backbone.

18. All are examples of primitive chordates *except* the   *a.* sea squirt   *b.* lancelet   *c.* flounder   *d.* acorn worm.

19. Which fish is least related to the others?   *a.* skate   *b.* ray   *c.* shark   *d.* barracuda

20. Which of the following structures is *not* present in all fish species?   *a.* two-chambered heart   *b.* closed circulatory system   *c.* gills   *d.* scales

21. A fish breathes by   *a.* taking in water through its mouth and passing it over its gills   *b.* taking in water through its gills and passing it out of its mouth   *c.* taking in water through the gills on one side of its body and passing it through the gills on the other side   *d.* fanning water into its gills with its pectoral fins.

22. Which of the following is a cartilaginous fish?   *a.* manta ray   *b.* mola   *c.* barracuda   *d.* sea horse

23. Which of the following is a jawless fish? *a.* shark
b. hagfish   *c.* mola   *d.* flounder

24. Why is the coelacanth an important fish to scientists?
a. It was the ancestor of all sharks.   *b.* It is an evolutionary
link between vertebrates and invertebrates.   *c.* It resembles a
fossil fish that was ancestral to amphibians.   *d.* It is extinct.

Research/Activity

■ Now that you know that fish can hear, find out how they can
"speak"—they actually make a variety of sounds. Prepare a
report about communication in bony fishes.

■ Find out about the ancient Japanese art of fish printing (*gyotaku*)
by searching the Internet to learn how it is done.

13 Marine Reptiles and Birds

When you have completed this chapter, you should be able to:

LIST the types of reptiles and birds that are found in the marine environment.

DISCUSS the special adaptations reptiles have for living in the sea.

DISCUSS the special adaptations seabirds have for an oceanic life.

13.1
Marine Reptiles

13.2
Marine Birds

The marine environment is home to a variety of birds and reptiles. At first glance, reptiles and birds may appear to be very different from one another. However, on closer examination, you will notice some important similarities.

Reptiles—such as the marine lizard shown above—have a scaly skin, and so do birds; but a bird's scaly skin is limited to its legs and feet. Reptiles and birds also are both vertebrates, animals with backbones. In addition, all birds and many reptiles lay eggs. These similarities are just part of the evidence showing that birds evolved from reptilian ancestors.

Marine reptiles are found mainly in tropical and subtropical habitats. Marine birds have a much wider range; they are found from polar seas to tropical shores. In this chapter, you will learn about some of the similarities and differences between these two classes of vertebrates, and about their adaptations for living in various marine habitats.

13.1 MARINE REPTILES

You have learned that fish are well adapted to life in water. Many millions of years ago, animals evolved from fish that were well adapted to life both on land and in the water—the amphibians. In time, reptiles evolved from the amphibians. Today, there are marine reptiles, but there are no truly marine amphibians. Thin-skinned animals such as frogs, toads, and salamanders cannot survive in salt water.

All reptiles share characteristics that originally evolved to suit life on land, but many have returned to a life in the water. There are four main groups of reptiles alive today. Each group has its marine representatives: sea turtles, sea snakes, marine lizards (iguanas), and saltwater crocodiles. Reptiles, which belong to the class Reptilia, have a dry scaly skin that protects against water loss. All reptiles live in warm or temperate climates because, as ectothermic (cold-blooded) animals, they become sluggish in cold temperatures. Most of a reptile's activities are dictated by the amount of warmth it receives from the sun. For example, if their bodies get too cold, aquatic reptiles such as crocodiles bask in the sun. And if their bodies get too warm from the sun, they cool off in the water (or, as in the case of sea snakes and sea turtles, swim to cooler waters). In some cases, ectotherms attain body temperatures that are actually higher than the temperature of their environment.

Adaptations of Reptiles

Marine reptiles have many characteristics for a terrestrial lifestyle that have been adapted to allow a life in the sea. Rather than gills, however, aquatic reptiles possess lungs for breathing. This means that they have to return to the water's surface periodically to gulp air, since they cannot obtain oxygen directly from the water as fish do. An important terrestrial adaptation first seen in the reptiles is the **amniotic egg**, which contains a large yolk to nourish the developing embryo and is enclosed in a leathery egg case to prevent it from losing water and drying out. Fertilization in reptiles is internal. Most aquatic reptiles, such as turtles and crocodilians, have to return to land to lay their eggs. Some sea snakes live-bear their young in the ocean. All reptiles have a three-chambered heart—an evolutionary "step up" from the fish's two-chambered heart—except for the **crocodilians** (alligators and crocodiles), which have a four-chambered heart. (A four-chambered heart enables the separation of oxygenated and deoxygenated blood.)

Marine reptiles also have adaptations that are specific to their life in the ocean. Because they live in a saltwater environment, sea turtles and marine lizards need to get rid of excess salt and to conserve freshwater. One of these adaptations is **salt glands**, which are positioned above the animal's eyes. The salt glands secrete great quantities of salty tears; this enables marine reptiles to live without access to freshwater. The salty secretions also wash sand from the turtles' and lizards' eyes when they move about on land. Another adaptation that marine reptiles have for conserving freshwater is the ability to excrete very concentrated urine by reabsorbing most of the water from it.

Saltwater Crocodiles

A large, predatory marine reptile is the crocodile, which belongs to the order Crocodilia. (See Figure 13-1.) Crocodiles often hunt by remaining just below the water's surface, with only their eyes and nostrils above the water, waiting to catch a meal by surprise. Whereas alligators are freshwater species, there are a dozen freshwater and saltwater species of crocodiles, found in Africa, Asia, Australia, and the Americas. (Alligators and crocodiles resemble each other, but crocodiles have a narrower snout.) The American saltwater crocodile (*Crocodylus acutus*) lives only in the Florida Keys. This surprisingly nonaggressive and shy reptile can grow up to 5 meters in length. Like other crocodiles, the American crocodile is endangered, since there are only about 500 to 1200 individuals left. Extensive development along Florida's coast has reduced the num-

Figure 13-1 The saltwater crocodile, an endangered marine reptile.

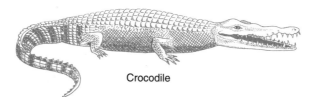

Crocodile

ber of crocodiles close to the point of extinction. Now these reptiles are protected by law, so their numbers are slowly increasing.

As mentioned above, unlike other reptiles, crocodilians have a four-chambered heart. Also, unlike most reptiles that lay eggs and abandon them, some crocodilians make a nest for their eggs, which they guard until the eggs hatch. The mother provides limited care for the new offspring as well.

Sea Snakes

Another marine reptile that is potentially dangerous to humans is the **sea snake**. There are about 55 species of sea snakes. All of them are venomous, although they are not particularly aggressive. Sea snakes usually range in size from 1 to 2 meters in length, and most species prey on small fish.

Various species of sea snakes inhabit the tropical Atlantic, Pacific, and Indian oceans. The yellow-bellied sea snake (*Pelamis platurus*) is a species found in Pacific waters, from California to Ecuador. (See Figure 13-2.) This sea snake hunts near the ocean surface, where it ambushes small tropical fish. The turtlehead sea snake is a more docile species; it eats fish eggs, which it scrapes off corals.

All snakes (and lizards) belong to the order Squamata. Sea snakes do, in fact, resemble their relatives on land, such as the coral snake. However, their special adaptations enable them to survive in the ocean. The flatter body (side-to-side) and paddlelike tail help the sea snake swim more efficiently. The presence of salt glands in

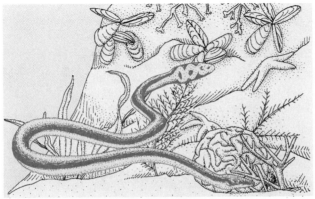

Sea snake

Figure 13-2 The yellow-bellied sea snake, a venomous marine reptile.

the mouth enables the sea snake to get rid of excess salts and thereby maintain a normal water balance in the body. Another adaptation in the sea snake is the special flap of tissue that covers the nostrils. During dives, this flap of tissue prevents water from entering the lungs. The lungs of a sea snake can inflate to three-quarters of its body length. This enhanced lung capacity lets the sea snake stay underwater for as long as two hours on a single breath. As mentioned before, some sea snakes bear their young live in the sea; others come ashore to lay their eggs.

Marine Lizards

Another type of reptile, the lizard, is mostly a land-dwelling animal. However, one species, the **marine iguana** (*Amblyrhynchus subcristatus*), swims and feeds in the ocean. (See Figure 13-3.) Marine iguanas live on the Galápagos Islands, which are located nearly 900 km off the coast of Ecuador in South America. In his book *The Voyage of the Beagle,* Charles Darwin described this marine lizard as a "...hideous looking creature, of a dirty black color, stupid and sluggish of movement...." Although the marine iguana may look menacing, it is harmless; and its movements underwater are graceful. In fact, the marine iguana is vulnerable to harm by humans; in January 2001, thousands of Galápagos iguanas died due to the effects of an oil spill.

Figure 13-3 The Galápagos marine iguana, the only lizard that swims and feeds underwater.

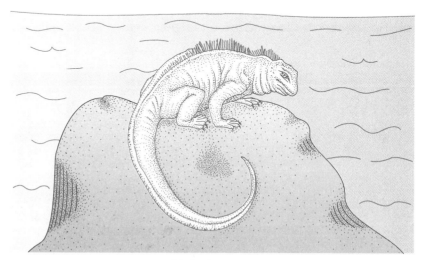

Marine iguana

Like the sea snake, the marine iguana has a flattened tail to aid in swimming. Marine iguanas dive into the ocean to graze on the seaweed and algae that grow on rocks in the subtidal zone. When their bodies get too cold from the water, the iguanas return to the land, where they warm themselves in the sun by lying across the heated lava boulders. Marine iguanas breed (lay their eggs) on land.

Sea Turtles

Of all the marine reptiles, the **sea turtle** is the most widely distributed. Sea turtles inhabit tropical and warm temperate oceans around the world. Along with freshwater turtles and land-dwelling tortoises, sea turtles belong to the order Chelonia. There are six species of marine turtles, and all are endangered: the hawksbill (*Eretmochelys imbricata*), leatherback (*Dermochelys coriacea*), loggerhead (*Caretta caretta*), Kemp's ridley (*Lepidochelys kempi*), Pacific, or olive, ridley (*Lepidochelys olivacea*), and green sea turtle (*Chelonia mydas*). The different sea turtle species can be distinguished from one another by their size and by the pattern of scales on their top shell, or carapace. All marine turtles have a hard carapace, except for the leatherback, whose top shell, as its name implies, is leathery in texture. In addition, the leatherback is the only sea turtle whose carapace is not fused to its backbone.

The largest marine turtle is the leatherback. An average-sized leatherback measures 2 meters in length and can weigh up to 450 kg. The smallest marine turtle is the Kemp's ridley, which is about 60 cm in length and weighs about 35 kg. (See Figure 13-4.) Some

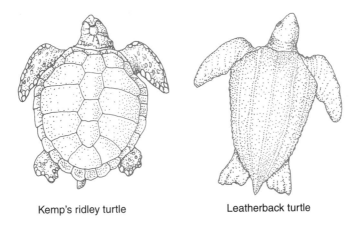

Kemp's ridley turtle

Leatherback turtle

Figure 13-4 The Kemp's ridley is the smallest sea turtle; the leatherback is the largest. (Not drawn to scale.)

sea turtles can live as long as 200 years—if they can survive their early years when they are most vulnerable to predators.

Sea turtles are well adapted to a marine environment. In fact, a sea turtle's body is smooth and streamlined for ease of movement in the water. The limbs (especially the forelimbs) have evolved into flippers, and their rapid movements enable turtles such as the leatherback to swim at speeds of up to 32 km per hour. These air-breathing animals have fatty deposits and lightweight bones for added buoyancy. However, they can stay underwater for up to 40 minutes on a single breath.

Feeding in Sea Turtles

Most sea turtles prefer coastal waters where food is most plentiful. Although turtles have no teeth, they do have strong jaws (and, in some cases, birdlike beaks) that they use either for breaking open the shells of crabs, clams, and other shelled animals, or for eating underwater vegetation. Mollusks, crustaceans, fish, and jellyfish make up the diet of the loggerhead turtle. The hawksbill turtle eats mollusks and crustaceans in addition to jellyfish and algae. The green sea turtle prefers to graze on the turtle grass and eel grass that grow in shallow waters, whereas the Pacific ridley eats invertebrates that live in eel grass. The leatherback ventures far offshore to feed on jellyfish. Sea turtles that eat jellyfish sometimes die when they accidentally ingest floating garbage such as plastic bags, which resemble their prey.

Reproduction and Development in Sea Turtles

Sea turtles are born on land but spend their lives at sea. The mature sea turtles return every few years to the beaches on which they were born to mate and lay their eggs. Mating occurs in shallow offshore waters. (It is thought that the female stores the sperm from these matings to fertilize the eggs she will lay two or three years later.) After mating, the female turtle swims to the shore and, during the night, emerges onto the beach. The flippers, which are adapted for swimming, cannot support the turtle's weight. So, using her hind limbs and forelimbs, the sea turtle drags herself up the slope of the sandy beach to find a nesting site above the high tide mark.

TECHNOLOGY

Satellite Tracking and Sea Turtles

Sea turtles are long-distance swimmers. For years, the journey of young sea turtles—particularly during their first year of life—has been a mystery to science. Now, through the use of a technology called satellite telemetry, scientists can track the movements of sea turtles across the open ocean. A small radio transmitter with a flexible antenna is glued to a sea turtle's top shell, or carapace. Each time the sea turtle surfaces to breathe, the unit transmits a signal. The signals are sent to four orbiting satellites that are operated by NOAA to monitor global weather patterns and to track radio-tagged animals.

The satellites make a total of six to eight passes each day over the tropical regions where most sea turtles are found. Special instruments in the satellites are designed to listen for transmissions from the turtles. However, to detect a specific turtle, the satellite must be overhead when that turtle surfaces to breathe. After a satellite receives a signal from a turtle, it sends the signal to a location in France, which then relays the data to a facility in the United States. There, a computer accesses the data and converts the information into readings of latitude and longitude. By plotting these continuous readings of latitude and longitude, scientists can track the route of a sea turtle.

Satellite telemetry has uncovered a trans-Pacific migratory route taken by juvenile logger-head turtles, from their nesting beaches in Japan to their feeding sites in Baja, Mexico. This information is critical for the protection of sea turtles. Now researchers can tell fisheries personnel about previously unknown sea turtle locations, so they can avoid catching these animals in their fishing gear. Satellite telemetry has also identified foraging zones in the open ocean where sea turtles congregate. One particular area, called the Transition Zone Chlorophyll Front (TZCF), is a region in the Pacific Ocean that is rich in sea life.

Scientists have obtained important data about the life cycle and whereabouts of sea turtles from satellite technology. However, coastal development, poaching, pollution, and trawling still take their toll on sea turtle populations. Today, all sea turtle species are considered endangered and are protected by law. Therefore, the valuable information that is gathered in space must be used to protect turtles in the sea.

QUESTIONS

1. How is satellite telemetry used to track sea turtles? What is done with the data?

2. Why is it important to know about the migratory routes of sea turtles?

3. Describe two discoveries made by satellite telemetry about sea turtle locations.

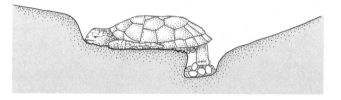

Figure 13-5 Sea turtles come ashore to lay their eggs.

At the nesting site, the turtle first uses all four limbs to dig a depression in the sand; then she uses her hind limbs to scoop out a hole about 30 to 60 cm deep. Into this hole the sea turtle lays about 100 fertilized leathery eggs that look like ping-pong balls. The female fills in the hole with sand and then heads back down the beach to the water, where she mates again. Over the next few weeks, the female will come ashore four or five more times to lay several hundred more eggs. (See Figure 13-5.)

While buried in the sand, the eggs are kept warm and moist and protected from such predators as raccoons, gulls, and rats. The embryos develop inside the leathery eggs for about two months. Interestingly, the sex of a sea turtle is determined by the position, and resulting temperature, of its egg within the nest. Sea turtle eggs that develop at about 28°C and below become males; those that develop at 30°C and above become females. After their development is completed, the baby turtles, called **hatchlings**, break through their shells and dig their way to the surface. Hatchlings usually emerge before dawn and are unprotected; they must quickly wiggle down to the sea. At this stage, they are vulnerable to predation by gulls and large fish.

Where do the hatchlings go after they enter the ocean? Marine biologists have been studying the migrations of sea turtles for a number of years. The green sea turtle, which lives in both the Atlantic and Pacific oceans, breeds every 2 to 3 years. (See Figure 13-6.) One of the breeding grounds is Ascension Island. This tiny island, just 8 km long, is located in the middle of the Atlantic Ocean, midway between South America and Africa. In one turtle study, young hatchlings were tagged on Ascension Island and then recaptured in the coastal waters of Brazil. Apparently, the young turtles were able to make this long journey by hitching a ride on clumps of seaweed that float on the South Equatorial Current, a large ocean current that moves across the Atlantic Ocean from east

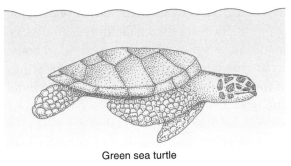

Figure 13-6 The green sea turtle migrates long distances between its breeding and feeding grounds.

Green sea turtle

to west. In addition, it is thought that sea turtles may be sensitive to magnetic fields and rely on them during their migrations.

The green sea turtles had migrated to Brazil to feed on the turtle and eel grasses that grow abundantly in the shallow waters. Mature turtles were also tagged in Brazil, and some of these turtles were later recovered on Ascension Island, where they nest. Marine biologists are not sure of the exact route taken by these turtles on their return to Ascension Island. More accurate methods are now being used to track these animals. Radio transmitters, mounted on the backs of some turtles, emit signals that can pinpoint their exact location. (See this chapter's feature on satellite tracking of sea turtles, page 315.)

13.1 SECTION REVIEW

1. What adaptations do sea snakes have for life in the ocean?

2. How does the marine iguana regulate its body temperature?

3. What kinds of food do the different sea turtle species eat?

13.2 MARINE BIRDS

The sea and the shore provide a haven for many species of birds. Birds that depend on the ocean for their survival are commonly referred to as **seabirds**. There are nearly 9000 species of birds; all are in the class Aves. Not all birds fly, but they all do share several important characteristics.

All birds have feathers, which are attached to the skin. There are two main types of feathers: down feathers and contour feathers. Down feathers are the small, fluffy feathers closest to the skin. These small feathers trap warm air and hold in body heat. Contour feathers are the larger feathers that cover the wings and the body; some of these are used for flight. Some aquatic birds have **powder feathers**, which repel water to protect the underlying down feathers. Some birds also have a special gland near their tail that produces a waterproof oil. The birds use their beak to spread this oil through their feathers when they groom, or **preen**, themselves.

Most birds have lightweight hollow bones, an adaptation for flight. However, some diving marine birds, such as the penguin, have denser bones. These heavier bones are adaptive for birds that spend time swimming underwater in pursuit of fish. Like most marine reptiles, all marine birds have to return to land (or, in the case of the penguin, to ice) to breed. All birds lay eggs, which are encased in a hard calcium-rich shell. Often both parents tend the eggs, keeping them warm and protecting them from predators. Marine birds nest in a variety of habitats: in tree branches, on cliff ledges along a rocky coast, on patches of vegetation, among pebbles and sand on a beach, or on a few stones out on the ice. Offspring are fed by the parent birds, who sometimes fly out to sea for many days to obtain enough fish or plankton to feed them.

Adaptations of Marine Birds

Seabirds have a variety of feeding methods. Some seabirds forage in the sand and mud along the shoreline. Others make short trips to the sea, diving several times a day for their catch of fish and invertebrates. The more oceanic species spend extended periods out at sea, gliding above the waves and diving into the sea for food.

Marine birds have some physical adaptations that are unique to their lifestyle. Seabird species that spend much of their time in and on the water have webbed feet for paddling and swimming. Like marine reptiles, marine birds have to get rid of excess salt and conserve fresh water. So, like the reptiles, seabirds have salt glands; they are special nasal glands that secrete a salty solution from the nostrils. In addition, seabirds conserve water by excreting a concentrated uric acid.

Common Shorebirds

Various species of marine birds search either in shallow water or in the sand along the water's edge for their food. You can often tell what a bird eats by looking at its beak, or bill. Among the marine birds, beak size and shape show great variation. Look, for example, at the three shorebirds shown in Figure 13-7.

The small **sandpiper** (*Calidris*) has a narrow pointed bill for poking in the sand for small invertebrates, such as worms and insects. Sandpipers are common shorebirds that move in small flocks in the intertidal zone, poking in the wet sand while managing to stay just ahead of the incoming waves. Various species of plovers, which are slightly larger shorebirds, also forage along sandy beaches.

The **oystercatcher** (*Haematopus*) uses its long, red knifelike beak to catch and eat various types of mollusks. The oystercatcher is a large bird, about the size of a hen, that inhabits the marshes and sandy beaches along parts of the Atlantic and Gulf coasts.

The **snowy egret** (*Egretta thula*) inhabits salt marshes along the Atlantic and Pacific coasts. How is the egret adapted for food-getting? Notice the long flexible neck and the pointed bill that it uses to quickly grab small fish that dart about in the shallow water. The long stiltlike legs give egrets the advantage of height in being able to spot fish. In the same family as egrets are various species of herons that hunt for fish and invertebrates in coastal marshes. In tropical areas, flamingos wade in the shallows, using their hooked beaks to capture and strain small invertebrates from the water.

Many aquatic birds obtain their food along the shore where

Figure 13-7 Three examples of common shorebirds found along Atlantic, Gulf, and Pacific coasts.

Snowy egret

Sandpiper

Oystercatcher

Sea duck
(merganser)

Figure 13-8 Sea ducks, such as this merganser, dive under the ocean surface to feed.

Black-backed gull

Figure 13-9 The sea gull is a common shorebird that scavenges along coasts.

plant communities thrive. Some species of ducks (*Anas*) are well adapted for life in coastal marshes. They use their webbed feet like paddles for moving through the water. Layers of fat, in addition to down and waterproof feathers, keep the birds warm and dry. Marsh ducks feed primarily on tiny aquatic plants, which they strain out of the water with their flat bills. Marshes are also used by ducks for nesting. Various types of hardy **sea ducks**, such as eiders (*Somateria*) and mergansers (*Mergus*), dive into the ocean to feed on mollusks, crustaceans, and fish. (See Figure 13-8.) Other aquatic birds that may feed in coastal areas include grebes and loons.

The **sea gull** is probably the bird most identified with the ocean. There are several different species of gulls. The herring gull (*Larus argentatus*) and the greater black-backed gull (*Larus marinus*) are the most common. (See Figure 13-9.) Gulls are largely scavengers that feed on dead marine animals that are carried ashore by the tides. Sea gulls also eat crabs on the beach and garbage at landfills. (They sometimes drop clams on hard surfaces, such as parking lots, to crack them open. Obviously, humans have had an impact on gulls' feeding choices and methods!) Although they will fly to mainland beaches to scavenge for food, sea gulls prefer to nest in isolated places on offshore islands.

Diving Shorebirds

For seabirds, flight is often necessary for food-getting. The **cormorant** (*Phalacrocorax*) is a common shorebird that dives from the sky for its food. When it spots a fish, the cormorant folds its wings, tucks in its feet, and dives into the water, where it catches the fish with its hooked beak. (See Figure 13-10.) The **common tern** (*Sterna hirundo*), a small shorebird that nests on sandy beaches, makes spectacular aerial maneuvers. The tern can hover over the water before it dives to catch a small fish. Since the tern lays its eggs in sand, the nesting sites are often disturbed by humans. To protect tern colonies, public access has been restricted and the use of recreational vehicles has been banned along many beachfronts.

Another diving bird is the **brown pelican** (*Pelecanus occidentalis*), shown in Figure 13-11. Pelicans are large birds that live along the Florida, Gulf, and California coasts. Pelicans literally "make a splash" when they dive from the sky into the water. This once-

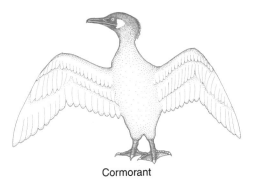

Cormorant

Figure 13-10 The cormorant dives into the water to catch its meal of fish.

Figure 13-11 The brown pelican scoops fish out of the water with its huge pouch.

endangered bird uses its large throat pouch like a net to scoop up fish. Water is squeezed out of the pouch and the fish are then swallowed headfirst to prevent their dorsal spines from getting stuck in the bird's throat.

An unusual shorebird that fishes on the wing but does not get its feet wet is the **black skimmer** (*Rynchops nigra*). The skimmer flies low over the water, with the tip of its lower jaw just beneath the surface. By skimming the surface in this way, the black skimmer eventually makes contact with a fish, which it swallows while still in flight. (See Figure 13-12.)

Another coastal bird that dives for its meal is the **osprey** (*Pandion haliaetus*), or fish hawk. The osprey swoops down from its nest on top of a tree to grab fish right out of the water. (See Figure 13-13.) Keen vision enables the fish hawk to spot its prey from high up. Like other hawks and eagles, the osprey has strong curved claws, called **talons**, which it uses to grab and hold onto a fish. The long, powerful wings of the hawk provide the lifting power needed to carry the bird and its prey back to the nest. The osprey, which feeds only on fish, lives along the Atlantic, Pacific, and Gulf coasts.

Black skimmer

Figure 13-12 While in flight, the skimmer uses its lower jaw to scoop fish from the water.

Osprey (fish hawk)

Figure 13-13 The osprey swoops down to grab fish with its sharp talons.

Diving Pelagic Birds

Many seabirds that nest on islands and along coastlines actually spend most of their lives at sea. The pelagic, or open ocean, diving seabirds include a variety of types, from sparrow-sized storm petrels

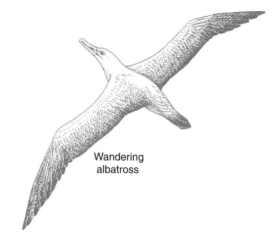

Figure 13-14 One of the largest seabirds, the albatross may spend years at sea before returning to land to breed.

Wandering
albatross

to auks, gannets, puffins, petrels, shearwaters, guillemots, and eagle-sized albatrosses. Some of these birds, such as the shearwaters and storm petrels, migrate thousands of kilometers each year as they follow schools of fish or drifting plankton. Probably the most oceanic of all the seabirds is the **wandering albatross** (*Diomedea exulans*) of the South Pacific. (See Figure 13-14.) The albatross, among the largest of all seabirds with a wingspan of about 3 meters, is adept at gliding effortlessly on air currents over the ocean. The albatross may spend 3 or 4 years at sea before returning to its birth island to breed. During this time, the bird rarely stops flying or gliding and may actually circle the entire globe.

Penguins

Penguins are the most aquatic of all seabirds. There are about 15 species of penguins; all but one species live in the southern hemisphere. They vary in height from about one-third of a meter to more than 1 meter tall. Penguins have no flight feathers and are completely flightless. However, they are excellent swimmers and divers. The smaller wings of a penguin function as flippers that can propel the bird through the water at speeds of up to 24 km per hour. Penguins have dense bones, which help give them the weight necessary for deep dives. (See Figure 13-15.)

On their dives, penguins catch a varied diet of fish, krill, squid, and shellfish. Like other birds that migrate in the open ocean in

search of food, some penguins go to sea for 2 years before returning to land or ice to nest. However, unlike the pelagic birds, which
go out to sea and fly over the ocean waves (sometimes diving in or
resting on the surface), penguins go out to sea under the waves,
swimming in the ocean. When they do return to breed, both males
and females share the duties of warming the eggs and feeding the
chicks after they hatch.

Penguins can survive the cold air and waters of the Antarctic
because they have a thick layer of fat under their skin and densely
packed soft down feathers for insulation. These features help to prevent loss of body heat in air temperatures that are often well below
zero. Like all other birds, penguins are **endothermic** (warm-
blooded). They can generate their own body heat, an ability that
enables them to live in regions with cold temperatures.

Chinstrap
penguin

Figure 13-15 Penguins
are flightless seabirds that
are excellent divers and
swimmers.

13.2 SECTION REVIEW

1. Describe three different feeding methods of marine birds.

2. What three features of seabirds are specifically adaptive for life
at sea?

3. How is the penguin adapted to live in the cold Antarctic?

Laboratory Investigation **13**

Adaptive Features of Marine Reptiles and Birds

PROBLEM: What are some features of birds and reptiles that show adaptations to the marine environment?

SKILLS: Observing and identifying adaptive features.

MATERIALS: Collection of photographs, illustrations, and/or plastic models of various marine birds and reptiles. You may also refer to the figures in this chapter.

PROCEDURE

1. Observe the features of various representative marine reptiles. In your notebook, make a copy of Table 13-1. Then list one or more features for each animal that you think may be specialized for its life in the water. What function does the particular feature serve? That is, how is it adaptive for an animal that spends more time in the water than on land? (You can note the same feature more than once.)

TABLE 13-1 ADAPTIVE FEATURES OF MARINE REPTILES

Reptile	Features/Structures	Purpose/Function
Crocodile		
Sea turtle		
Sea snake		
Marine iguana		

2. Observe the features of various representative marine (or aquatic) birds. In your notebook, make a copy of Table 13-2. Then list the features for each bird that you think enable it to function well in a marine environment (salt marsh or ocean), or list the particular function or purpose that the adaptive feature serves for the bird (depending on which space is left blank). You can add other features or functions that are not listed in the table but that are also important for survival in the bird's particular habitat.

TABLE 13-2 ADAPTIVE FEATURES OF MARINE (AQUATIC) BIRDS

Bird	Features/Structures	Purpose/Function
Skimmer	Longer lower bill	
Osprey		Catching fish (in flight)
Penguin		Swimming underwater
Oystercatcher	Knifelike bill	
Sea duck	Webbed feet	
Sandpiper		Feeding along shore
Heron (egret)	Long legs and bill	
Albatross		Long-distance flight
Pelican	Large pouch	
Cormorant		Fishing underwater

OBSERVATIONS AND ANALYSES

1. Describe three features of marine reptiles that are adaptive for living in an aquatic environment.

2. Describe five features of seabirds that are adaptive for feeding and/or living in a marine environment.

3. What are some adaptive features that are similar in both marine reptiles and marine birds? List the particular birds and reptiles.

Answer the following questions on a separate sheet of paper.

Vocabulary

The following list contains all the boldface terms in this chapter.

amniotic egg, black skimmer, brown pelican, common tern, cormorant, crocodilians, endothermic, hatchlings, marine iguana, osprey, oystercatcher, powder feathers, preen, salt glands, sandpiper, seabirds, sea ducks, sea gull, sea snake, sea turtle, snowy egret, talons, wandering albatross

Fill In

Use one of the vocabulary terms listed above to complete each sentence.

1. Birds can regulate their body temperature; they are _____.

2. The _____ has venom, a paddlelike tail, and salt glands.

3. A lizard that swims and feeds in the ocean is the _____.

4. Seabirds spread an oil through their feathers when they _____.

5. The most widely distributed marine reptile is the _____.

Think and Write

Use the information in this chapter to respond to these items.

6. Marine turtles bury their fertilized eggs in the sand, leaving them to hatch on their own. Discuss the risks and benefits.

7. How do sea snakes resemble their relatives on land? How do they differ, as a result of living in the ocean?

8. Briefly describe the feeding methods of the pelican, osprey, sandpiper, and egret. How is each method suited for getting specific resources in and near the sea?

Inquiry

Base your answers to questions 9 through 11 on the graph below, which shows the approximate number of sea turtle nesting sites on a stretch of monitored beaches along Florida's Atlantic Coast.

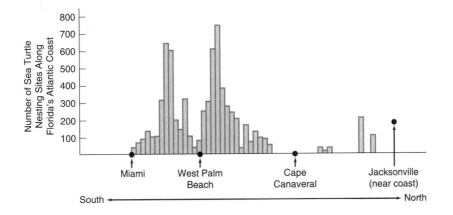

9. The beaches with the highest number of nesting sites are located between   *a.* Miami and West Palm Beach   *b.* West Palm Beach and Cape Canaveral   *c.* Cape Canaveral and Jacksonville   *d.* Miami and the Gulf Coast.

10. Which is an accurate statement regarding the data in the graph?   *a.* More nesting sites have been found in northern Florida than in southern Florida.   *b.* The largest number of nesting sites were found on the beaches around Miami. *c.* The nesting sites increase in number from east to west. *d.* More than 700 nesting sites were found on the coast north of West Palm Beach.

11. A tentative conclusion that can be drawn from the data is that   *a.* most suitable nesting sites are found on beaches within the big cities   *b.* most suitable nesting sites are found on beaches near, but not in, the big cities   *c.* there is an even distribution of nesting sites along Florida's Atlantic Coast *d.* there is an even distribution of nesting sites along Florida's Gulf Coast.

Multiple Choice

Choose the response that best completes the sentence.

12. The pattern of reproduction
shown in this reptile is
a. internal fertilization and
external development
b. internal fertilization and
internal development
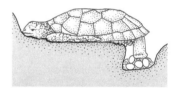
c. external fertilization and external development
d. external fertilization and internal development.

13. The marine reptile that guards its eggs in the nest until they
hatch is the   *a.* sea snake   *b.* green sea turtle   *c.* crocodile
d. iguana.

14. Marine reptiles may have any of the following features
except   *a.* a thick, scaly skin   *b.* salt glands   *c.* blubber
d. an amniotic egg.

15. Sea turtles leave the water and come up on land to
a. find food   *b.* lay eggs   *c.* find a mate   *d.* cool off.

16. Marine turtles have all of the following adaptations *except*
a. a smooth, streamlined body   *b.* flippers for forelimbs
c. the ability to stay underwater for a long time   *d.* gills.

17. The female sea turtle lays eggs that have a   *a.* leathery shell
b. hard calcium shell   *c.* jellylike coating   *d.* spiny shell.

18. An adaptive feature that this
shore bird has for food-getting is
a. a pointed bill for grabbing small
fish   *b.* sharp talons for holding
onto prey   *c.* webbed feet for
moving through water   *d.* a
streamlined body for diving.

19. Penguins have all of the following
adaptations *except*   *a.* soft down
feathers   *b.* a thick layer of
insulating fat   *c.* flipperlike wings
d. contour feathers for flight.

20. The sandpiper's beak, used to find tiny creatures in the sand, is   *a.* long and curved   *b.* broad and flat   *c.* short and dull   *d.* narrow and pointed.

21. The bird that has webbed feet for moving in the water is the   *a.* osprey   *b.* egret   *c.* oystercatcher   *d.* sea duck.

22. Marine reptiles are found primarily in   *a.* cold regions   *b.* tropical habitats   *c.* temperate regions   *d.* volcanic areas.

23. The osprey uses its sharp talons to   *a.* grab and hold onto fish   *b.* dig holes to find fiddler crabs   *c.* scrape barnacles off wood pilings   *d.* crack shells open.

Research/Activity

■ Sea turtles travel hundreds of kilometers between their birthplace and feeding grounds. What clues do sea turtles rely on to find their way between their feeding and breeding grounds? Imagine that you had to make such a journey every few years. Write about the clues you would rely on to find your way.

■ Collect some feathers near seabird nesting or perching sites. Use your library or the Internet to do research about feather structure; try to identify the species your feathers come from. Prepare a report; illustrate the feather's structure and/or attach the feathers. (Be sure to comply with regulations; do not enter restricted seabird nesting sites.)

14 Marine Mammals

When you have completed this chapter, you should be able to:

DESCRIBE the basic characteristics and behaviors of the cetaceans.

DISCUSS adaptations of pinnipeds and other marine mammals.

IDENTIFY unique diving response features of marine mammals.

According to the fossil record, about 50 million years ago some terrestrial mammals began to make the transition from a life on land to a life in the sea. In the vastness of the ocean, these animals found new sources of food and plenty of room. Over time, their bodies evolved into forms that were more suitable for swimming and diving than for walking on land.

One type of animal that made a successful and complete transition from land to water in the course of its evolution is the whale. The whale is an air-breathing animal that spends its entire life in the ocean. Like humans, whales are mammals. The humpback whale (*Megaptera novaeangliae*), for example, feeds her offspring milk from her mammary glands, a distinguishing characteristic of all mammals. Another characteristic shared by mammals is the four-chambered heart. All mammals are placed in the class Mammalia. Mammals that live in the ocean are called marine mammals. In this chapter, you will learn about the whale and other marine mammals that have adapted to life in the sea.

14.1 CETACEANS: WHALES AND DOLPHINS

Whales and dolphins belong to the order of mammals called Cetacea. There are about 80 different species of **cetaceans**. The largest cetaceans are the whales and the smallest cetaceans are the dolphins and porpoises, which are actually small whales that range from 2 to 4 meters in length. Dolphins and porpoises are extremely intelligent, social animals. In general, the difference between dolphins and porpoises is that dolphins have an elongated snout and can swim faster. Dolphins can swim so fast that they like to catch up with speeding boats to ride in their bow waves (although Dall porpoises are also known to do this). Dolphins and porpoises display a variety of acrobatic leaps, spins, and somersaults that take them out of the water and high into the air. (See Figure 14-1.)

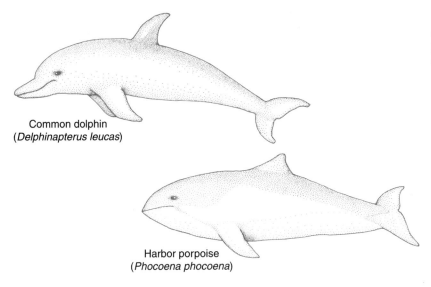

Common dolphin
(*Delphinapterus leucas*)

Harbor porpoise
(*Phocoena phocoena*)

Figure 14-1 A dolphin and a porpoise. Unlike the porpoise, the dolphin has an elongated snout.

Baleen Whales

The whales are classified into two main groups—the baleen whales and the toothed whales. The **baleen whales**, which belong to suborder Mysticeti, are filter feeders that eat plankton and small fish. They include such species as the blue, finback, humpback, right, and

Figure 14-2 Two examples of baleen whales, which are filter feeders.

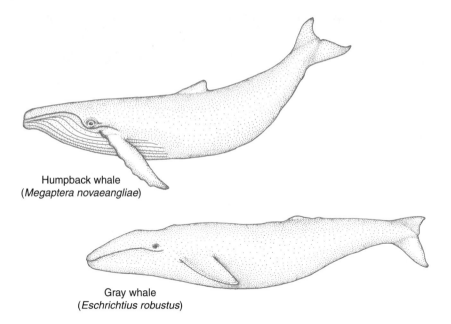

Humpback whale
(*Megaptera novaeangliae*)

Gray whale
(*Eschrichtius robustus*)

gray whales. (See Figure 14-2.) At 30 meters long (and weighing more than 150,000 kg), the blue whale (*Balaenoptera musculus*) is the biggest animal that has ever lived on Earth—even bigger than the largest dinosaurs that lived millions of years ago. (See Figure 14-3.)

While swimming through the water, a baleen whale opens its mouth to take in enormous quantities of water that contains zooplankton. The throat of some baleen whales—the rorquals—is pleated like an accordion to expand and hold the large volume of water. As the water is forced from the whale's mouth, it passes through overlapping plates of a fibrous protein material called **baleen**. The baleen plates, which look like giant combs, hang from the roof of the whale's mouth. They are strainers that filter small organisms from the water. The water is squirted out of the whale's mouth, and then the food is swallowed.

Figure 14-3 At 30 meters long, the blue whale is the largest animal on Earth.

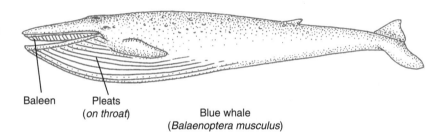

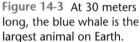

Baleen   Pleats
          (*on throat*)   Blue whale
                          (*Balaenoptera musculus*)

There are three types of feeding methods in the baleen whales. These different methods are reflected in differing shapes and sizes of the baleen plates. The rorquals take huge gulps of water to get krill and small fish. The right whales swim slowly through near-surface waters with their mouths held open to skim and strain the small zooplankton (copepods). Bottom feeders such as the gray whale suck up sediments on the seafloor and then filter out and eat small crustaceans and other invertebrates.

Toothed Whales

The **toothed whales**, which belong to suborder Odontoceti, include all other whales (such as the sperm, killer, pilot, and beluga), dolphins, and porpoises. These whales have peglike teeth on their jaws, with which they catch prey such as fish, seals, penguins, and squid. The toothed whales are active hunters. After seizing its prey, a toothed whale usually swallows it whole. Compartments in the stomach "chew" the food. The sperm whale (*Physeter macrocephalus*) is the largest of the toothed whales; it grows to about 15 meters in length. The smallest whale (not counting dolphins and porpoises) is the narwhal (*Monodon monoceros*), which grows to about 5 meters in length. (See Figure 14-4.) Each of these whales has unique teeth. The sperm whale has large cone-shaped teeth, but only on its long, narrow lower jaw. The male narwhal has an elongated front tooth that grows out of the left side of its upper jaw. This long spiral tusk, which was once thought to belong to the mythical unicorn, is used at breeding time to attract females and to fend off rival males.

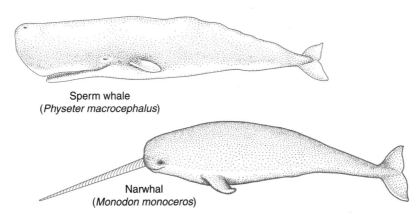

Sperm whale
(*Physeter macrocephalus*)

Narwhal
(*Monodon monoceros*)

Figure 14-4 Two examples of toothed whales, which are active hunters. The sperm whale, at 15 meters, is the largest toothed whale; the narwhal, at 5 meters, is the smallest. (Not drawn to scale.)

Reproduction in Whales

Cetaceans are fully aquatic animals; they do not return to the land to breed. Fertilization and development are internal. The period of embryonic development, or gestation period, may last from about 11 months in some species to as long as 18 months in the largest whales. Whales breed about once every 3 years and usually give birth to one calf at a time.

Mother whales invest a great deal of parental care in their offspring. As a result, a close bond develops between a whale and her calf. Whales are born tail-first. So that it will not drown, the newborn whale is pushed by its mother to the surface to get its first breath. The whale nurses her calf for about 6 to 10 months. The milk of whales is rich in protein and fat; this helps the newborn grow fast and add on layers of insulating fat. For example, a blue whale feeds her calf the equivalent of 400 glasses of milk a day. The milk contains about 50 percent fat. At this rate, the blue whale calf gains about 90 kg a day—a good start for the world's largest animal!

14.1 SECTION REVIEW

1. Compare the eating habits of baleen whales and toothed whales.

2. Describe how rorqual whales use their baleen to feed.

3. What do whale calves eat? How are they fed?

14.2 WHALE ADAPTATIONS AND BEHAVIORS

Whales possess numerous structures that enable them to survive in the ocean. You will learn about some of the important features and behaviors of whales (and dolphins) in this section.

Breathing in Whales

The whale breathes through an opening on the top of its head called the **blowhole**, which is its nose or nostrils. As early whales evolved and became more fully aquatic, their nostrils moved from the front (snout) to the top of the head. This position is more adaptive for an animal that lives in the water but must surface to breathe; the whale's body (including its eyes and mouth) can remain underwater while it breathes.

Toothed whales have one nostril; baleen whales have two nostrils. A whale breathes in air at the surface through its blowhole; the air is then carried to the lungs. The whale also breathes out through its blowhole. Water that is present in the exhaled air and in the blowhole produces a visible plume of water vapor, thus giving rise to the sailors' expression, "There she blows."

Swimming in Whales

Perhaps you have seen a killer whale performing in an aquarium show and have been impressed with its swimming and leaping. Whales are powerful swimmers. The killer whale, or orca (*Orcinus orca*), is the fastest of all the marine mammals, having been clocked at 55 km per hour. (See Figure 14-5.) Vigorous contractions of its body muscles cause the up-and-down movements of the powerful hind flippers, or **tail flukes**, which propel the animal through the water. The dorsal fin, which varies from 2 meters tall in the killer whale to very small in the baleen whales (and nonexistent in the narwhal), is used for staying on course. The pectoral fins, which range from the small, stubby flippers of the narwhal to the 5-meter-long winglike flippers of the humpback, are used for steering, braking, and balance. Although the pectoral fins of a whale or dolphin may resemble the fins of a shark, they have a different internal

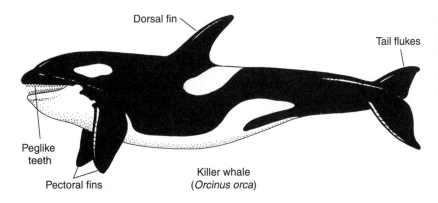

Dorsal fin

Tail flukes

Peglike teeth

Pectoral fins

Killer whale (*Orcinus orca*)

Figure 14-5 The killer whale, or orca, is the fastest of the marine mammals; its powerful tail flukes help propel it through the water.

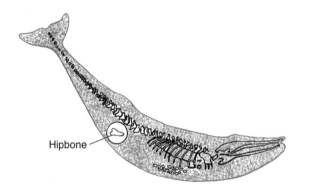

Figure 14-6 The floating hipbone is all that remains of the whale's ancestral hind limbs.

Hipbone

structure. Inside the pectoral fins of cetaceans are bones that are similar in structure to those of their land-dwelling ancestors; in fact, they resemble the bones of a human hand. *Note:* Rear limb buds in whale embryos, and floating hipbones in the adults, are all that remain of the ancestral hind limbs. (See Figure 14-6.)

Other Movements in Whales

Baleen and toothed whales often move and behave in spectacular ways. For example, the right whale (*Eubalaena glacialis*) is very acrobatic, as shown in Figure 14-7. (The right whale was so named because early whalers found this whale to be the easy or "right" whale to hunt, since it floats at the surface when killed.) Besides swimming, diving is the most common activity of whales, usually done in pursuit of food. The whale uses its pectoral fins to change position from horizontal to vertical. Then it submerges headfirst beneath the surface. The tail flukes are the last part of the body to enter the water, as they push the whale downward. You will learn more about the diving ability of whales and other marine mammals in the last section of this chapter.

Some baleen whales, such as the humpbacks, capture their prey by diving down around them, while producing a wall of bubbles that confuses and "entraps" them. The whales then swim up to the surface and engulf the small organisms in a feeding process called *lunging*.

Sometimes a whale waves its tail in the air and then smashes it on the ocean surface, producing a sound that can be heard several kilometers away. This behavior is called **lobtailing**. Lobtailing is not

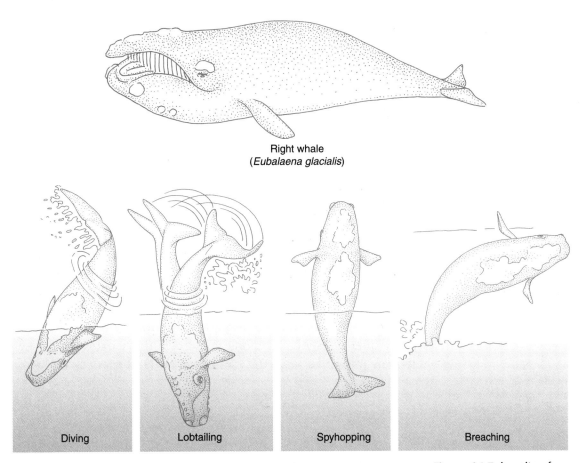

Right whale
(*Eubalaena glacialis*)

Diving

Lobtailing

Spyhopping

Breaching

Figure 14-7 In spite of its large size, the right whale is very acrobatic. Like several other whales, it exhibits a variety of behaviors and movements.

fully understood by marine scientists. Some believe that it is a sign of aggression. Others think that lobtailing is just the whale's way of announcing its presence.

Another common behavior in whales is **spyhopping**. When spyhopping, a whale raises its head above the water's surface to look around for a few seconds. It appears that whales are curious about the world above the ocean. Scientists do not yet fully understand the purpose of this behavior.

The most spectacular kind of whale behavior—and one that is often seen in humpback whales—is **breaching**. When a whale breaches, it leaps almost completely out of the water and then crashes back down, creating a huge, loud splash. Scientists are not sure why whales breach. There are several possible reasons: it might be a means of dislodging skin parasites, another form of communication, or just a playful type of behavior.

Whale Migrations

Whales are great long-distance swimmers. Over the course of a year, whales—traveling in their extended family group, or **pod**—can cover a course thousands of kilometers long. It is thought that whales find their way by locating geological features along the seafloor and by sensing changes in ocean currents, water chemistry, Earth's magnetic field, and the position of the sun. Look at the migratory route of the California gray whale (*Eschrichtius robustus*), shown in Figure 14-8, which is the longest of any whale species. Gray whales migrate in a north-south direction, from Alaska to Baja California, then back to Alaska again, a round-trip distance of more than 12,000 km.

Why do whales migrate? Many whales migrate between their feeding and breeding grounds, just as many species of birds do. Gray whales spend their summers feeding in the cold waters of the Bering Sea and Arctic Ocean. These nutrient-rich Arctic waters produce great quantities of plankton, on which the whales feed. They migrate to overwinter in the warm, shallow (and relatively secluded)

Figure 14-8 Migratory routes of two whale species—the humpback and the gray whale.

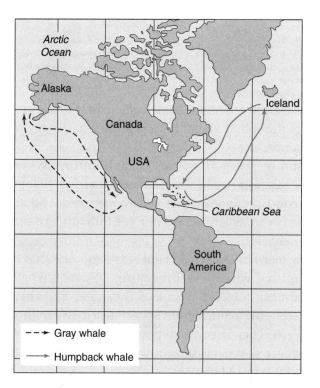

waters around Baja California, where they bear their young and breed. Whale babies lack the thick fat layers of the adults, so it is most adaptive for them to be born in the warmer water. The adult gray whales do not feed at this time; they live off the fat they accumulate during the summer in the Arctic.

Another baleen whale that migrates between feeding and breeding grounds is the humpback whale (*Megaptera novaeangliae*). Trace the migratory route of the Atlantic humpback whale, also shown in Figure 14-8. The Atlantic population of humpback whales migrates each year between Iceland and the Caribbean Sea. The Pacific populations of humpbacks (not shown in the figure) include those that migrate between the Bering Sea and southern Mexico (alongside California) and those that migrate between the Aleutian Islands and Hawaii.

Keeping Warm in Cold Waters

How are the migratory whales adapted to live in both polar and tropical seas? Whales are endothermic (warm-blooded) animals. Their body temperature remains the same in spite of the external water temperature. Whales are able to keep warm in the frigid polar seas because (unlike land mammals that have fur) they possess thick layers of fat, called **blubber**, under their skin. The blubber, which may be more than 60 cm thick in some whales, traps and prevents body heat from being lost through the skin.

Even in tropical waters, whales need the insulation. Warm bodies naturally lose heat to water. In addition, some whales have to dive to great depths for food, where the water is cold. The deepest diving whale is the sperm whale, which dives in pursuit of its preferred prey, the giant squid. In 1955, off the coast of Peru, a sperm whale was found entangled in an electric cable at a depth of 1127 meters. However, they may actually dive to more than three times that depth. (The sperm whale got its name from the fact that it has white waxy oil in its head called *spermaceti*, which plays a part in the whale's diving ability. Sperm whales increase the density of the oil before they dive and decrease its density when they ascend. Spermaceti was used many years ago in the manufacture of candles and as a lubricant for watch parts and delicate machinery. As a result of overhunting for this oil, the sperm whale population is now much smaller than it used to be.) Sperm whales also migrate through the world's oceans in search of rich feeding grounds, from the North Pole to south of the equator.

Communication and Echolocation in Whales

Cetaceans have relatively large, well-developed brains and are considered to be very intelligent. Part of that intelligence is shown in their ability to communicate among themselves, often in complex ways. Dolphins are known to communicate through a series of clicks and other sounds. These sounds are produced in the dolphin's airway and then focused or directed by a fatty bump in its forehead, called the **melon**. (Refer to Figure 14-1 on page 331.) For years, worldwide, dolphins have been observed hunting cooperatively, herding schools of fish along a shoreline—clearly, some form of communication between pod members is involved in this process.

Dolphins and whales (such as sperm whales) also use sound waves to sense objects in their environment and to locate prey. This natural form of sonar, in which sound produced by the cetacean is bounced off an object, is called **echolocation**. Sperm whales are also known to communicate through complex patterns of clicks referred to as codas. It is thought that the huge chamber of spermaceti in their foreheads is involved in the production of sounds and in echolocation. Sound waves emitted by dolphins and whales can also be used to stun both prey and enemies alike.

One of the most interesting aspects of whale communication is their ability to produce songs. Belugas are known to produce a great variety of sounds when they vocalize. (See the feature on sound in the sea in Chapter 19.) Perhaps most impressive of all, however, are the now-famous songs of the humpback whales. These have been recorded and analyzed for decades, and it has become apparent to scientists that humpbacks not only sing specific songs but also modify these songs over the years. It appears that the singing is done primarily by the breeding males and is related to competition for mates. Each population of humpbacks has its own variety of these songs, which changes slightly over time.

A puzzling and unfortunate kind of behavior that is sometimes seen in cetaceans is **stranding** on beaches. There have been many recorded cases of groups of whales (both toothed and baleen species) swimming into shallow waters and becoming beached, or stranded, when the tide goes out. Lying on the beach, the whales die, their internal organs crushed by their own body weight. Attempts to tow them back out to sea are not always successful. Many whales just swim back to the beach and strand themselves again.

The beaching of whales is still a mystery that marine scientists are trying to unravel. Several reasons for cetacean strandings have been hypothesized. Since whales rely on echolocation to sense their surroundings, it is possible that disease, parasites, or an infection, particularly in the leader of a pod, may cause the whales to become disoriented along a shoreline. In fact, toxins from water pollution, such as PCB and DDT, as well as natural toxins (called biotoxins) produced by red-tide dinoflagellates, have been found in stranded dolphins' tissues. It appears that these contaminants impair the overall health of the cetaceans. This makes them susceptible to bacterial and viral infections and, as a result, more likely to strand.

14.2 SECTION REVIEW

1. What are the two main reasons for the migrations of whales?
2. Describe three different kinds of whale behaviors. What is the purpose of each behavior?
3. How do whales and dolphins communicate among themselves?

14.3 SEALS AND OTHER MARINE MAMMALS

You can usually tell what kind of food an animal eats by looking at its teeth or mouthparts. For example, the baleen in a blue whale's mouth shows that this whale filter feeds on small organisms. The pointed teeth of dolphins and toothed whales show that these animals catch larger, more mobile prey. Likewise, marine mammals such as seals and sea otters have sharp, pointed teeth that are used in hunting. The canine teeth in front are used to seize prey, while the molars in back are used for cutting and tearing its flesh. Flesh-eating animals, which usually have sharp teeth, are known as

carnivores. The order of mammals called Carnivora includes both land and aquatic mammals. The seal, sea lion, walrus, sea otter, and polar bear are marine mammals that belong to this order. Animals that eat only vegetation are called **herbivores**. The manatees and dugongs are marine mammals that are herbivores.

Seals and Sea Lions

To move efficiently through water, another group of marine mammals has paddlelike appendages, or flippers, and a torpedo-shaped body. These fin-footed carnivorous marine mammals are classified in the suborder Pinnipedia (meaning "wing-foot" or "feather-foot"), which includes the seals, sea lions (and fur seals), and walruses. The **pinnipeds**, as they are called, are found throughout the world. There are about 30 different species. Seals and sea lions have the widest distribution and inhabit all the oceans. Walruses are found only in the polar seas of the north. The pinnipeds' main sources of food are fish and squid, although some will eat mollusks, crustaceans, or much larger prey.

At first glance, sea lions and seals look very much alike. However, there are some important structural differences. (See Figure 14-9.) Notice that the forelimbs of the sea lion are longer and more developed than those of the seal. The sea lion uses its forelimbs to move through the water. It can also prop itself up and move quickly on land, whereas the seal can only drag its body awkwardly along the beach. In contrast, the seal propels itself swiftly through the water using its stronger hind flippers. Another difference between the two groups is that sea lions (also called *eared seals*) have external ear flaps, whereas true seals (*Phoca*) do not. Seals and sea lions both have stiff facial whiskers, which help them locate their food.

Figure 14-9 The seal and the sea lion are examples of two similar types of pinnipeds. Unlike the seal, which is an agile swimmer, the sea lion has longer forelimbs (for quicker movement on land) and external ear flaps.

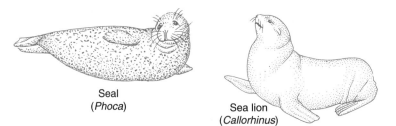

Seal
(*Phoca*)

Sea lion
(*Callorhinus*)

Reproduction in Pinnipeds

Unlike whales and dolphins, which are fully aquatic and can breed in the water, pinnipeds have to return to land to mate and give birth. During the breeding season, pinnipeds swim onto the shore, where they often congregate by the thousands. Females come ashore to give birth and to nurse their young. Mature males fight for access to the females, and the dominant males acquire "harems" of females with which they mate. Interestingly, seals and sea lions have evolved the ability to delay the development of their embryos, so that the birth of their single offspring occurs exactly 12 months after mating—when they come ashore again to breed.

The Walrus

The walrus (*Odobenus rosmarus*) inhabits the frigid Arctic and sub-Arctic waters. Notice the tusks of the walrus, shown in Figure 14-10. The tusks are overly developed canines that are used for digging up mollusks from the seafloor and for hauling the walrus up onto the ice. Tusks are found in both male and female walruses. In males, the tusks are longer and are used to establish dominance. Walruses use their sensitive, stiff facial whiskers to locate clams and mussels on the sandy seafloor. They can dive more than 90 meters deep to find their food.

Walrus
(*Odobenus rosmarus*)

Figure 14-10 The walrus is a large pinniped that inhabits the North Polar seas. It uses its tusks for digging up mollusks on the seafloor and for hauling itself onto the ice.

Adaptations of Pinnipeds

Although seals, sea lions, and walruses are not as fully aquatic as whales, they have many characteristics that indicate a life spent much of the time in the sea. Pinnipeds are intelligent, social animals. Recent research has shown that, like whales, pinnipeds communicate with one another by using a variety of sounds. It is possible that pinnipeds also use echolocation, since the sounds they produce are similar to those of the whales. Pinnipeds also have the ability to dive to great depths in search of food. (See Section 14.4.) The deepest diver of all the pinnipeds may well be the elephant seal; one was recorded diving more than 1500 meters below the surface.

All pinnipeds have a layer of blubber that insulates against the cold air and water. In addition, both seals and sea lions are covered by dense fur. However, the true seals are adapted to live in colder waters than sea lions are, because seals have denser fur and thicker skin. Walruses do not have much fur, but they have a thick layer of blubber under their skin to keep them warm in the Arctic.

For many years, people have hunted seals and sea lions for their thick beautiful furs, causing some populations to become endangered. In addition, people have killed these animals to reduce competition for commercially valuable fish. Walruses are also hunted —mostly illegally—for their ivory tusks. At present, most marine mammals are protected by law, but fatalities still occur. Many seals and sea lions are accidentally caught and drowned in huge drift nets, which are set out to catch entire schools of fish. Pollution, disease, loss of habitat, and declining food sources (due to overfishing by people) also threaten some pinniped populations.

The Sea Otter

Compared to cetaceans and pinnipeds, the **sea otter** (*Enhydra lutris*) only recently became adapted to living in the ocean (about three million years ago). Sea otters are closely related to the land-dwelling weasels and minks. They are the smallest of the marine mammals, growing to only about a meter in length. The three main populations live in the Pacific—along the coasts of California, Alaska and the Aleutian Islands, and various Russian islands. (See Figure 14-11.)

Figure 14-11 The sea otter eats shellfish as it floats on its back in kelp beds along the rocky Pacific coast.

Sea otter
(*Enhydra lutris*)

Sea otters are commonly found in the giant kelp forests along the rocky California coast. Kelp is the huge seaweed that grows from the seafloor to the ocean surface. As they float at the surface, sea otters hold onto the kelp, often wrapping it over their bodies to help anchor themselves in the choppy water. With their webbed hind feet, flattened tail, and streamlined body, sea otters are efficient swimmers and divers. On a typical dive, a sea otter swims to the bottom, locates a mussel, crab, abalone, or sea urchin, and returns to the surface where it rolls over to float on its back. Taking a rock that it also picked up from the bottom and placing it on its chest, the sea otter cracks open the shellfish by smashing it against the rock. After several whacks, the shell is opened and the sea otter consumes the contents. Sea otters have to repeat this process many times each day to supply their energy needs. The otters also eat fish and snails that live on the kelp. By eating snails and sea urchins, which graze on the kelp, sea otters help to maintain and promote growth of the seaweed—a great benefit to the kelp harvesting industry.

Sea otters spend most of their time in the ocean—they eat, sleep, mate, and rear their young in the water. Only on rare occasions does a sea otter venture on land at low tide to forage for mollusks among the rocks. Sea otters move awkwardly on land and quickly return to the water. Unlike the cetaceans and pinnipeds, sea otters have no blubber under their skin. In the chilly waters along the Pacific Coast, the sea otter can retain its body heat because it has incredibly thick fur, which it constantly grooms with oils from its skin. As the otter grooms its fur, air bubbles are trapped within it, thus greatly increasing its insulating ability.

Threats to Sea Otters

Oil spills from tankers are particularly devastating to sea otters because the oil coats their fur, which the animals then cannot clean and groom. The otters quickly freeze due to the loss of insulation. The soft, thick fur that sea otters depend on was so prized by fur trappers in the 1800s that, by the early 1900s, the California sea otter was almost hunted into extinction. Today, sea otters are protected, so they are making a comeback. But the California population, which numbers fewer than 2500 otters, is increasing very slowly. The Russian and Alaskan populations of sea otters are in better shape, with about 20,000 and 120,000 otters, respectively. Sea otters are also at risk of predation by great white sharks and killer whales, although they are not their main prey item.

Another type of otter that is sometimes found in the sea is the river otter, populations of which are found in Europe and in North and South America. Although they live mainly in freshwater, as their name implies, river otters are sometimes found on coasts, where they scramble along the rocks and venture into the ocean for food. River otters, which are closely related to sea otters, need protection too. Their numbers are dwindling due to loss of habitat, pollution, and the hunting of them for their fur.

Manatees and Dugongs

A docile marine mammal, the **manatee** (*Trichechus manatus*) lives in the warm, shallow waters of Florida. The manatee lives underwater, feeding on vegetation that grows in the rivers and waters along Florida's Gulf and Atlantic coasts. (See Figure 14-12.) About every 15 minutes, the manatee surfaces for a breath of air and then quickly submerges. Manatees are social animals that communicate with one another using a variety of high-pitched squeaks and whistles. During the winter, manatees swim up into the warmer coastal rivers that feed into the ocean. As the summer approaches and the ocean warms, the manatees leave the inland waters and migrate along the coasts. The manatee is also found in the shallow waters of the Caribbean (and so is often referred to as the West Indian manatee), in the Amazon River, and along the Atlantic coasts of South America and West Africa.

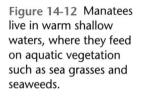

Figure 14-12 Manatees live in warm shallow waters, where they feed on aquatic vegetation such as sea grasses and seaweeds.

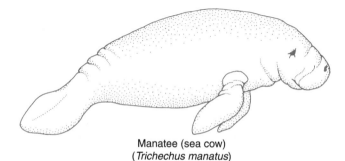

Manatee (sea cow)
(*Trichechus manatus*)

RESEARCH
The Migrating Manatee

A 3-meter-long marine mammal is on the loose and possibly heading your way! If you see him, report it to the authorities. Do not try to capture him. There is a $20,000 fine for harassment because this mobile animal is a manatee, a type of endangered marine mammal. Many bulletins like this were sent out by officials trying to track the whereabouts of a migrating manatee, one that usually inhabits the warm coastal waters and rivers of Florida. Why? Because sometimes, in the summer months, manatees leave the Florida waters and migrate north along the East Coast.

In October 1994, a migrating manatee was spotted in Chesapeake Bay, Virginia, earning him the nickname Chessie. At first, the manatee was thought to be lost. So he was caught and flown back to Florida, where he was returned to the wild. However, before Chessie was released, researchers attached a radio transmitter to his large, paddlelike tail in order to track his movements. The next time Chessie migrated, he swam as far north as Rhode Island.

Why would Chessie migrate more than 2000 km north? Well, Chessie is male; and male manatees tend to wander, whereas female manatees tend to stay within their home range. Scientists are not sure why Chessie swam so far north. It may have been the unusually warm ocean temperatures that year that caused him to swim farther than the Carolinas, the typical northernmost range.

During one of his yearly migrations, Chessie lost his satellite tag, making it difficult to track him. Rather than retag the manatee, the researchers decided to rely on the unique pattern of scars on his back (caused by impacts with boats) as a means of identification. Over the past few years, people have both sighted and photographed Chessie along his migratory route. As of this writing, the last photo of Chessie was taken in Virginia waters on September 26, 2001. Apparently, this West Indian manatee was still alive and well as he continued his north and south migrations along the eastern seaboard!

QUESTIONS

1. Why was Chessie's behavior considered so unusual?

2. By what means have scientists been tracking this manatee?

3. Why is the public interested in this marine mammal?

The manatee moves slowly through the water, propelled by a gentle up-and-down movement of its wide, paddlelike tail. The manatee has no hind limbs, and its forelimbs are used mainly for holding the aquatic plants on which it feeds. Adult manatees may consume up to 45 kg of vegetation a day (grinding it up with their molars) to maintain their body weight, which can reach over 900 kg. As a result, this grazing aquatic mammal is also commonly referred to as the *sea cow.*

Along with the **dugong** (*Dugong dugon*), which is its close relative, the manatee is classified in the order of mammals called Sirenia. These **sirenians**—named for the mythical female sirens that lured sailors to their doom—are actually distant relatives of the elephant, which is also a herbivore. The manatee uses its large upper lip, called a prehensile (meaning "handlike") lip, to grasp vegetation in a manner similar to that of the elephant using its trunk to pull up shrubs.

Dugongs are found in the tropical Pacific and off the east coast of Africa. Both manatees and dugongs have a cylindrical body shape similar to that of the pinnipeds. However, the dugong's tail is triangular rather than rounded like that of the manatee. Like the manatee, the dugong feeds on vegetation in shallow waters, concentrating on various types of sea grasses and seaweeds.

Threats to Manatees

Manatees and dugongs are endangered species, vulnerable to hunting, loss of habitat, and pollution. Unrestricted use of powerboats and continued development along Florida's coastal waterways are both responsible for a decline in the manatee population. As of the mid-1990s, there were about 2500 manatees left in Florida waters. Although protected by law, dozens of manatees are accidentally killed each year by powerboats' hulls and propellers when they surface to breathe. The impact of the hulls can kill them; and propeller blades cause deep wounds in the backs of the animals, either killing them directly or causing them to die from infection. Manatees that survive are left with permanent scars. In addition, a potent toxin from red tide algal blooms caused numerous manatee deaths in Florida in the mid-1990s.

Certain areas have been designated as manatee sanctuaries to help ensure their survival. (See Chapter 23.) It is hoped that manatees and dugongs will fare better than the Steller's sea cow, which was both discovered and hunted into extinction in the 1700s. This 9-meter-long sea cow grazed on seaweeds in the shallow waters of the Bering Sea, where it was easy prey for hungry sailors.

The Polar Bear

The marine mammal that is the most terrestrial is the polar bear (*Ursus maritimus*), which lives on ice floes and along the shore in the North Polar region. (See Figure 14-13.) Its dense fur and thick layer of blubber keep out the Arctic cold and retain body heat. (In addition, the hairs are hollow, which further helps to insulate the bear.) Although the polar bear is mainly adapted for life on land, it has a streamlined head and body, which helps it to be an excellent swimmer. The polar bear uses its powerful forelimbs to paddle from one ice floe to another. However, the polar bear cannot swim fast enough to catch a seal, its preferred food, in the water. Since its white coat enables it to blend in with the Arctic landscape, the polar bear can successfully stalk a seal that is sunning itself on the ice. Polar bears also wait at a hole in the ice to seize a seal when it comes up for a breath of air. Depending on the season, polar bears will also eat fish, birds, and plants. Except for a mother with her cubs, polar bears are mainly solitary animals.

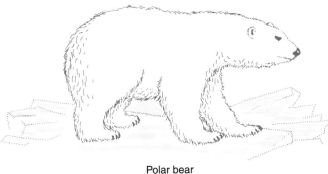

Polar bear
(*Ursus maritimus*)

Figure 14-13 The polar bear is the most terrestrial of all marine mammals, but it is also an excellent swimmer. It is an active predator both on land and at sea.

14.3 SECTION REVIEW

1. What are some physical adaptations of seals and sea lions to their aquatic way of life?

2. Describe the feeding habits of the California sea otter. Explain how the sea otter population benefits the growth of kelp.

3. What type of food do manatees and dugongs eat? Where?

14.4 THE DIVING RESPONSE

Many marine mammals have adaptations for diving deeply in pursuit of food. Marine mammals that dive very deep, such as whales, may need to hold their breath for as long as one-and-a-half hours. (Marine mammals automatically shut their nostrils when they are underwater.) How are these air-breathing mammals able to stay below the surface of the water for such a long period of time?

Diving marine mammals, such as whales and seals, can dive to great depths on a single breath, as shown in Table 14-1, because they have adaptations that increase the oxygen-carrying capacity of their bodies. These structures and behaviors make possible a group of responses that are collectively called the **diving response** (or diving reflex). One diving response is the detouring, or shunting, of blood that contains nutrients (glucose) and oxygen from the rest of the body to the vital organs—the brain, heart, lungs, and muscles. Whales oxygenate their blood by taking in a large volume of air at the surface. Just before a dive, they empty the air from their lungs.

TABLE 14-1 DEPTH AND TIME OF BREATH-HOLD IN DIVING MARINE MAMMALS

Mammal	Depth (meters)	Maximum Breath Hold (minutes)
Sea otter	55	5
Porpoise	305	6
Sea lion	168	30
True seal	575	73
Sperm whale	2200	90

The concentration of oxygenated blood and glucose in their important organs allows diving mammals to maintain their vital functions during the critical period of breath-hold. The cells of diving mammals can also increase energy production during breath-hold by carrying out anaerobic respiration. In this way, glucose can be metabolized in the absence of oxygen to produce small amounts of chemical energy (ATP).

Other Diving Response Adaptations

Another part of the diving response is the ability of marine mammals to inhale and exhale quickly, and nearly completely, between dives. Elastic tissue in their lungs and chest permits greater expansion during inhalation. The recoil action of elastic tissue in the lungs (along with the push of powerful chest muscles) allows the lungs to empty more quickly during exhalation.

Perhaps most important, diving mammals have a higher blood volume and a greater concentration of oxygen-binding red blood cells than nondiving mammals. The protein molecule hemoglobin, which is present in red blood cells, holds onto the oxygen. Diving mammals also possess another oxygen-binding protein, called **myoglobin**, which is located in their muscles. Together, the hemoglobin and myoglobin increase the oxygen-carrying capacity of their bodies during a dive.

Another important component of the diving response in marine mammals is **bradycardia**, the ability to slow the heart rate. For example, when the northern elephant seal dives below the surface, its heart rate drops from 85 to 12 beats per minute. Since oxygen-rich blood is shunted to just the vital organs during a dive, there is less blood circulating around the animal's body. Consequently, heart action slows, which helps to conserve the animal's energy.

Interestingly, bradycardia is not limited to marine mammals. In 1973, a Frenchman, Jacques Mayol, had a drop in heart rate from 72 to 36 beats per minute while setting a world's record for a single-breath-hold dive of 85 meters. What is the mechanism that causes bradycardia and the shunting of blood during deep dives? Scientists discovered that both of these diving responses involve reflex actions of the nervous system. Water pressure acts on pressure receptors located in the head of the diving mammal, causing impulses to be sent to the brain. The brain, in turn, sends impulses back to the heart, telling it to regulate its rate. Impulses from the brain are also sent to the blood vessels, causing them to constrict or dilate, thus influencing the direction of blood flow.

Can bradycardia be demonstrated in humans under laboratory conditions? You can find out by doing the lab investigation at the end of this chapter.

14.4 SECTION REVIEW

1. What are some features of marine mammals that enable them to make deep dives?

2. What is bradycardia? How does it work along with other features of the diving response?

3. Which three marine mammals dive the deepest (according to Table 14-1 on page 350)? Why do they dive so deep?

Laboratory Investigation **14**

Diving Response in Humans

PROBLEM: Can a diving response be demonstrated in humans?

SKILL: Measuring pulse rates.

MATERIALS: Basin or bucket of warm water (about 25 to 30°C), catch basin, stopwatch, ear plugs, towel, swim cap.

PROCEDURE

1. Work with a partner. Locate the pulse in your partner's wrist. (The pulse rate is the rate at which the heart beats.) Take your partner's pulse while he/she holds his/her breath for 15 seconds. (Either you or your partner can look at the stopwatch.) Multiply by 4 to get the pulse rate per minute. In your notebook, record the result in a copy of Table 14-2. (See page 354.)

2. Switch roles and have your partner take your pulse for 15 seconds, while you hold your breath. Again, multiply by 4 to get the pulse rate per minute. Record the result in your table. If time permits, repeat this procedure for a total of six trials (three per partner). Calculate the average by adding the results for each trial and dividing by the number of trials. These data serve as the control group, since the experimental factor (water pressure) was not involved.

3. Fill a basin or bucket to the top with warm water and place it in a container large enough to catch the overflow. Put on a swim cap and ear plugs. Now submerge your face in the water for 15 seconds, while your partner takes your pulse. (In this case, your partner will watch the time!) Multiply by 4 to get the pulse rate per minute. (If 15 seconds feels too long, submerge and time the pulse for 10 seconds, then multiply by 6 to get the pulse rate per minute.) Record the result in your table.

4. Repeat for a total of six trials (with partners taking three turns each, alternating at submerging and timing) and calculate the average. These data represent the experimental group, with facial submersion in water (water pressure) being the experimental factor. Compare with above (control group) results to see if there is any difference.

5. Write all students' results on the chalkboard. Compute a class average to see if the overall results appear to be significant.

TABLE 14-2 COMPARING PULSE RATES OF NOT-SUBMERGED AND FACIALLY SUBMERGED PEOPLE

Trial	Not-Submerged Pulse Rate (beats per minute)	Facially Submerged Pulse Rate (beats per minute)
1		
2		
3		
4		
5		
6		
Total		
Average		

OBSERVATIONS AND ANALYSES

1. What was your average pulse rate for breath-hold in the air? In the water? What does your pulse rate actually represent?

2. What differences, if any, did you observe in the two sets of trials? Why did you hold your breath for the control group trials?

3. Based on your results, can bradycardia occur in humans? Does it begin with facial submersion? Explain.

Answer the following questions on a separate sheet of paper.

Vocabulary

The following list contains all the boldface terms in this chapter.

baleen, baleen whales, blowhole, blubber, bradycardia, breaching, carnivores, cetaceans, diving response, dugong, echolocation, herbivores, lobtailing, manatee, melon, myoglobin, pinnipeds, pod, sea otter, sirenians, spyhopping, stranding, tail flukes, toothed whales

Fill In

Use one of the vocabulary terms listed above to complete each sentence.

1. The term _____ refers to the seals, walruses, and sea lions.

2. The _____ have special "plates" to filter feed on plankton.

3. Humpbacks may be seen _____, or leaping out of the water.

4. Layers of fat, or _____, enable whales to live in cold waters.

5. Whales travel in an extended family group, or _____.

Think and Write

Use the information in this chapter to respond to these items.

6. List some characteristics that marine mammals and humans have in common.

7. How are pinnipeds adapted for food-getting underwater?

8. In what kinds of environments are the sirenians now found?

Inquiry

Base your answers to questions 9 through 12 on the graph on page 356 and on your knowledge of marine science.

A student hypothesized that a human underwater would show a similar diving reflex (bradycardia—the slowing of the heart rate) to that of a whale or a seal. She tested her hypothesis by recording her pulse rate at different depths in a swimming pool of uniform temperature and light intensity. The results of her experiment are shown in the following graph.

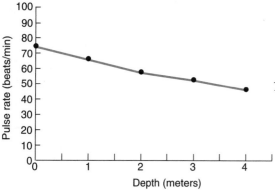

Pulse rate (beats/min) vs Depth (meters)

9. The results of the experiment show that, as depth increases, the human pulse rate *a.* increases *b.* increases then decreases *c.* decreases *d.* remains the same.

10. Based on the data in the graph, which statement is most correct? *a.* The hypothesis is supported by the data. *b.* The hypothesis is not supported by the data. *c.* The hypothesis could not be tested. *d.* There appears to be no relationship between depth and pulse rate.

11. A tentative conclusion that can be drawn from this experiment is that *a.* humans have nothing in common with seals and whales *b.* people do not have any diving reflex *c.* bradycardia probably does occur in humans *d.* people are not affected by changes in water depth.

12. Which of the following would be an appropriate title for this experiment? *a.* The effect of water temperature on pulse rate in humans. *b.* The effect of chlorinated water on pulse rate in humans. *c.* The effect of water depth on pulse rate in marine mammals. *d.* The effect of water depth on pulse rate in humans.

Multiple Choice

Choose the response that best completes the sentence or answers the question.

13. How is the whale adapted for breathing in the ocean? *a.* Air enters the blowhole and exits the mouth. *b.* Air enters and exits through the mouth. *c.* Air enters and exits through the blowhole. *d.* Air enters the mouth and exits the blowhole.

14. Of the following, which marine mammals have been heavily hunted for their fur? *a.* seals and gray whales *b.* seals and sea otters *c.* dolphins and manatees *d.* gray whales and manatees

15. Which of the following is a filter-feeding marine mammal? *a.* sea otter *b.* sperm whale *c.* humpback whale *d.* manatee

16. The pinniped whose forelimbs are better developed for moving on land is the *a.* true seal *b.* walrus *c.* sea lion *d.* sea otter.

17. The marine mammal that cracks shells open with a rock is the *a.* sea lion *b.* sea otter *c.* river otter *d.* manatee.

18. Which of the following is part of the diving response? *a.* the shunting of blood away from the vital organs *b.* the speeding up of the heart rate *c.* a slowing of the heart rate *d.* a decrease in the oxygen-carrying capacity

19. Evidence of the whale's terrestrial ancestors is clearly seen in the internal structure of its *a.* teeth *b.* forelimbs *c.* tail flukes *d.* forehead.

20. Which of the following is *not* a characteristic of all marine mammals? *a.* being warm-blooded *b.* having a blowhole *c.* having a four-chambered heart *d.* nursing their young

21. Which of the following is a toothed whale? *a.* blue whale *b.* humpback whale *c.* gray whale *d.* sperm whale

22. Bradycardia, a type of diving response, occurs as a result of *a.* water temperature *b.* water pressure *c.* hunger *d.* fatigue.

23. It is thought that whales strand themselves because they *a.* follow surfers ashore *b.* ride the bow waves of boats *c.* are disoriented due to disease or infection *d.* are blind.

24. The following marine mammal is classified as a *a.* pinniped *b.* large porpoise *c.* toothed whale *d.* baleen whale.

Research/Activity

- Choose one baleen whale and one toothed whale, or one true seal and one sea lion, and prepare a chart that illustrates the distinguishing features of each group.

- Investigate the breathing rate of a dolphin or whale in an aquarium. Record its breathing rate in several trials at different times of the day to see if daylight is a factor. Prepare a report based on your results and additional research.

UNIT 5

THE WATER PLANET

Most people who look at a globe or world map will see it as land separated by oceans. Yet, people who spend their lives at sea will see oceans separated by land. Which perception is correct? Actually, both perceptions are valid — they just depend on one's point of view. However, the oceans do cover over 70 percent of the planet's surface, so it is accurate to call Earth a water planet.

Scientists who study the ocean know water has special properties that affect the organisms living in it. Water is the very basis of life. In the first chapter of this unit, you will learn about the chemistry of water and why it is so important for the support of life on Earth.

In the chapter on geology of the ocean, you will learn about the origin of the oceans and continents, and how they have changed over time. The ocean influences temperature, precipitation, and the formation of storms. The last chapter in this unit discusses the ocean's effect on climate.

359

15 The World of Water

When you have completed this chapter, you should be able to:

DISCUSS Earth's water cycle and water budget.

EXPLAIN the basic chemistry of water as a solvent.

DESCRIBE the importance of pH to water chemistry.

DISCUSS the sources of, and variations in, the ocean's salinity.

All day and all night—every day and every night—the ocean keeps up a steady rhythm as it advances and then retreats from the shore, propelled by tidal forces. Moving in, the ocean waves break along the sloping shore; and just as quickly, the waves recede.

In some places, shallow tidal pools are left behind as the waves move away from the shore. Scientists think that several billion years ago, life on Earth began in the ocean—perhaps in a tiny pool of water left behind by the unending movement of the waves. Most likely, life probably began in the sea because ocean water contains important substances needed for life processes.

The chemical composition of ocean water is the main topic of this chapter. Knowledge about water and the substances it contains will provide you with a better understanding of how life exists in the sea.

15.1 THE WATER PLANET

Since so much of Earth is covered by water, it is sometimes called the "water planet." If you look at a world map in an atlas, you will see that there is more water than land. In fact, more than 70 percent of Earth's surface is water, with most of that water in the ocean. The ocean surrounds landmasses known as continents, which are the remaining 29 percent of Earth's surface. The continents divide the ocean into four major parts: the Atlantic Ocean, Pacific Ocean, Indian Ocean, and Arctic Ocean. Often, where two continents lie close together, a smaller part of an ocean called a **sea** is formed. Locate the Caribbean Sea in Figure 15-1. Notice that it connects to the Atlantic Ocean. In fact, all of Earth's oceans and seas flow into one another, forming one continuous body of water.

Latitude and Longitude

To locate geographical areas with precision, people draw a grid, or series of lines, over maps and globes. The equator is one of the most important lines on this grid; it divides the world in half. The

Figure 15-1 The oceans and seas of the world.

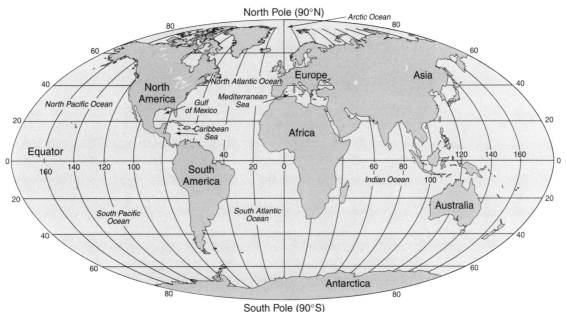

northern hemisphere is the part of Earth located north of the equator; and the southern hemisphere is the part of Earth located south of the equator. The southern hemisphere has 20 percent more ocean than the northern hemisphere. Lines that run parallel to the equator are called lines of **latitude**. Distances north or south of the equator are measured in degrees. The latitude of the equator is 0 degrees. The latitude of the North Pole is 90 degrees north, and the latitude of the South Pole is 90 degrees south. All other latitudes fall between 0 and 90 degrees north or south of the equator.

Try to locate the Galápagos Islands on the map in Figure 15-1. These islands are represented by a tiny spec right on the equator, at 0 degrees latitude, west of South America. If you follow east along the 0 degree line, you will see that many other land areas also lie on the equator. For example, Ecuador, Brazil, and several countries in Africa lie on the equator.

How are geographical areas located at the same latitude distinguished from one another? Another measurement, called **longitude**, when combined with latitude measurements can pinpoint geographical locations with great precision. On a globe, you can see that lines of longitude run from the North Pole to the South Pole, intersecting the lines of latitude. Longitude is measured in degrees east or west of the prime meridian. The prime meridian runs through Greenwich, England, and it is assigned the starting point of 0 degrees longitude. Earth is divided into 24 meridians; meridians are 15 degrees apart. The meridians cover a distance 180 degrees east and 180 degrees west of the prime meridian.

The precise location of a geographical area can be given by specifying the latitude and longitude lines that pass through the area. Look again at Figure 15-1 and find the exact location of the Galápagos Islands; they are located at about 85 degrees west longitude. In contrast, the mouth of the Amazon River in Brazil (not labeled on the map) is located at 50 degrees west longitude. However, both places are located at 0 degrees latitude; that is, both are at the equator.

15.1 SECTION REVIEW

1. Why is Earth often called the "water planet"?

2. How are geographical places on Earth located with accuracy?

3. What is the precise difference between an ocean and a sea?

ENVIRONMENT
A Beacon on a Vanishing Beach

For centuries, sailors have used measurements of latitude and longitude to help them find their way. In addition, seafarers have relied on specific landmarks to guide their ships safely into port. Lighthouses are among the most important of these landmarks, helping sailors steer their way around offshore hazards.

The famous lighthouse in Cape Hatteras, North Carolina, built on a barrier island in 1870, is one such landmark that has been given a second life. At risk of toppling into the surf only 50 meters away, this historic structure was hoisted up by special lifts, placed onto platforms, and moved on rollers 400 meters inland. Now it stands 60 meters high, a beacon in the night to the countless ships that sail in these turbulent coastal waters—an area often referred to as the "graveyard of the Atlantic." But how long will it remain standing on this vanishing barrier beach? Scientists estimate that, if beach erosion continues at its present rate, by the year 2090 the lighthouse will have to be moved inland again or it will collapse into the sea.

Building on a barrier beach is like trying to build a castle in the sand. Sand is a very unstable substrate. Both wind and water can easily shift and move the sand about. Climatic conditions—in the form of coastal storms, heavy winds, tidal surges, and wave action—constantly change and erode the landscape of these barrier beaches, which run parallel to the main coastline.

But people like to live near the sea. Barrier islands along the Atlantic and Gulf coasts continue to attract developers, in spite of the hazards of beach erosion. Local governments have built a variety of barriers—including sea walls parallel to the beach and rock walls perpendicular to the beach—to keep the sand from washing away. Yet these efforts have been largely unsuccessful, since nature continues to reshape the barrier islands through erosion and deposition.

Dredging sand from the seafloor offshore and dumping it onto the beach is the latest attempt by coastal communities to keep erosion at bay. But the people of Cape Hatteras did not choose this method to save their lighthouse. After all, it would only be a short period of time before erosion would claim more land. Instead, they retreated inland with their lighthouse, which continues to send out signals in the night, but perhaps with a more symbolic warning for developers—don't build on a vanishing beach.

QUESTIONS

1. Why was the Cape Hatteras lighthouse moved 400 meters inland?

2. Describe the climate conditions that can contribute to beach erosion.

3. Compare the pros and cons of building on a barrier beach with building inland.

15.2 THE WATER BUDGET

Most people have to plan carefully how they spend their money. Very few people have unlimited amounts of money to spend. The amount of money needed for expenses over a given period is called a budget.

Earth also has a budget, a **water budget**. Earth's water budget is the total amount of water contained in and on the planet. As you can see in pie graph A in Figure 15-2, the oceans contain about 97 percent of all the water on Earth. Which ocean has the largest volume of water? The Pacific is the largest ocean, followed by the Atlantic, Indian, and Arctic oceans. (See pie graph B in Figure 15-2.) Only 3 percent of all the water on Earth is freshwater. Most of that 3 percent is frozen in glaciers and in ice found near the polar regions. Only about one-third of all freshwater on Earth is found as liquid in rivers and lakes, and in underground sources. Compared to the amount of salt water, the amount of freshwater on Earth is very small indeed.

The amount of water in the ocean determines sea level. The **sea level** is the point at which the ocean surface touches the shoreline. Over Earth's long history, the sea level has changed. About 12,000 years ago, during the last great ice age, the sea level was lower than it is today, perhaps as much as 100 meters lower. At that time, the edge of the sea was at the continental shelf. Why was the sea level so much lower then than it is today? During that period, Earth's climate was colder, snowfall increased, and as a result, much of the world's water was frozen—locked up in the form of glaciers and polar ice caps. When water freezes, less is available for the oceans,

Figure 15–2 Pie graphs: (A) Earth's total water budget and (B) the volume of water in Earth's oceans.

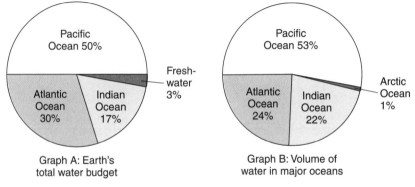

Graph A: Earth's total water budget

Graph B: Volume of water in major oceans

and the sea level drops. Since the Ice Age, the climate of Earth has warmed, causing much of the ice to melt. This melting ice added water to the ocean (through rainfall and runoff), and the sea level has risen.

Will the warming trend continue? Will the level of the sea continue to rise? The past offers some clues. Several ice ages have occurred during Earth's history. And each ice age was followed by a warming trend. If the past provides hints about the future, there will most certainly be another period of global cooling, perhaps followed by another ice age. Some scientists are almost certain of this; it is only the timing of these events that remains uncertain. For now, the sea level still appears to be rising, very slightly, each year.

The Water Cycle

Water falls to the ground as rain, snow, sleet, and hail. These forms of moisture that fall to the earth are called **precipitation**, and they become the streams and rivers that eventually flow into the oceans. The land area through which water passes on its way to the ocean is called a **watershed**. Rain even falls directly over the ocean. You may wonder, then, why the sea level doesn't keep rising indefinitely. If you look at Figure 15-3, you will see that water also moves from the ocean to the atmosphere and back again to the land. This continuous movement of water is called the **water cycle**. The water cycle is responsible for the reuse, or recycling, of this most important of natural resources.

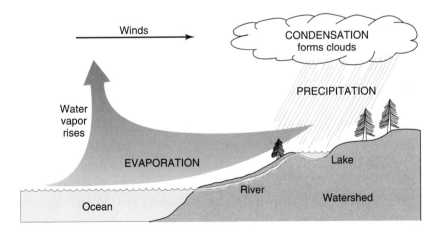

Figure 15-3 Earth's water cycle occurs through the processes of evaporation, condensation, and precipitation.

Follow the pathway of water. One of the stages in the water cycle is **evaporation**. Evaporation is the process by which liquid water changes to a gas. During evaporation, water molecules at the water's surface move into the air as water vapor, a gas. You cannot observe water molecules moving from a liquid to a gas. The water molecules are too small and too far apart to be seen. But if weather conditions are just right, you can observe water vapor in the air. If the temperature is cold enough, water vapor forms a cloud. Fog is a cloud of water vapor that forms close to the ground.

The process of cloud formation is called **condensation**. Condensation occurs when molecules of water vapor come close enough together to form a cloud. Condensation is another stage in the water cycle. Clouds are moist because they contain tiny droplets of water. If the droplets are heavy enough, they fall to the earth as rain or another form of precipitation. Thus, precipitation is an important part of the water cycle. As a result of the water cycle, freshwater is returned to the land to be used by plants and animals. Eventually, this water returns to the ocean.

15.2 SECTION REVIEW

1. Why was the sea level lower during the last ice age? Why is the sea level higher today?

2. What is Earth's water budget? How much of it is freshwater?

3. Why doesn't the sea level rise continuously as water from rivers and rainfall enters the ocean?

15.3 WATER AS A SOLVENT

The next time it rains, observe some puddles of water. Note that, after they dry up, a residue of substances is left behind. Even tap water can leave a stain when it evaporates from a drinking glass or tabletop. The stain, or residue, includes chemical substances that were dissolved in the water. A liquid, such as water, that contains dissolved substances is called a **solvent**. Water is a good solvent because it can dissolve many different substances, including salt. In fact, water is so good at dissolving substances that it is often called the universal solvent. Ocean water contains many dissolved substances. The study of the chemical substances that are found in ocean water is a branch of marine science called *chemical oceanography.*

Substances in Sea Water

Ocean water is made up of about 96.5 percent water molecules. The remaining 3.5 percent is mostly salt, which is dissolved in the water. Any substance that water holds in a dissolved state is called a **solute**. In ocean water, salt is a solute because it is dissolved in water, the solvent. Salt mixed with water forms salt water, a **solution**. The relationship among solute, solvent, and solution can be seen in the following equation.

$$\text{Salt} \quad + \quad \text{Water} \quad \longrightarrow \quad \text{Salt water}$$
(solute)   *(solvent)*   *(solution)*

Many solutions, like ocean water, are also mixtures. A **mixture** contains two or more substances that can be separated by ordinary physical means. For example, seawater contains salts and water. You can separate the salt from the water by putting some ocean water in a beaker. After a short time, the water will evaporate, leaving a residue of salt behind. You can hasten this process by heating the beaker, causing the water to evaporate more quickly.

The salts dissolved in ocean water are called *sea salts*. A sea salt is a type of compound. A **compound** is a substance that contains two or more kinds of atoms that are chemically joined, or bonded, together. The compound sodium chloride is the most common of the sea salts. You already know it by its common name, table salt. Sodium chloride is made up of the elements sodium (Na) and chlorine (Cl). (See Table 15-1, which lists the sea salts, on page 368.)

Ocean water can be prepared by dissolving sea salts in tap water. When a salt is added to water, the elements joined in the compound break apart, or ionize, to form ions. An *ion* is an atom that is not electrically neutral and thus has a charge. Ordinary table salt (NaCl), when added to water, ionizes into the positively charged sodium ion (Na^+) and the negatively charged chloride ion (Cl^-).

Another compound found in the ocean is calcium carbonate ($CaCO_3$). Calcium carbonate is also called *limestone*. It is the main component of seashells and of the limestone that makes up coral

TABLE 15-1 SEA SALTS FOUND IN OCEAN WATER (PERCENT)

Type of Sea Salt	Sea Salt Content
Sodium chloride	67.0
Magnesium chloride	14.6
Sodium sulfate	11.6
Calcium chloride	3.5
Potassium chloride	2.2
Other sea salts	1.1

reefs. Chalk, which was originally formed in the ocean, is made of calcium carbonate. The wearing down of seashells and coral reefs releases calcium carbonate into the seawater. Again, in this case, its ions are released: the positively charged ion Ca^{2+} and the negatively charged ion CO_3^{2-}. Note that each of these ions has a charge of 2.

The charge on each ion results from the presence or absence of the negatively charged particles called *electrons*. An excess of electrons produces an ion with a negative charge; a lack of electrons produces an ion with a positive charge.

Ions with opposite charges are attracted to each other by a force known as *electrostatic force*. The elements in both sodium chloride (NaCl) and calcium chloride ($CaCl_2$) are held together by electrostatic forces when these compounds exist as solids (not in solution). However, when they are in solution, these solids break into ions because the electrostatic forces in the water pull them apart.

The Water Molecule

How does salt dissolve in water? We have to focus on the structure of a molecule of water before we are able to answer this question. Look at Figure 15-4. The chemical, or molecular, formula for water is H_2O. A water molecule is formed from two atoms of hydrogen and one atom of oxygen. A *molecule* is defined as the smallest quantity of an element or a compound that can exist without losing the properties of that substance.

Notice in Figure 15-4 that the oxygen atom is larger than the hydrogen atom. An atom of oxygen contains more atomic particles than does a hydrogen atom. Each oxygen atom has eight protons and eight neutrons in its nucleus. Each of these atomic particles is assigned a weight of one, so an atom of oxygen has an atomic weight of 16. Each atom of oxygen also has eight electrons. The electrons are located in an area surrounding the atom's nucleus. However, the weight of an electron is so small that it does not contribute to the atomic weight. Hydrogen has one proton and no neutrons in its nucleus. It also has one electron. Because it contains only one proton, the hydrogen atom is assigned a weight of one.

Now let's focus on a single molecule of water. The oxygen atom has more electrons than do the two hydrogen atoms. Therefore the oxygen side of a water molecule is more negative than the hydrogen side. The hydrogen side of a water molecule is less negative, or more positive, than the oxygen side. The opposite charges on the two sides of all water molecules cause them to be attracted to each other, forming weak hydrogen bonds.

When you sprinkle table salt (NaCl) into tap water, the salt dissolves because the opposite charges on the water molecules attract ions from the salt. The oppositely charged water molecule, which is called a polar molecule, behaves like a magnet. When sodium chloride is added to water, the Na^+ is attracted to the negative end of the water molecule, and the Cl^- is attracted to the positive side. The sodium and chloride ions separate and become surrounded by water molecules. The result is that salt dissolves in water.

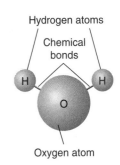

Figure 15-4 The water molecule; hydrogen and oxygen atoms are held together by chemical bonds.

pH

Not all water molecules exist as the polar molecule shown in Figure 15-4. A small number of water molecules separate into positive and negative ions, as shown in the following equation:

$$H_2O \longrightarrow H^+ + OH^-$$

Water   *Hydrogen ion*   *Hydroxyl ion*

The ion with the positive charge is called the *hydrogen ion,* and the ion with the negative charge is called the *hydroxyl ion.*

A solution that contains a larger number of hydrogen (rather than hydroxyl) ions is called an **acid**. A solution that has a larger number of hydroxyl (rather than hydrogen) ions (OH^-) is called a **base**, or alkaline solution. The degree of acidity or alkalinity of a solution is called its **pH**. The term *pH* refers to the concentration (power) of hydrogen ions in a solution. The pH of a solution is measured on a scale that runs from 0 to 14.0. An acid has a pH that lies between the numbers 0 and 6.9 on the pH scale. One characteristic of an acid is its sour taste. Imagine biting into a lemon or a lime. The sour taste of these two fruits is due to the presence of citric acid ($C_6H_8O_7$), a substance that in water produces a large number of hydrogen ions.

A basic solution has a pH that lies between 7.1 and 14.0 on the pH scale. Bases include sodium hydroxide (NaOH) and potassium hydroxide (KOH). Soaps and detergents are bases. Some common medicines that are used to neutralize stomach acid are bases. If a substance contains equal numbers of hydrogen ions and hydroxyl ions it is called **neutral**. A neutral solution has a pH of 7.0. Find the pH of ocean water on the pH scale shown in Figure 15-5. As you can see, the pH of ocean water is approximately 8.0. Is ocean water acidic or basic?

Sometimes bodies of water can become acidic due to acid rain, also called **acid precipitation**. Acid rain is formed when chemicals released by the burning of fossil fuels are absorbed by moisture in the air. Acid rain has changed the pH of hundreds of lakes and ponds. In extreme cases, the pH of these bodies of water may fall as low as 4 or 5. This high acidity has been responsible for a drop in fish populations in many freshwater environments.

Ocean water has a more stable pH and is affected less by acid rain than is a smaller body of water such as a lake or pond. When it rains into the ocean, the large volume of ocean water dilutes the acid rain. Also, chemicals in ocean water, called **buffers**, help main-

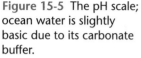

Figure 15-5 The pH scale; ocean water is slightly basic due to its carbonate buffer.

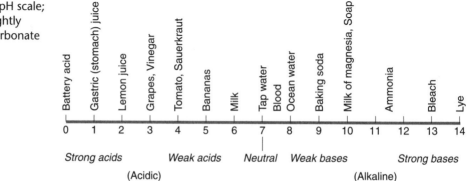

tain a stable pH. A buffer is a substance that lessens the tendency of a solution to become too acidic or too basic. One of the buffers present in ocean water is the carbonate buffer (CO_3^{2-}). The carbonate buffer can accept hydrogen ions, causing the water to become less acidic or more basic. The carbonate buffer can also release hydrogen ions, making the water more acidic and less basic. The actions of this buffer can be seen in the following reaction:

Carbon dioxide + Water $\longrightarrow$ Carbonic acid $\longrightarrow$ Bicarbonate ion + Hydrogen $\longrightarrow$ Carbonate ion + Hydrogen ions

Acidity increases (an increase in hydrogen ions)

$\longrightarrow$

$$CO_2 + H_2O \longrightarrow H_2CO_3 \longrightarrow HCO_3^- + H^+ \longrightarrow CO_3^{2-} + 2H^+$$

$\longleftarrow$

Acidity decreases (hydrogen ions are removed)

In ocean water, the buffering action goes to the left, producing a slightly basic pH that lies between 8 and 9. However, the pH of ocean water varies slightly within this range over each 24-hour period, as shown in the graph in Figure 15-6. During the day, as marine algae carry out photosynthesis (the manufacture of simple carbohydrates), CO_2 is removed from the water. The removal of dissolved CO_2 forces the buffering reaction to move to the left, which removes hydrogen ions. As hydrogen ions are removed, the water becomes less acidic or more basic. The pH of ocean water rises. At night, photosynthesis does not occur—but respiration (the release of stored energy found in nutrients) does. Respiration adds CO_2 to the water. The CO_2 causes the buffering reaction to move to the right. The number of hydrogen ions is increased, and the pH of ocean

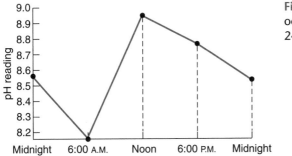

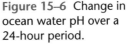

Figure 15–6 Change in ocean water pH over a 24-hour period.

water is lowered. The opposite effects of photosynthesis and respiration cause the slight changes in ocean water pH that occur over each 24-hour period.

Oxygen in the Water

If you try to hold your breath, your body's need for air will overcome your attempt not to inhale within a minute or two. You need air to survive—specifically, the oxygen in the air, which is the gas produced by plants during photosynthesis. It may surprise you to learn that a great deal of Earth's supply of oxygen is the result of photosynthesis carried out by algae and plants that live in water.

The oxygen produced in photosynthesis is in the form of molecular oxygen (O_2). After oxygen leaves the plant, some of it dissolves in the water. The rest is released into the atmosphere at the water's surface (some of which goes back into the sea). The oxygen you take in with every breath includes some that comes from marine plants and algae!

The oxygen that dissolves in ocean water is called **dissolved oxygen**, or **DO** for short. Some marine organisms take in oxygen through pores and moist cell membranes on their body surface; other organisms, such as fish, obtain their oxygen through membranes in special structures called gills.

However, oxygen is not very soluble in water. The quantity is fairly small, so scientists measure the DO in parts per million (ppm). Ocean water can hold from 1 to 12 ppm of dissolved oxygen (depending on the water's temperature). This amount is much less than the amount of oxygen in air, which is about 200 ppm.

How is oxygen distributed in ocean water? Figure 15-7 shows the distribution of oxygen from the surface of the ocean to the bottom. You can see in the graph that, as depth increases to about 1000 meters below the surface, the DO in the water decreases. Notice that the lowest amount of oxygen, called the **O_2 minimum zone**, is located at 1000 meters and not at the bottom of the ocean. The bottom contains a slightly higher level of DO than the O_2 minimum zone. This increase in DO on the bottom is a result of the temperature and pressure of the water at this depth. Cold, dense water contains a higher concentration of oxygen than does warm, less dense water.

Why is there more DO at the surface of the ocean than in the

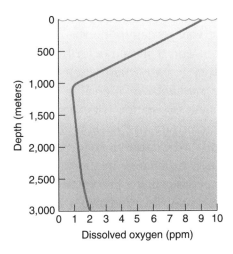

Figure 15-7 Dissolved oxygen levels at different ocean depths.

depths? Recall that much of the oxygen in ocean water is produced by marine plants and algae within the sunlit zone, which extends down to a depth of about 60 meters. Since photosynthesizing plants cannot live in the darkness below this depth, the level of dissolved oxygen found there is much less.

You may wonder why there is any oxygen at all dissolved in the deepest parts of the ocean. Actually, the ocean is not a calm, unmoving body of water. Waves and currents circulate the water throughout the ocean, causing some oxygen from surface waters to reach the lowest depths.

15.3 Section Review

1. How does salt (sodium chloride) dissolve in water?

2. Why is the ocean more basic (alkaline) than a freshwater pond?

3. What is the relationship between DO and ocean depth?

15.4 THE SALTY SEA

Fill two beakers, one with clean ocean water and the other with tap water. Ask a classmate to distinguish between them. It won't be easy, because they both look alike; yet the beaker of ocean water contains 3.5 percent salt. How can you tell them apart? Of course, you (or your classmate) could taste a sample. However, this method of testing is not the best way to determine the chemical makeup of an unknown substance—and it should *never* be used to test any substances in a laboratory (or anywhere else)! A much safer way is to use an *indicator solution*. An indicator solution changes in the presence of a particular substance. For example, you can use an indicator called silver nitrate ($AgNO_3$). When silver nitrate is added to salt water, a milky white substance forms. When silver nitrate is added to freshwater, no change in the solution is observed. The milky white substance is called a **precipitate**. A precipitate is a solid substance that may be produced when two liquids are mixed. In this case, the precipitate in the beaker of salt water is the compound silver chloride ($AgCl$). Silver nitrate will produce a precipitate only when it is added to a solution that contains chloride ions. This chemical reaction is summarized in the following equation:

$$NaCl \quad + \quad AgNO_3 \quad \longrightarrow \quad AgCl \quad + \quad NaNO_3$$

sodium chloride   silver nitrate   silver chloride   sodium nitrate

The silver nitrate test for chloride ions is a qualitative test. A *qualitative test* is used to reveal the presence or absence of a substance. In this case, it tests whether there are chloride ions in solution. A test that measures the actual amount of a substance (numerically) is called a *quantitative test*. For example, the hydrometer described in Chapter 2 is used to measure the salinity of water in units of specific gravity. Thus, you can use a hydrometer to quantitatively test the salt content of your aquarium tank.

Salinity Variations

Does the salinity of ocean water vary around the world? Look at the salinity chart of surface waters shown in Figure 15-8. As you can see, the salinity, expressed in parts per thousand (ppt), does not vary greatly from place to place, but there are differences. For example, the middle of the Pacific shows a salinity of over 35 ppt, whereas the middle of the Atlantic shows a salinity of over 37 ppt. Readings from the Red Sea and the Mediterranean Sea can be even higher. How can the high salinity in these two seas be explained? Both seas

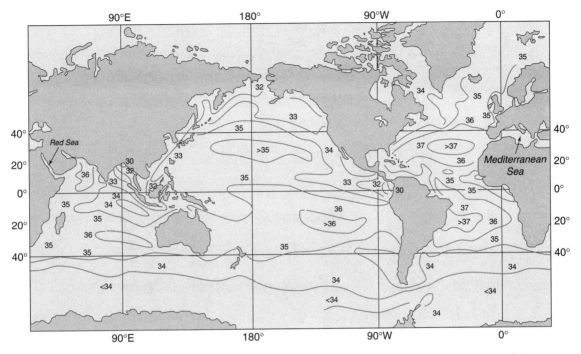

Figure 15-8 Salinity ranges of surface ocean waters.

are in hot areas with high levels of evaporation and less precipitation than open oceans at the same latitude. When water evaporates from the ocean, salt is left behind. Thus, evaporation increases the salinity of an ocean, especially if it is a small, enclosed sea such as the Mediterranean. (See Figure 15-9 on page 376.)

Salinity also varies with latitude. Look again at Figure 15-8. Notice that at 20 degrees north latitude and 20 degrees south latitude, salinity is approximately 36 ppt, higher than at the equator. The salinity is lower at the equator because it rains more there. Rain dilutes the ocean, making it less salty. At the 20 degrees north and south latitudes, there is less precipitation and more evaporation. The lower rainfall and greater evaporation cause the ocean water in this area to be slightly saltier than the water at the equator.

You may live near the coast where a river or stream enters the ocean. Freshwater from these sources lowers the salinity of the nearby ocean. Rainwater runoff from the land also lowers the salinity of ocean water near the coast.

Salinity also varies with the depth of the ocean. Look at the graph in Figure 15-10, on page 376. The salinity at the bottom of

Figure 15-9 The Mediterranean, which is a small, enclosed sea, has a higher salinity than that of the Atlantic Ocean.

the ocean is greater than at the surface. However, the change in salinity that occurs with increasing depth does not occur at a uniform rate. There is a layer of water, located at a depth between 100 and 200 meters, called the **halocline**. The halocline layer shows a rapid change (increase) in salinity. Why does salinity increase with depth? It has to do with the temperature of the water. Water is much colder at great depths than at the surface. You may recall that

Figure 15-10 Salinity levels at different ocean depths; note the halocline.

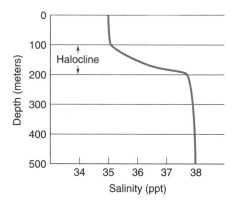

cold water contains molecules that are closer together. Cold water also causes salt ions to move closer together, thus increasing the salinity. In warmer surface water, the ions are farther apart, making the water less salty.

Why the Sea Is Salty

The salty taste of ocean water is mainly due to the presence of sodium chloride, one of various salts found in ocean water. The ocean gets its salts from several different sources. Rivers and streams that flow into the ocean are not just an important source of freshwater. Salts eroded from streambeds and riverbeds, and from adjoining land areas, also flow toward the sea. These salts are important to life in the ocean in various ways. Some marine organisms use salts directly. For example, to make their shells of calcium carbonate (a type of salt), mollusks remove calcium from ocean water. Silica is absorbed by diatoms (unicellular organisms) and used to make up their glassy cell walls. The sodium and chloride ions in NaCl are not removed in large amounts from the water, so these ions accumulate in the sea.

Another source of salt is the ocean floor itself. In ancient times, the Norse people believed that a giant "salt mill" ground salt from rocks on the ocean floor. Actually, this folktale was not too far from the scientific truth. Recently, oceanographers have discovered hot water spewing from hydrothermal vents on the seafloor. This hot water contains minerals dissolved from deposits found beneath the ocean floor. These minerals include sodium and chloride, the main components of sea salts.

15.4 SECTION REVIEW

1. How does the salinity of ocean water vary with depth?
2. How does the salinity of the ocean vary with latitude?
3. Describe some sources for the salt found in the sea.

Laboratory Investigation **15**

Determining Seawater Salinity

PROBLEM: How much salt is dissolved in ocean water?

SKILLS: Using a graduated cylinder and a triple-beam balance; calculating percentages.

MATERIALS: Triple-beam balance, 100-mL beakers, graduated cylinder, salt water, Bunsen burner or hot plate, beaker cover, tongs, cooling pad.

PROCEDURE

1. Balance your scale at the zero point. Determine the mass of an empty 100-mL beaker. In your notebook, record the amount in a copy of Table 15-2.

2. Measure out 20 mL of salt water into a graduated cylinder. Pour the 20 mL into the 100-mL beaker. Use the balance to find the mass of the beaker containing the water. Record the amount in the table.

3. Place the beaker of water over the burner or hot plate and heat it until all the water is boiled off. Place a cover on the beaker when most of the water is boiled off to prevent the salt from splattering out. (See Figure 15-11.)

4. Use the tongs to remove the beaker from the hot plate. Place the beaker on a cooling pad for a few minutes; then place it on the balance to find its mass. Record the mass of the beaker plus salt in your copy of Table 15-2.

5. Calculate the mass of the salt and the mass of the water (subtract the mass of the empty beaker from that amount plus salt and that amount plus water, respectively) and write the amounts in the table. To calculate percent salt, divide the results for #3 into the results for #5 and multiply by 100.

TABLE 15-2 SALINITY DETERMINATION

1. Mass of empty beaker	_____ grams
2. Mass of beaker plus water	_____ grams
3. Mass of water (subtract #1 from #2)	_____ grams
4. Mass of beaker plus salt	_____ grams
5. Mass of salt (subtract #1 from #4)	_____ grams

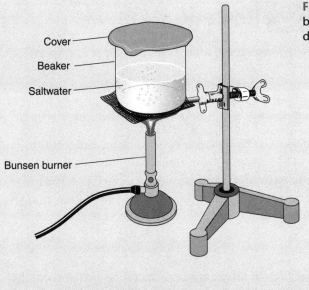

Figure 15-11 Heating a beaker of salt water to determine the salinity.

Cover

Beaker

Saltwater

Bunsen burner

OBSERVATIONS AND ANALYSES

1. What is meant by salinity? What is the salinity of your water sample?

2. Give two reasons why students using the same water samples may determine different salinities.

3. Which location would have a higher salinity, an estuary or the open ocean? Explain your answer.

Chapter 15 | Review

Answer the following questions on a separate sheet of paper.

Vocabulary

The following list contains all the boldface terms in this chapter.

acid, acid precipitation, base, buffers, compound, condensation, dissolved oxygen (DO), evaporation, halocline, latitude, longitude, mixture, neutral, O$_2$ minimum zone, pH, precipitate, precipitation, sea, sea level, solute, solution, solvent, water budget, water cycle, watershed

Fill In

Use one of the vocabulary terms listed above to complete each sentence.

1. The acidity or alkalinity of a substance is called its _____.

2. Forms of moisture that fall to the earth are called _____.

3. Liquid water changes to a gas in the process of _____.

4. Molecules of water vapor form a cloud when _____ occurs.

5. The total amount of water on Earth makes up its _____.

Think and Write

Use the information in this chapter to respond to these items.

6. Why does water at the ocean bottom have more dissolved oxygen than seawater at a depth of about 1000 meters?

7. Explain why a small sea may have a higher salinity than the open ocean at the same latitude.

8. How do the processes of evaporation and condensation contribute to the water cycle?

Inquiry

Base your answers to questions 9 through 11 on the graph below, which shows fluctuations in the pH of ocean water over a 24-hour period, and on your knowledge of marine science.

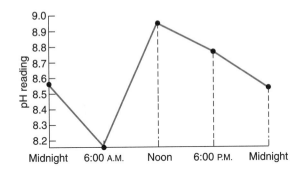

9. At what time of the day is ocean water the least alkaline?
   a. noon   *b.* 6:00 A.M.   *c.* midnight   *d.* 6:00 P.M.

10. Which statement is accurate, based on data in the graph?
    a. The pH varies from very acidic to very basic over 24 hours.
    b. From midnight to 6:00 A.M., there is a decrease in pH.
    c. From 6:00 A.M. to noon, there is a decrease in pH.
    d. From noon to 6:00 P.M., there is an increase in pH.

11. What conclusion can be drawn from the information in the graph?   *a.* The pH of ocean water varies during a 24-hour period.   *b.* The time of day has no impact on the pH of ocean water.   *c.* The average pH of ocean water is about 8.9.   *d.* The pH of ocean water falls within an acidic range.

Multiple Choice

Choose the response that best completes the sentence or answers the question.

12. The continuous movement of water between the ocean, atmosphere, and land is called the   *a.* water budget   *b.* water cycle   *c.* water level   *d.* watershed.

13. The layer of ocean water in which the salinity increases is called the   *a.* mixture   *b.* watershed   *c.* halocline   *d.* solvent.

Base your answers to questions 14 and 15 on the graph on page 373.

14. As the depth of the ocean increases, the amount of dissolved oxygen   *a.* increases only   *b.* decreases only   *c.* increases and then decreases   *d.* decreases and then increases slightly.

15. At about what depth does the O_2 minimum zone appear?
    a. 100 meters   *b.* 500 meters   *c.* 700 meters
    d. 1000 meters

16. A likely reason for any decrease in dissolved oxygen below 60 meters would be   *a.* the respiration of marine animals   *b.* a decrease in pH   *c.* a decrease in photosynthesis   *d.* an increase in salinity.

17. The pH of ocean water is about   *a.* 1   *b.* 3   *c.* 5   *d.* 8.

18. Water is an excellent solvent mainly because   *a.* it is a liquid   *b.* it can dissolve other substances   *c.* water molecules are uncharged   *d.* objects can float in it.

19. Of the following items, which one is probably *not* a source of O_2 in seawater?   *a.* algae and plants   *b.* waves   *c.* the atmosphere   *d.* water molecules

20. If a sample of ocean water is tested and found to have a pH of 6, it means that   *a.* OH ions outnumber H ions   *b.* H ions outnumber OH ions   *c.* H and OH ions are equal in number   *d.* the water is slightly alkaline.

21. The correct sequence of stages in the water cycle is
    a. precipitation, condensation, evaporation   *b.* evaporation, precipitation, condensation   *c.* condensation, evaporation, precipitation   *d.* evaporation, condensation, precipitation.

Base your answer to question 22 on the graph on page 376.

22. The graph illustrates that   *a.* as depth increases, salinity tends to decrease   *b.* as depth increases, salinity tends to increase   *c.* as depth increases, pressure decreases   *d.* salinity does not change with depth.

23. The diagram at right represents a molecule of   *a.* water   *b.* sodium chloride   *c.* carbon dioxide   *d.* hydrogen peroxide.

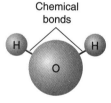

Chemical bonds

24. The following equation, $H_2O \longrightarrow A + OH^-$, shows the dissociation of water into two substances. What missing substance does the letter A represent? *a.* H^- *b.* OH *c.* H_2 *d.* H^+

Research/Activity

Does ocean ice that melts cause the sea level to rise? Find out for yourself. Add a few ice cubes to a glass of water; be sure to fill the glass to the top (without overflowing). Place the glass on a dry paper towel. Let the water sit at room temperature until the ice cubes have melted. Did the water overflow? Report your results. What is the relationship between melting ice and sea level? Does it make a difference if the ice is located on land or in the sea? Explain.

16 Geology of the Ocean

When you have completed this chapter, you should be able to:

RELATE the theories of continental drift and plate tectonics to the formation of the continents and oceans.

EXPLAIN the development of seafloor topographic features.

DESCRIBE the formation of coastal features and reef types.

16.1
Origin of the Ocean and the Continents

16.2
The Theory of Plate Tectonics

16.3
Ocean Floor Topography

16.4
Coasts and Reefs in Profile

The ocean is like a blanket of water that covers the seabed, or ocean floor. But there is no true "rest" on the seabed, and Earth is not a sleeping giant. There are many dynamic forces operating within Earth's solid part, or lithosphere, which influence and affect the characteristics of the liquid part, or **hydrosphere**. For example, the sudden emergence of a volcanic island in the middle of the ocean (such as Surtsey, shown above) is evidence of the forces at work thousands of meters below the surface of the sea.

The study of the development and physical characteristics of our planet's seafloor and continents—and of the forces that have shaped them—makes up the field of science called **geology**. Scientists who specialize in the study of geological features of the ocean are called *marine geologists*. You will begin your study of the geology of Earth's oceans and coastlines by going back in time to learn about the beginning of Earth itself.

16.1 ORIGIN OF THE OCEAN AND THE CONTINENTS

If you could go back before the beginning of time—back almost 20 billion years—you could observe the start of our universe and indeed all the galaxies that make up the vast array of space. Today, the space telescope has permitted humans on Earth to get a glimpse back in time to the very beginnings of matter. Scientists think that our universe formed as a result of a gigantic explosion, called the *big bang.*

The Earth Forms

At the time of the big bang, all the matter in the universe was contained in one sphere. Extremely hot and dense, the sphere exploded, sending out matter in all directions in a kind of giant cloud. As the cloud moved out from the explosion, some of the matter came together and formed clumps that eventually became galaxies—including the Milky Way, the home galaxy of planet Earth. In time, further clumping of the matter caused the formation of stars and planets. Scientists think that Earth, and all the other planets in our solar system, formed about 4.6 billion years ago, about 15 billion years after the big bang.

Shortly after its formation (geologically speaking), Earth was a hot molten mass, too hot for solid rocks to exist, too hot for water to exist as a liquid, and much, much too hot for life to exist at all. Evidence shows that at about four billion years ago, Earth had cooled enough for liquid rock to become solid at Earth's surface. But this early Earth was not a quiet place. For many millions of years, the solid surface of Earth was disturbed by volcanic activity that occurred over the whole planet.

An atmosphere (the layer of gases that surrounds a planet) began to form on Earth about 3.5 billion years ago. Earth's first atmosphere was very different from the atmosphere that exists today. Billions of years ago, the atmosphere probably contained some water vapor (water as a gas), carbon monoxide, hydrogen sulfide (a gas that smells like rotten eggs), nitrogen (the gas that makes up most of today's atmosphere), and hydrogen cyanide (a deadly gas).

The Ocean Forms

Remember that oceans could not exist on early Earth because of the high temperatures. But by about 4 billion years ago, Earth became cool enough for water vapor within the mantle to cool. This eventually formed liquid water on the surface. As Earth cooled still more, thunderclouds began to form. For many thousands of years, thunderstorms occurred and covered Earth with water, filling in the low spots that were to become the early ocean.

Some of the ocean's water came from the activity of volcanoes, which spewed great quantities of water vapor into the atmosphere. The impact of many meteors, which heated the surface of Earth, is also thought to have caused the release of water vapor. In addition, heated water in the crust boiled up to the surface and formed hot springs. Some of the hot water emerged from the surface as a geyser, or spray. In some places on Earth today, hot springs and geysers still exist, which are evidence that the area beneath them is hot. (See Figure 16-1.) Scientists call this heat (a form of energy that forms within Earth) *geothermal energy*. Iceland is an island that derives much of its energy from geothermal sources. In fact, in places in Iceland where this underground energy comes to the surface in the form of hot springs and steam, you can get a small glimpse of what vast areas of early Earth must have looked like. People in Iceland heat their homes by tapping these underground sources of heat.

Ocean water also came from molecules of water that were bound up in compounds within Earth's crust and released due to

Figure 16-1 Heated water in the crust forms hot springs and geysers.

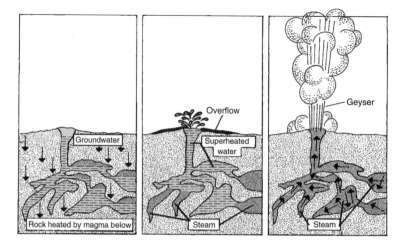

heating. Compounds that contain water are called hydrated compounds. An example of a hydrated compound is copper sulfate. Hydrated copper sulfate has the formula $CuSO_4 \cdot 5H_2O$. Notice that the formula shows five water molecules attached to a molecule of copper sulfate. When hydrated copper sulfate is heated, water is given off (and anhydrous copper sulfate is produced), as shown in the following reaction:

$$CuSO_4 \cdot 5H_2O \quad + \quad Heat \quad \longrightarrow \quad CuSO_4 \quad + \quad 5H_2O$$

| Hydrated copper sulfate | | Anhydrous copper sulfate | Water |

You can measure water loss from a hydrated compound by performing the lab investigation at the end of this chapter.

Origin of the Continents

In 1912, Alfred Wegener, a German meteorologist, proposed a hypothesis that caused a great deal of controversy in the scientific community. He suggested that the continents were not always located in their present positions, that over time they had moved. Wegener had noticed that the continents fit together like the pieces of a jigsaw puzzle. (See Figure 16-2 on page 388.) He suggested that about 200 million years ago the present continents formed one large landmass he called *Pangaea*, surrounded by a single huge ocean. At that time, Pangaea began to break up into smaller continents that moved over the surface of Earth, ultimately reaching their present positions. What is the evidence that the present-day continents originated in the single landmass Pangaea? Wegener cited the similarities of fossils and rock formations on different continents (especially on either side of the Atlantic) along with other technical evidence to support his hypothesis, which he called the theory of **continental drift**. His theory was not well received by most geologists at that time. In fact, scientists were quick to point out that Wegener was not even a geologist; he was a meteorologist—a scientist who normally studies the weather. However, studies of the seafloor that were conducted in the 1950s provided more evidence to confirm Wegener's theory, and it is now generally accepted by the scientific community.

What caused the single landmass to break apart into several continents? And what forces caused the continents to drift apart?

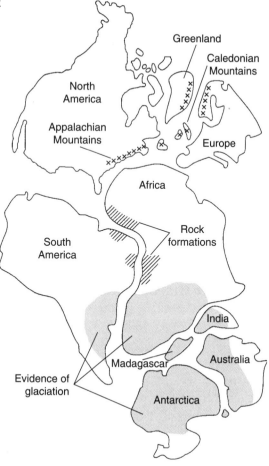

Figure 16-2 Evidence that the continents were once joined.

Greenland

Caledonian Mountains

North America

Appalachian Mountains

Europe

Africa

Rock formations

South America

India

Madagascar

Australia

Evidence of glaciation

Antarctica

Wegener was ridiculed because he was not able to provide an exact mechanism for the movement of the continents. (See Figure 16-3.) However, we now know that powerful forces inside Earth's interior caused the breakup of the continents. Look at the diagram of Earth's interior, shown in Figure 16-4. The interior of our planet is composed of several layers. At Earth's center is the inner core, surrounded by the outer core, the mantle, and then the crust. Much of Earth's interior is in a hot, molten state. The inner core has the highest temperatures, with a range of about 6200 to 6600°C. The **mantle**, which is a region of geologic activity between the core and the crust, has a temperature range of about 1200 to 5000°C.

The high temperatures inside Earth are hot enough to melt rock. In fact, Earth's interior—from the inner core to the upper mantle—

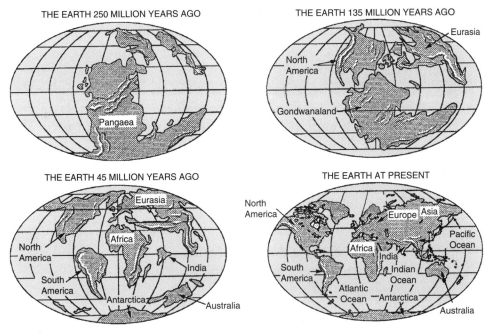

THE EARTH 250 MILLION YEARS AGO

Pangaea

THE EARTH 135 MILLION YEARS AGO

Eurasia

North
America

Gondwanaland

THE EARTH 45 MILLION YEARS AGO

Eurasia

Africa

North
America

South
America

India

Antarctica

Australia

THE EARTH AT PRESENT

North
America

Europe Asia

Pacific
Ocean

Africa

India

South
America

Indian
Ocean

Atlantic
Ocean

Antarctica

Australia

Figure 16-3 The breakup of Pangaea and continental drift over time.

consists of hot, molten material. This molten material within the mantle is called **magma**. The churning of the magma creates a force that generates great pressures upward into Earth's surface layer, or **crust**. Earth's crust is only about 40 km thick. If enough force is generated, it cracks. Then the ground trembles and moves, producing an earthquake. Sometimes magma flows out of a crack in the crust,

Figure 16-4 The layers of Earth's interior.

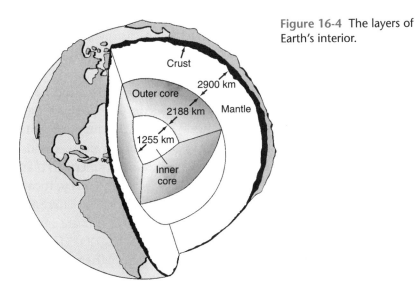

Crust

2900 km

Outer core

2188 km

Mantle

1255 km

Inner
core

producing a volcanic eruption. Magma that flows out of the crust onto Earth's surface is called *lava*. Disturbances (vibrations) in Earth's crust, such as volcanic eruptions and earthquakes, are examples of *seismic activity*.

Seismic activity was involved in the breakup of the supercontinent Pangaea. In the next section, you will learn about the mechanism that is responsible for making the continents drift apart.

16.1 SECTION REVIEW

1. How did the stars and planets come into being?
2. What were the sources of the ocean's water?
3. Describe some of the evidence that indicates the continents originated from a single landmass.

16.2 THE THEORY OF PLATE TECTONICS

Satellite photos show that the continents are moving at a rate of approximately one centimeter per year. The Atlantic Ocean is getting wider and the Pacific Ocean is getting narrower. What exactly is causing the continents to drift? Research on Earth's interior has revealed that its crust is divided into segments called **plates**, which float like rafts on the molten interior layer. The continents ride on top of these plates. This idea that Earth's crust is divided into segments that drift about has been developed into the theory of **plate tectonics**.

To understand the theory of plate tectonics, look at the world map in Figure 16-5. Earth's crust is divided into seven major plates and about a dozen minor plates. Locate the North American plate. The eastern border lies in the middle of the Atlantic Ocean, and the western border runs through California. The North American plate is drifting westward. To understand what causes the plates to move, look at the profile of the mantle shown in Figure 16-6. There is a big difference in temperature between the upper mantle (about 1400°C) and the lower mantle (about 2600°C). Scientists think that this difference in temperature creates convection currents. A **convection current** is a transfer of heat in a liquid or gas that causes

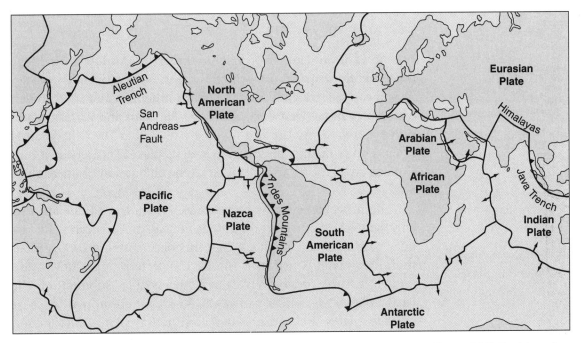

Figure 16-5 Earth's major crustal plates.

the molten magma to rise up through the mantle and into the crust, forming an *oceanic ridge.* Off the east coast of North and South America, it is the Mid-Atlantic Ridge that is formed; this ridge runs the entire length of the Atlantic Ocean, dividing it in half. Magma that breaks through Earth's oceanic crust as lava can also accumulate to form mid-ocean islands.

Figure 16-6 Convection currents in the mantle cause seafloor spreading.

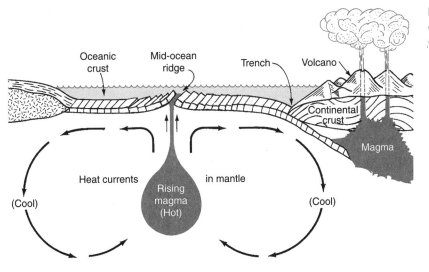

Seafloor Spreading

The upward movement of magma under the Mid-Atlantic Ridge causes **seafloor spreading**, which is the moving apart of the plates. As you can see in Figure 16-5, the North American and South American plates move westward, while the Eurasian and African plates move eastward. As the hot magma rises under the Mid-Atlantic Ridge, cooler magma moves in to take its place. This sets up a continuous circulation pattern similar to the circulation of warm air in a room.

The movement of the plates causes them to collide with one another. Notice, in Figure 16-5, that the North American plate collides with the Pacific plate. When this occurs, one plate overrides the other plate. Crust from the Pacific plate plunges downward under the North American plate in a process called **subduction**. The Pacific plate slides under part of the North American plate because oceanic plates are denser than continental plates. Subduction destroys old plates as the crust descends into the mantle to become molten magma. This process occurs in several areas around the world (at subduction zones) and forms **trenches**, the deepest and steepest depressions found on the ocean floor. (See Figure 16-7.)

Figure 16-7 Deep-sea trenches are formed at subduction zones.

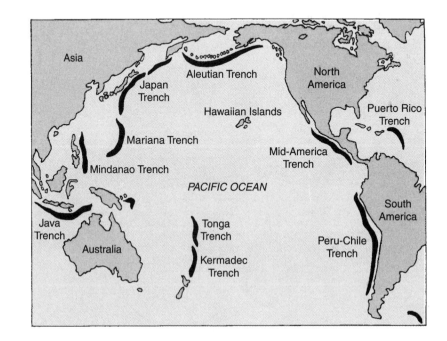

The theory of plate tectonics helps to explain various geological phenomena. At the margin of the plates, a crack occurs in Earth's crust that is called a **fault**. For example, the San Andreas fault, which cuts through California, forms the boundary between the North American and Pacific plates. Earthquakes and volcanic activity tend to occur along the margins of plates, where there is movement of, and friction between, the adjoining plates.

Ocean Floor Formation

Plate tectonics also explains how the ocean floor was formed. As magma continued to rise up to form the Mid-Atlantic Ridge, the North American and Eurasian plates moved farther apart, creating a new ocean floor. This process, which has been occurring for many millions of years, is recorded in the symmetrical, parallel bands of basalt (the volcanic rock that makes up the ocean floor) that spread out along either side of the ridge. The spreading apart of the Atlantic seafloor is an ongoing process.

Further evidence that the seafloor is spreading comes from the study of rock samples taken from both sides of the mid-ocean ridge. (See Figure 16-8.) The youngest rocks are closer to the ridge and the oldest rocks are farther away. Scientists discovered that the magnetic properties of rocks on both sides of the ridge were as symmetrical as

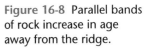

Figure 16-8 Parallel bands of rock increase in age away from the ridge.

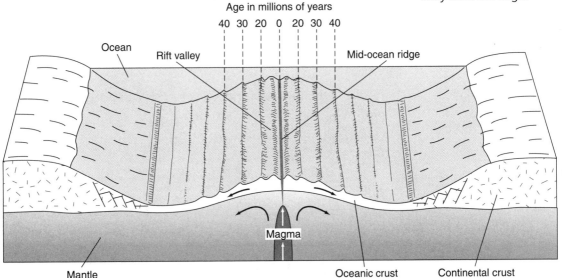

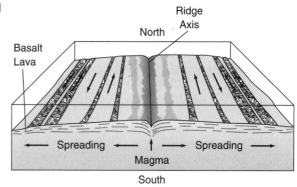

Figure 16-9 Identical magnetic bands are seen on each side of the ridge.

the bands of rocks. When the magma hardened, the magnetic minerals in the rock aligned in the direction of Earth's magnetic field. During Earth's history, the magnetic poles have reversed several times. The pattern of polarity on either side of the ridge was identical and reflected these pole reversals. Scientists concluded that, as the magma hardened, half moved to one side of the oceanic ridge and the other half moved to the other side. This pattern of magnetic bands provided strong evidence for ocean floor spreading. (See Figure 16-9.)

The theory of plate tectonics is called a unifying theory because it explains the origin of, and connections between, such phenomena as earthquakes, volcanic activity, faults, continental drift, and seafloor spreading. A knowledge of plate tectonics is also helpful in explaining how some structures on the ocean floor were formed. (See Figure 16-10, which shows a profile of the Atlantic Ocean floor.)

16.2 SECTION REVIEW

1. What process causes the continents to drift apart? How?

2. How was the Mid-Atlantic Ridge formed? Explain how the process is related to seafloor spreading.

3. Why is the theory of plate tectonics called a unifying theory?

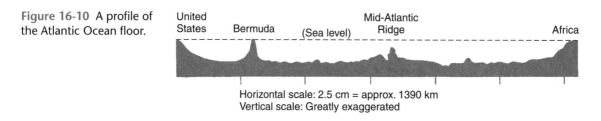

Figure 16-10 A profile of the Atlantic Ocean floor.

Horizontal scale: 2.5 cm = approx. 1390 km
Vertical scale: Greatly exaggerated

16.3 OCEAN FLOOR TOPOGRAPHY

What do the features of the ocean bottom look like? The average depth of the ocean is about 3636 meters—much too deep for scuba divers to explore. However, oceanographers can obtain a profile of the ocean floor (without submerging in an underwater vehicle) by using sonar. Modern ships are equipped with sonar. A ship's sonar device beams a continuous sound signal downward. After the sound wave hits the bottom, the returning signal, called an *echo*, is received by a depth recorder in the ship. This produces a line tracing of the ocean floor. (See Figure 16-11.) Notice that the depth recording of the ocean floor shows a bottom that varies from fairly smooth to jagged, or irregular. The irregular part could be debris dumped on the ocean floor, a sunken ship, or a natural feature. Recall that the *Titanic* and other sunken ships have been located by using sonar. Modern fishing boats also use sonar to locate schools of fish. Sonar

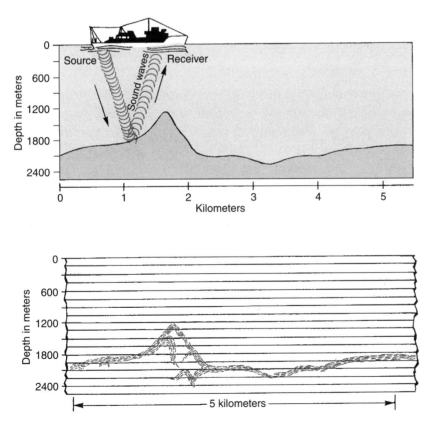

Figure 16-11 Sonar is used to obtain a profile of the ocean floor.

is very useful to help ships navigate in shallow waters. For example, a reef may be located only several meters below the surface—close enough to make ships cautious when they pass by.

Sonar, Ocean Depth, and Topography

Ocean depth is calculated automatically by sonar. Two pieces of information are needed to calculate the depth—the speed of sound in water (1454 meters per second) and the time it takes for the signal to reach the bottom. If a signal takes one second (after being sent) to return to the ship, then it takes one-half second to travel to the bottom. Since sound travels 1454 meters per second underwater, it travels 1454 divided by 2 in one-half second. Therefore, the depth of the water is 727 meters. The following formula is useful in calculating ocean depth using sonar:

Depth (D) = 1454 meters per second × time (t) ÷ 2

In this example,

$D = 1454 \times 1 \div 2$

$D = 727$ meters

Ships equipped with sonar have been crisscrossing the oceans for a number of years in an attempt to map as much of the ocean floor as possible. Thousands of sonar tracings have been made. The data from the ships' sonar maps have been combined to create a single map that shows all seafloor elevations and depressions. The study of Earth's surface features, such as elevations and depressions, on the land and the ocean floor is called **topography**. Seafloor topography includes some very dramatic depressions and elevations. (See Figure 16-12.)

Figure 16-12 Features of the ocean floor.

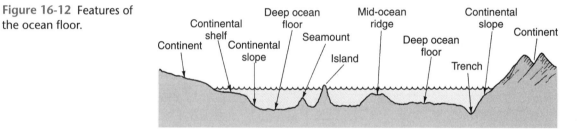

Seafloor Features

Recently, the U.S. Navy released a treasure trove of formerly classified data on the oceans that were collected during the Cold War. These images of the ocean floor were obtained via satellite readings. The satellites *Seasat* and *Geosat* used radar to measure sea level. They measured bumps and depressions on the ocean's surface that reflected the pull of gravity (on the water) exerted by seafloor objects. A variety of features were uncovered. High ridges off the coast of Oregon were formed when plates collided. Off the west coast of Florida, the seafloor has steep-walled elevations 2 km high. Off New Jersey's coast, there is a continental slope with deep canyons that were most likely produced by submarine avalanches. A **continental slope** is the area where the seafloor drops steeply at the outer edge of the continental shelf. (See Figure 16-13.) In the Gulf of Mexico, off the Louisiana coast, the images show a moonlike, pockmarked surface, formed when the gulf was dry and filled with evaporated sea salt. Sediments from the Mississippi River covered the salt. When the sea level rose, the weight of ocean water on these sediments created the odd shapes.

Cutting through the continental shelf and slope are steep V-shaped depressions called **submarine canyons**. Some of these canyons are as large as the Grand Canyon. Two well-known submarine canyons along the U.S. coasts are the Monterey Canyon off California and the Hudson (River) Canyon off New York. (See Figures 16-14a and 16-14b on page 398.) How did these huge canyons form? Many submarine canyons are extensions of sunken river valleys from the adjoining continent. During the last ice age, when the sea level was much lower, many canyons existed as river valleys. When the sea level rose at the end of the ice age, these valleys were submerged, creating the submarine canyons. The deeper parts of

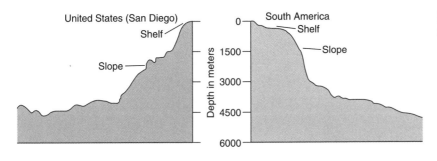

Figure 16-13 Continental shelves and slopes.

Figure 16-14a The Hudson (River) Canyon.

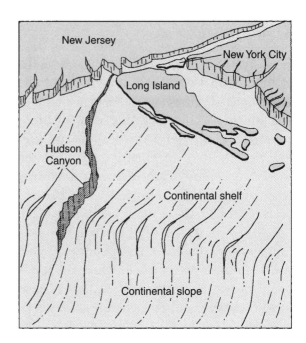

submarine canyons are formed by swift undersea currents. Smaller valleys that cut through a slope are formed as a result of erosion by mudslides, which move from the shelf out to the ocean basin. The accumulation of mudslide sediments at the base of a slope creates a slightly elevated region called the **continental rise**.

Sonar maps of the ocean floor often reveal small submarine mountains called **seamounts** (shown in Figure 16-12). Seamounts

Figure 16-14b A sonar scan of the Hudson (River) Canyon.

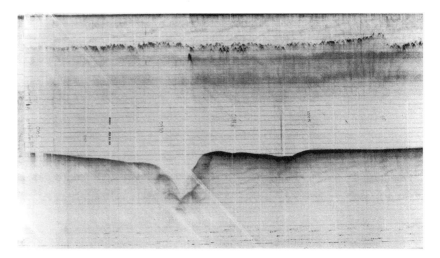

are produced in regions of intense volcanic activity, where magma (lava) pushes through the crustal plate and piles up on the seafloor. If the lava breaks the surface of the ocean, then a volcanic island is formed. The Hawaiian Islands were formed from a chain of seamounts. The main island of Hawaii, at the eastern end of the chain, is the youngest (formed about 800,000 years ago) and most geologically active of the five Hawaiian Islands. The progressively older seamounts stretch in an arc to the northwest. (The oldest Hawaiian Island is about 4 to 6 million years old.) Such a chain of islands is formed when a crustal plate moves over an area of intense activity in the mantle, called a **hot spot**. The area over the hot spot develops a seamount as lava pours through its crust. As the plate moves along, the hot spot breaks through the next area of crust, forming a new seamount. (See Figure 16-15.)

Some seamounts actually may be former islands that have sunk beneath the surface. Erosion by waves and currents can cause the tops of seamounts to become flattened, forming structures called **guyots** (pronounced GEE-ohz), shown in Figure 16-15. Trenches, as mentioned above, are another topographical feature of the ocean floor. Recall that trenches are found at the margins (subduction zones) of crustal plates, where one plate descends into the mantle below the other plate. (See again Figure 16-12.) The deepest ocean trench is the Mariana Trench, located in the western Pacific Ocean. The Mariana Trench is 10,958 meters deep, which is deep enough to contain Mt. Everest, the tallest terrestrial mountain on Earth.

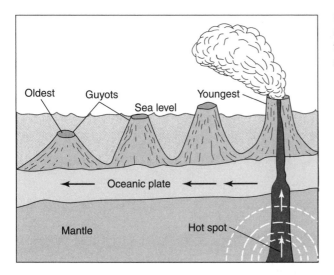

Figure 16-15 A chain of islands forms over a hot spot.

TECHNOLOGY

Loihi on the Rise

The inhabitants of the main island of Hawaii must be among the bravest people in the world. These islanders live about 30 km from the most active volcano on Earth. This volcano, named Loihi (meaning "long one" in Hawaiian) is so active that it produced nearly 1500 earthquakes in just one week! Even though the volcano is more than 3 km high and about 40 km long, local Hawaiians cannot see it. This is because Loihi rises from the ocean floor, with its peak about 1 km below the ocean surface. Marine geologists predict that Loihi will be the next Hawaiian island to emerge from the ocean–about 50,000 to 100,000 years from now.

Loihi has been erupting continuously for about 20 years. Most of the earthquakes it produces are relatively weak. However, they seem to be increasing in magnitude and frequency, causing local authorities to be concerned for the safety of the islanders. Some recent quakes have been recorded at magnitudes between 4 and 5 on the Richter scale. (Either the Richter or the magnitude scale is used to measure earthquake intensity.) Seismologists (scientists who study earthquakes) fear that an underwater earthquake with a magnitude of 6.8 might produce a huge wave a that would reach the big island in just 15 minutes—not enough time for people along the coast to prepare for emergency evacuation to higher ground.

Several federal agencies are monitoring the volcano's activity. Local officials have decided to set up an early warning system that will give island residents more time to evacuate in the event of a serious earthquake. The Hawaiian Undersea Research Laboratory at the University of Hawaii is using submersibles and robots to monitor and videotape the eruptions underwater, thus giving scientists the opportunity to observe and record never-before-seen events going on inside the crater.

QUESTIONS

1. Explain why the inhabitants of Hawaii cannot see Loihi's eruptions.

2. Why is Loihi's seismic activity potentially dangerous for Hawaiians?

3. Describe the technology scientists are using to monitor Loihi's activity.

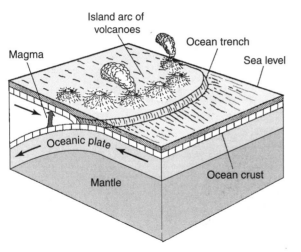

Island arc of
volcanoes

Magma

Ocean trench

Sea level

Oceanic plate

Mantle

Ocean crust

Figure 16-16 Volcanic island arcs form over trenches.

Associated with the trenches are groups of volcanic islands that form an arc in the ocean, called **island arcs**. (See Figure 16-16.) Most trenches, and their island arcs, are located on the periphery of the Pacific Ocean (that is, along the west coasts of North and South America and the east coast of Asia). Many of these islands are still volcanically active, so the area bordering the Pacific is called the Ring of Fire. Frequent earthquakes also occur along the Ring of Fire, due to the movement of subducting plates.

The Mid-Ocean Ridge

When magma rises up from the mantle through the oceanic crust, it forms ridges. The prominent **mid-ocean ridge** is the continuous undersea volcanic mountain range that encircles the globe, marking the boundaries of several crustal plates. As noted above, the part of the ridge that runs through the middle of the Atlantic Ocean is called the Mid-Atlantic Ridge. (See Figure 16-17 on page 402.)

The Mid-Atlantic Ridge rises about 3030 meters above the ocean floor; its highest underwater ridge lies about 900 meters below the surface of the ocean. Iceland, which rises above the surface, is a volcanic island formed by lava that poured out of the Mid-Atlantic Ridge. Other smaller volcanic islands have formed near Iceland, such as Surtsey, which was "born" in 1963. Eruptions continued on this island for several years, causing Surtsey to increase in size. (See photograph on page 384.)

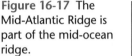
Figure 16-17 The Mid-Atlantic Ridge is part of the mid-ocean ridge.

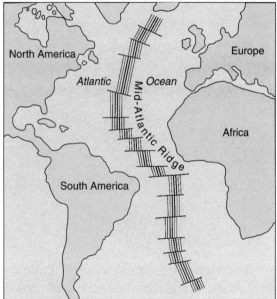

A depression called the central **rift valley** runs along the crest of the mid-ocean ridge (including the Mid-Atlantic Ridge). The rift is a crack in the seafloor through which molten rock from the mantle is expelled. The rift zone is of special interest to oceanographers because that is the place where new seafloor is continually being created. When the hot magma pours out on the ocean floor, it cools and solidifies to form new ocean crust. Lateral movements on both sides of the rift then cause the seafloor to spread apart. About 250 million years ago, all the continents were joined together at the Mid-Atlantic Ridge. Over geologic time, hot molten matter rising up through the mantle and into the crust caused the continents to split apart. The continuing flow of magma pushed the continents farther apart and created a space between them, called an *ocean basin*. This ocean basin filled up with water and became the Atlantic Ocean.

Features of the Rift Zone

The oceanic crust is very porous in the area of a rift zone. Ocean water seeps down through cracks in the crust and gets heated from the hot magma below. The hot water rises up through the crust, dis-

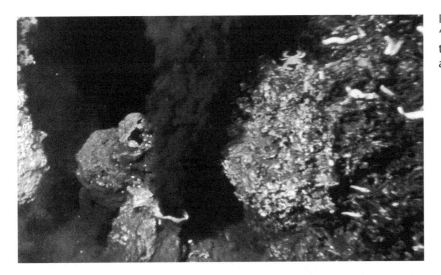

Figure 16-18 Mineral-rich "black smokers" (such as the one shown here) form at hydrothermal vents.

solving minerals out of the rock as it flows. When the hot water emerges from the seafloor, it makes contact with the cold ocean water. Then the minerals dissolved in the hot water form a cloud that looks like smoke coming from a chimney. Oceanographers call these springs of mineral-laden waters "black smokers." (See Figure 16-18.)

The area in the rift zone where these hot springs emerge is called a **hydrothermal vent**. Submersibles have visited the hydrothermal vents, some more than 2400 meters deep. But the submersibles cannot get too close to a vent. The temperature can be as high as 371°C. (Recall that water boils at 100°C.) The water was so hot at one vent that it melted the first thermometer used to record the temperature.

In 1977, scientists aboard the *Alvin,* an American submersible, made an amazing discovery while investigating some hydrothermal vents. Parts of the seafloor near the vents were carpeted by a thick growth of living things. There were clusters of giant tube worms, large clams, albino crabs, deep-sea fishes, and other organisms. Oceanographers wondered how so much life could exist at such a great depth, where there is no light. Since 1977, the *Alvin* and other submersibles have made many more dives to study the vents and to bring back water samples and specimens. The water was found to be very rich in minerals, particularly the compound hydrogen sulfide (H_2S). Researchers also found that the water had a high concentration of bacteria. It turns out that these bacteria use the hydrogen sulfide to produce their food.

Some types of bacteria share the food they make in a symbiotic relationship with giant tube worms and other deep-sea creatures. You may recall that this form of food making by bacteria, which is not based on photosynthesis, is called *chemosynthesis*. (Refer to Chapter 8 for a review of how chemosynthetic bacteria use hydrogen sulfide to make their food.) Marine scientists continue to collect water samples from the areas around hydrothermal vents. They recently discovered large masses of heat-loving bacteria, called *thermophilic bacteria*, which live on the outside of the vents in water at or near the boiling point! In addition, many new and unique species of crabs, mussels, octopus, shrimp, and fish have been discovered near the vents.

16.3 Section Review

1. Calculate the depth of a sunken ship if a sonar signal takes two seconds to return to the research ship after it is emitted.

2. How were the Hawaiian Islands formed? Are they all the same age?

3. What important geological process occurs in a rift zone?

16.4 COASTS AND REEFS IN PROFILE

Approximately 70 percent of the U.S. population lives within 80 km of a coast. The coast, or shore, is the boundary between land and sea. As you have learned in previous chapters, some coasts are rocky, while others have sandy beaches. A beach is a part of the shore that contains loose sediments eroded from the land.

Beaches

Our coastal states have an abundance of sandy beaches and rocky shores. In Hawaii, there are both sandy beaches and rocky coasts of volcanic origin. Volcanic rock originates from the molten lava that pours out of volcanoes and flows down to the sea. When the lava

Figure 16-19 A typical sandy beach is produced by erosion of rocks and is mildly sloped.

reaches the ocean, it boils the water into steam, and the lava hardens into rock. Sand on a beach is mainly the product of erosion from rocks along the shore. Waves pound on the rocky shores, and pieces of rock break off and fall into the surf. Tides move the rocks back and forth, wearing them down into pebbles. Over time, the pebbles are ground into sand by rubbing against one another as they are tossed by the waves. (See Figure 16-19.)

Beach sand may also come from eroded rocks from mountains located hundreds of km away. Rivers and streams wear down the rocks. Sediments from the rocks are transported downstream to the ocean, where they are deposited as sand on the beach. (You already examined sand grains under the microscope to determine their origin in the lab investigation in an earlier chapter.) The erosion of volcanic rock produced the black sands found on some Hawaiian beaches. The white and pink sandy beaches of many other tropical islands are largely composed of fine sediments eroded from offshore coral reefs. Sand may also contain shell or bone fragments, fish scales, and other debris from marine animals. (See Figure 16-20 on page 406.)

Coasts have differently shaped profiles. On sandy shores with heavy surf, the crashing waves erode the sand, forming a steeply sloped beach. On beaches where large rivers empty into a calm sea,

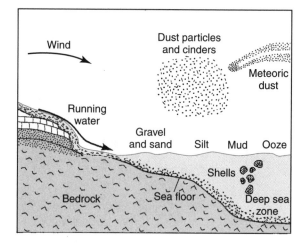

Figure 16-20 Beach sand and seafloor sediments come from eroded rock and other debris.

sediments carried by the river are deposited along the shore, producing a fan-shaped feature called a **delta**. (See Figure 16-21.) The Nile River, which flows into the Mediterranean, and the Mississippi River, which empties into the Gulf of Mexico, both form deltas. (See Figure 16-22.)

Rocky Coasts

Compared to sandy beaches, rocky coasts are often very steep. (See Figure 16-23.) How are they formed? The rocky coast of Maine was formed 12,000 years ago, toward the end of the last Ice Age. As the climate warmed, the glacier that covered much of North America retreated, carving out troughs, or valleys, which later became river valleys. When the glaciers melted, the sea level rose. The ocean

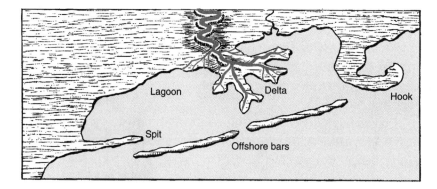

Figure 16-21 Typical shoreline features, including a delta.

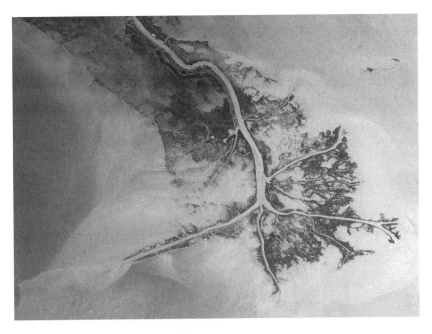

Figure 16-22 The fan-shaped Mississippi Delta extends into the Gulf of Mexico.

invaded the land, filling in the eroded valleys left by the retreating glaciers. Coasts eroded by glaciers are found in such places as Alaska, Chile, Greenland, Norway, and Scotland, where they are known as fjords. A **fjord** (pronounced FEE-yord) is a narrow inlet from the sea that is both steep and deep. For example, one of the fjords in Chile is more than 1210 meters deep. (See Figure 16-24 on page 408.)

Figure 16-23 A typical rocky coastline is usually quite steep.

Figure 16-24 Fjords, such as this one, are very deep and steep; they are found along coasts eroded by glaciers.

Types of Reefs

You may recall that a coral reef is a limestone structure that is built by coral polyps that live on the reef's surface. There are three kinds of reefs: fringing reefs, barrier reefs, and atolls. (See Figure 16-25.) A **fringing reef** lies a few kilometers offshore and is parallel to the mainland. On the shore side of the reef, the water is shallow; on the ocean side, it is deep. A fringing reef grows most rapidly on the ocean side of the reef because there is greater water circulation, which brings more food and oxygen to the coral. Fringing reefs are typically found in the Florida Keys and in the Caribbean.

A reef that grows farther offshore is called a **barrier reef**. Most barrier reefs lie approximately 25 km offshore and are separated from the island by a channel. The world's most famous barrier reef is the Great Barrier Reef, which is actually a series of reefs that lie between 16 and 160 km off the northeast coast of Australia. The Great Barrier Reef is 2000 km long.

Fringing reefs and barrier reefs grow right up to the surface of the ocean. At low tide, the tops may extend above the water. Waves

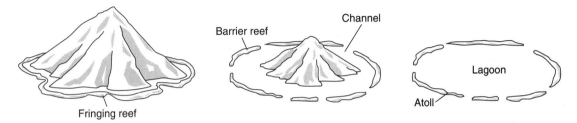

Fringing reef   Barrier reef   Channel   Lagoon   Atoll

Figure 16-25 The three types of coral reef structures also represent the main stages in reef evolution.

and currents break off pieces of the coral; these chunks of coral stone accumulate on the seafloor. If enough coral piles up, small islands called **keys** or **cays** are formed. The Florida Keys and the Cayman Islands are formed from coral stone.

Scattered throughout the South Pacific and the Indian Ocean are coral structures called atolls. An **atoll** is a string of coral islands that forms a circle. In the middle is a shallow lagoon that may vary in width from 1 to 12 km. In 1837, the naturalist Charles Darwin observed these islands as he sailed on the research vessel H.M.S. *Beagle,* and he wondered why the islands form a circle. He hypothesized that the circular shape represents the last stage in reef evolution, which is associated with the sinking of a volcanic island. According to Darwin, a fringing reef appears first along the shoreline of a volcanic island. While the island begins slowly to sink or erode, the fringing reef continues to grow upward and outward to form a barrier reef. Finally, the island sinks completely below the surface, leaving only a circular fringe of reefs, that is, the atoll. Scientists have confirmed Darwin's hypothesis by drilling through the coral limestone and discovering a foundation of volcanic rock beneath the reef. There are many well-known coral atoll islands in the Pacific, such as Wake, Midway, Bikini, and Eniwetok.

16.4 SECTION REVIEW

1. What are three common sources of beach sand?

2. What are deltas and how are they formed?

3. Why are some rocky coasts (with fjords) so steep?

Laboratory Investigation 16

Getting Water from a "Stone"

PROBLEM: How can we show that Earth's crust contains water?

SKILLS: Heating and measuring a chemical compound; calculating weight (mass) and percentages.

MATERIALS: Safety glasses, Bunsen burner or hot plate, porcelain evaporating dish, hydrated copper sulfate, spatula, tongs, triple-beam balance, cooling pad.

PROCEDURE

1. Put on the safety glasses. Heat the evaporating dish over a Bunsen burner or on a hot plate for a minute to evaporate possible moisture from the dish.

2. Use tongs to transfer the dish to a cooling pad for a few minutes.

3. After it cools, transfer the dish to the balance to be weighed. In your notebook, record the weight (mass) in a copy of Table 16-1.

4. Use a spatula to measure out 2 grams of copper sulfate and put it into the evaporating dish. Record the weight (mass) of the dish plus the copper sulfate in the table.

5. Place the evaporating dish that contains the copper sulfate onto a hot plate or over the Bunsen burner. Heat gently for five minutes, until the blue color of the copper sulfate disappears.

6. Use tongs to transfer the dish to a cooling pad and wait a minute for it to cool. Place the dish on the balance and record the weight (mass) in the table.

TABLE 16-1 WEIGHTS (MASSES) OF COPPER SULFATE

Evaporating dish (empty):	_____ grams
Evaporating dish plus copper sulfate (before heating):	_____ grams
Evaporating dish plus copper sulfate (after heating):	_____ grams

CALCULATIONS

1. To find the weight (mass) of the copper sulfate, subtract the weight (mass) obtained in step 3 from that of step 4.

2. To find the weight (mass) of the water, subtract the weight (mass) obtained in step 6 from that of step 4 (line 3 from line 2 in the table).

3. To calculate the percentage of water in the hydrate, use the equation

   Percentage of water = weight of water ÷ weight of hydrate × 100.

4. You can calculate the number of water molecules in the hydrate by using the equation

   Number of water molecules = weight of hydrate ÷ weight of water.

OBSERVATIONS AND ANALYSES

1. Compare your answer for calculation 4 with those of the other students. Your answers may vary. How can you explain these differences? (The correct number of water molecules is five.)

2. Describe what happened—physically and chemically—to the copper sulfate hydrate when it was heated.

3. What is the important difference between the copper sulfate before it was heated and the copper sulfate after it was heated?

Chapter 16 | Review

Answer the following questions on a separate sheet of paper.

Vocabulary

The following list contains all the boldface terms in this chapter.

atoll, barrier reef, continental drift, continental rise, continental slope, convection current, crust, delta, fault, fjord, fringing reef, guyots, hot spot, hydrothermal vent, island arcs, keys (cays), magma, mantle, mid-ocean ridge, plate tectonics, plates, rift valley, seafloor spreading, seamounts, subduction, submarine canyons, topography, trenches

Fill In

Use one of the vocabulary terms listed above to complete each sentence.

1. The theory of _____ explains continental drift.
2. A _____ is an area of intense activity in the mantle.
3. The crust of one plate plunges below another during _____.
4. Molten material within the mantle is called _____.
5. A _____ is where black smokers emerge in a rift zone.

Think and Write

Use the information in this chapter to respond to these items.

6. What forces caused the supercontinent Pangaea to split apart?
7. How do coral atolls form? How are they related to reefs?
8. Explain how seamounts are related to guyots and islands.

Inquiry

Base your answers to questions 9 through 11 on the information in Figure 16-8 on page 393, and on your knowledge of marine science.

9. Based on this profile of the ocean floor, which statement is correct?   *a.* The sides of the ridge are moving away from each other.   *b.* The sides of the ridge are moving toward each other.

c. The bands of rocks closest to the center of the ridge are the oldest.   *d.* A deep-sea trench is being formed at the mid-ocean ridge.

10. The magma that flows through the mid-ocean ridge comes up from the   *a.* continental crust   *b.* oceanic crust   *c.* mantle   *d.* rift valley.

11. What is an accurate statement regarding the data in this diagram?   *a.* Seafloor spreading began about 20 million years ago.   *b.* The ocean is deepest where it covers the rift valley.   *c.* The two sides of the ridge are very different from each other.   *d.* The topographic features on both sides of the ridge are very similar.

Multiple Choice

Choose the response that best completes the sentence or answers the question.

12. The fan-shaped feature that is formed by a river depositing sediments near the shore is a   *a.* barrier reef   *b.* continental rise   *c.* delta   *d.* plate.

13. The structure labeled "C" in the diagram is the   *a.* crust   *b.* mantle   *c.* inner core   *d.* outer core.

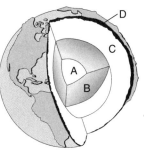

14. The largest area of ocean floor is the   *a.* ocean basin   *b.* continental slope   *c.* continental shelf   *d.* continental rise.

15. Deep-sea trenches are caused by   *a.* faulting   *b.* subduction   *c.* volcanic eruptions   *d.* turbidity currents.

16. A feature of the seafloor that provides evidence for the theory of plate tectonics is   *a.* sedimentary layers   *b.* coral reefs   *c.* small canyons   *d.* the mid-ocean ridge.

17. Which topographical features are in the proper sequence, going from the Mid-Atlantic Ridge to the North American continent?   *a.* ocean basin, slope, shelf   *b.* shelf, slope, ocean basin   *c.* trench, slope, basin   *d.* ocean basin, shelf, rise

18. Which of the following items is *not* a topographical feature?
    a. seamount   *b.* oil deposits   *c.* trenches   *d.* guyots

19. The scientist most responsible for formulating the theory of continental drift is   *a.* Alfred Wegener   *b.* Charles Darwin   *c.* Jacques Cousteau   *d.* Sir Charles Thompson.

20. Which is *not* an accurate statement about plate tectonics?
    a. Some continents are drifting apart.   *b.* The continents are fixed in position.   *c.* The continents ride on crustal plates.   *d.* There is seismic activity at the margins of crustal plates.

21. The original supercontinent that existed about 200 million years ago is called   *a.* Loihi   *b.* Atlantis   *c.* Antarctica   *d.* Pangaea.

22. If sound travels 1454 meters per second in water, how deep is the ocean floor if the echo of a ship's signal takes one second to return to the surface?   *a.* 1454 meters   *b.* 727 meters   *c.* 2181 meters   *d.* 484 meters

23. Darwin hypothesized that the last stage in coral reef evolution is the   *a.* fringing reef   *b.* barrier reef   *c.* coral atoll   *d.* key.

24. What natural process is illustrated in the following diagram?
    a. the after-effects of a tidal wave   *b.* a rise in sea level   *c.* ecological succession on a volcanic island   *d.* the evolution of a coral (reef) atoll.

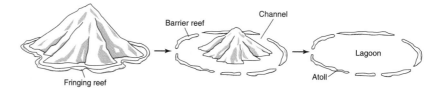

Barrier reef   Channel   Lagoon   Atoll   Fringing reef

Research/Activity

Use different colors of modeling clay to make a model of seafloor spreading, an oceanic ridge, a subduction zone, or Earth's plates. Label the structures that you have represented in your model.

17 Climate and the Ocean

When you have completed this chapter, you should be able to:

DESCRIBE how ocean temperatures affect weather and climate.

EXPLAIN how ocean moisture creates fogs and hurricanes.

DISCUSS warming events such as El Niño and the greenhouse effect.

Three major "layers" make up the planet Earth. In the last chapter, you learned about two layers: the land, or lithosphere, which is made up of the core, mantle, and crust, and the water, or hydrosphere, which covers much of the land. In addition, there is the layer of air, or atmosphere, which covers the entire hydrosphere and lithosphere. These different layers interact with one another to produce changes in each part. For example, the wind makes waves. The evaporation of water from the ocean forms clouds. This, in turn, may produce precipitation over nearby land.

Local, short-term conditions in the atmosphere, such as the humidity, temperature, and wind velocity, are called *weather*. Weather varies from place to place and from day to day in any given location. However, large geographical areas have long-term prevailing patterns of weather, called *climate*.

The ocean exerts a great influence on weather and climate, producing a variety of conditions, from mild sea breezes to violent hurricanes (such as the one shown above). In this chapter, you will see how interactions among the hydrosphere, lithosphere, and atmosphere produce weather and climate changes.

17.1 OCEAN TEMPERATURE AND WIND

As you know, Earth does not have a uniform climate. It is, generally, warmest at the equator and coldest at the poles. Why does the temperature vary over the surface of the planet? The answer has to do with the rays that Earth receives from the sun—the kinds of rays, the angle at which they strike the surface, what kind of surface they strike, how they are reflected or absorbed, the time of the year (seasons), and Earth's position in relation to the sun.

Energy from the sun, which is called **radiant energy**, travels through space as rays (also called radiation). The sun emits different kinds of radiation. Visible light is one of these forms of radiant energy.

Some radiant energy that strikes Earth's surface is absorbed and changed into heat. The rest is either absorbed by the atmosphere or reflected back into space. The amount of solar energy that reaches any part of the planet depends on the angle at which the rays strike Earth's surface, called the **angle of insolation**. The amount of energy that is absorbed as heat depends on conditions in the atmosphere, physical properties of the surface the rays reach, and the angle of insolation. (See Figure 17-1.)

Solar energy is most intense at the equator because the rays are direct, that is, they strike Earth at an angle of 90 degrees. Due to Earth's curvature and tilt on its axis, regions north and south of the equator receive the rays at an angle that is slanted relative to the surface. Slanted rays are spread out over a larger area than are direct rays (which are more concentrated), so slanted rays are less intense. (See Figure 17-2.)

Ocean Temperature

The uneven heating of Earth's surface causes the ocean's temperature to vary with latitude. The ocean is warmest at the equator. As the distance north and south of the equator increases, the temperature of the ocean's surface waters generally decreases (not counting effects of specific ocean currents). The ocean is coldest at the poles because those areas have the lowest angle of insolation and receive the least intense solar heat.

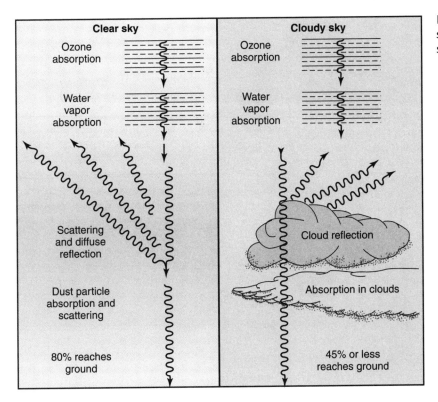

Figure 17-1 Absorption of solar radiation, in a clear sky and a cloudy sky.

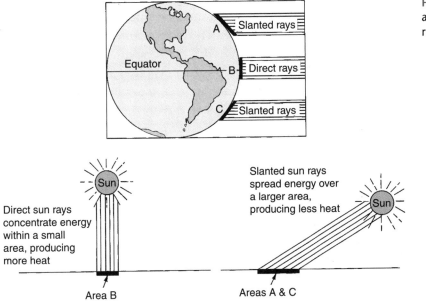

Figure 17-2 Slanted rays are less intense than direct rays.

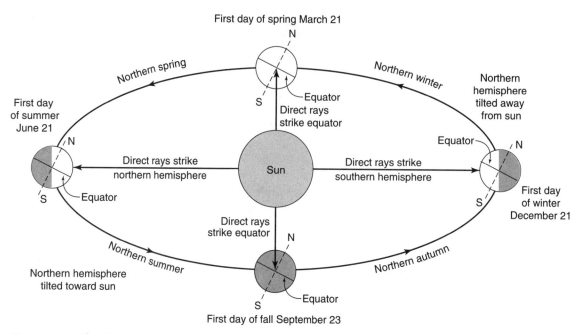

Figure 17-3 The tilt of Earth's axis causes the seasons.

Ocean temperature also varies by season. Look at Figure 17-3, which shows the sun's rays striking Earth at the start of each of the four seasons. The axis of Earth has a tilt, or angle of inclination, of 23½ degrees from the vertical in relation to the sun. In the northern hemisphere during the summer, which begins on June 21, Earth is tilted, or inclined, toward the sun. The northern oceans warm because the sun's rays strike Earth's surface directly at this time of the year, heating it most intensely. In the winter, which begins on December 21, the northern hemisphere is tilted away from the sun, so the rays strike Earth's surface at an angle, that is, slanted. Since slanted rays deliver less intense heat, the northern oceans are colder. (Because of Earth's tilt, summer and winter seasons are reversed in the Southern Hemisphere.)

The ocean heats more slowly and retains heat longer than land does because the sun's rays can penetrate farther into water (which is transparent) than into land. The depth to which the rays travel depends on their energy. Most solar radiation is absorbed within the top 60 meters of the ocean, although some light penetrates hundreds of meters into the water. (You will learn more about light in the sea in Chapter 19.)

In summary, the rays that strike Earth directly (at about 90 degrees, during the summer and at the equator) penetrate and heat the water more intensely than do rays that strike at a lower angle (slanted) during the winter and above or below the equator.

Ocean Wind

On a hot summer day, you may go to the beach to get some relief from the heat. You feel cooler as you approach the sea because a cool wind blows off the water. Often, the wind (which is a mass of moving air) is in the form of a gentle breeze, called a **sea breeze**. (See Figure 17-4.) The sea breeze is cool because air over the ocean is cooler than air over the adjacent land. Ocean air is cooler because water takes a longer time to heat from the sun's rays than land does. Consequently, on a summer day, the ocean releases less heat into the air above it than the land does. This cooler air moves in, as a sea breeze, from over the ocean to above the land.

The opposite occurs at night. Cooler air from the land moves seaward, producing what is called a **land breeze**. (See Figure 17-4.)

Figure 17-4 A sea breeze occurs during the day; a land breeze occurs at night.

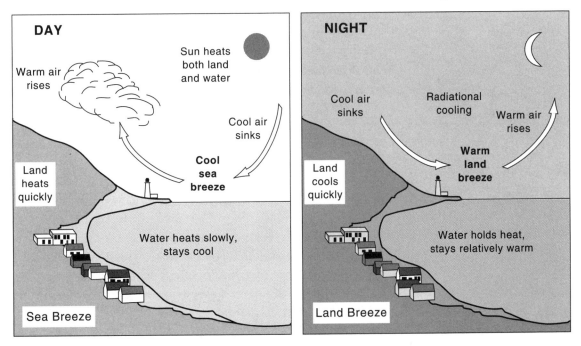

After the sun sets, the land (and the air above it) cools (that is, loses heat) more quickly than the ocean does. As warmer air rises over the ocean, cooler air from the land moves in to take its place. Why the difference in heat loss between the two surfaces? Only the top layers of the land are heated, so the ground loses its heat quickly. However, as mentioned above, water is heated to a greater depth than land is, so the ocean gives up its heat more slowly and retains more heat. In fact, water has a heat storage ability, or **specific heat**, that is about three times greater than that of land. Water's capacity to retain heat longer and lose heat more slowly provides marine organisms with a more moderate environment than that found on land, where temperatures can fluctuate more dramatically.

What causes a sea breeze to blow toward the land, and a land breeze to blow toward the sea? Look at Figure 17-5, which shows a demonstration of air flow in a smoking chimney. When the candle inside one chimney is lit and smoke paper is placed over the other chimney, the smoke descends into that chimney and goes up the chimney above the candle. (Smoke is used to make the flow of air visible.) The smoke behaves in this way because warm air rises—and the air in the first chimney is heated by the candle, causing it to rise. Cooler air flowing into the second chimney is more dense, and so it sinks and moves over to take the place of the rising air in the first chimney. This demonstration of how temperature differences cause air flow illustrates a *convection current*, which is the continual movement of a gas or a liquid in a cycle as the heated part rises and

Figure 17-5 An illustration of air flow in a chimney.

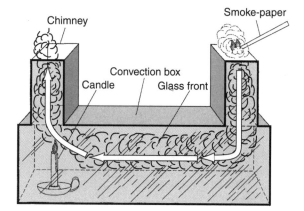

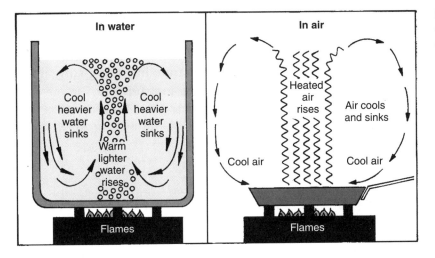

Figure 17-6 Convection currents in water and in air.

the cooler part sinks. (See Figure 17-6.) So the sea breeze and the land breeze are actually convection currents. During the day, the sun heats the sand, causing hot air to rise above it; cooler air over the ocean flows in to take its place, producing the sea breeze. And during the night, cooler air over the land flows out to take the place of warmer air that rises above the ocean.

The fact that the land and the sea gain and lose heat at different rates helps to explain, in part, why temperatures along a coast are generally more moderate than those of inland areas. During the summer, coastal cities are often a few degrees cooler than nearby inland cities (at the same latitude), due to the cooling effect of the ocean. And during the winter, coastal cities are often a few degrees warmer than nearby inland cities (at the same latitude), because they benefit from the warming effect of the ocean, which slowly releases heat that was absorbed from the sun.

17.1 SECTION REVIEW

1. Why is the ocean surface warmer at the equator than at the poles?

2. How does the temperature of the ocean vary with the seasons? Explain how that effect is related to Earth's angle of insolation.

3. Why is the air usually cooler at the beach than it is inland?

17.2 MOISTURE IN THE AIR

The ocean not only affects air temperature but also greatly influences the amount of water vapor, or moisture, in the air. The amount of moisture in the atmosphere is called **humidity**. Because they are near the ocean, coastal regions are often more humid than inland regions.

Humidity

How is humidity determined? Actually, scientists calculate relative humidity. **Relative humidity** is the amount of water vapor in the air compared to the maximum amount (saturation) of water vapor the air can hold at a given temperature. Saturation occurs when the air is completely filled with water vapor and cannot hold any more. Relative humidity is expressed as a percentage. (See Figure 17-7.) You can use the following formula to calculate relative humidity:

Relative humidity = amount of water vapor in air ÷ amount needed for saturation × 100

For example, on a summer day, when the air temperature is 26°C, the amount of moisture needed for the air to be saturated would be 25 grams per cubic meter (m³) of air. (Water vapor can be measured in different units, for example, in grains. In this case, it is measured in grams.) If the amount of moisture in the air is 22 grams per m³ of air, you would figure the relative humidity as follows:

Relative humidity = 22 grams ÷ 25 grams × 100

= 0.88 gram × 100

= 88 percent

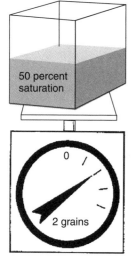

Figure 17-7 Relative humidity: 100 percent and 50 percent saturation. (Note: one gram is equal to about 17 grains.)

Relative humidity can also be determined using instruments called **hygrometers**. One type of hygrometer, called the sling psychrometer, is shown in Figure 17-8. It is composed of two thermometers: one is uncovered (dry bulb); the other has a wet cloth covering its bulb (wet bulb). By spinning the psychrometer in the air for a few seconds and then reading the temperature of each thermometer, you can use a relative humidity chart to determine the relative humidity. If the air is not saturated, there will be some evaporation of moisture from the wet cloth, so the wet-bulb thermometer will have a lower temperature than the dry-bulb thermometer. (The difference in degrees on the chart is based on the wet-bulb reading.) When the air is very dry, more water vapor evaporates, thus giving the wet bulb a much lower temperature than the dry bulb. For example, when the dry-bulb temperature is 20°C and the wet-bulb temperature is 19°C (a difference of only 1°C), the relative humidity is 91 percent; but when the dry-bulb temperature is 20°C and the wet-bulb temperature is 15°C (a difference of 5°C), the relative humidity is only 59 percent.

At each temperature, the air can hold a different quantity of water vapor. (See Figure 17-9.) Warm air can hold more water vapor than cold air can. In other words, warm air has a higher saturation point than cold air does. (Just think of a hot, steamy jungle!) When warm, moist air gets chilled by encountering a cold surface, the air may become saturated (because cold air has a lower saturation point), and visible forms of water vapor such as fog or dew appear.

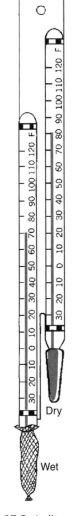

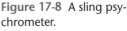

Figure 17-8 A sling psychrometer.

Fog

The air at sea level in many coastal regions is often filled with so much moisture that it looks like a cloud. Air that is saturated with moisture near the ground is called **fog**. Actually, a fog can be considered a ground-level cloud. Fog forms when warm, moist air makes contact with a cold surface. The mirror in your bathroom often "fogs up" when warm, moist air from the shower makes contact with the cold surface of the mirror. When water condenses (changes from a vapor to a liquid) on a solid surface, it is called **dew**, which you may have seen on cars and lawns on cool summer mornings. Dew is moisture that comes from the warm air and condenses on the

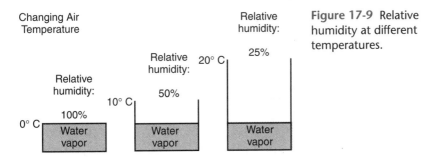

Figure 17-9 Relative humidity at different temperatures.

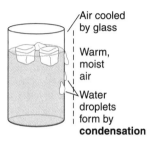

Air cooled by glass

Warm, moist air

Water droplets form by **condensation**

Figure 17-10 Condensation of water on a cool surface.

surface of solids that cool over night. Likewise, a glass of ice water appears to "sweat" because moisture from the warmer air condenses on its cool surface. (See Figure 17-10.) The temperature at which water vapor condenses as a cloud or fog is called the **dew point**, and it depends on the amount of water vapor, or humidity, in the air.

Coastal fog forms at the surface of the ocean when the dew point is reached. The San Francisco fog occurs when the cold California Current moving south along the coast meets warm, moist air carried in by prevailing westerly winds. The cold surface water cools the moist air above it to the point at which the air cannot hold all the water vapor. (Remember: cold air has a lower saturation point than warm air.) The excess water vapor appears as fog. The foggiest place in the world is the Newfoundland coast in Canada, which averages about 120 foggy days per year. There, fog is produced when the warm, moist air above the Gulf Stream meets the cold Labrador Current coming down from the north. The moisture condenses to form a thick fog that can be a serious hindrance to navigation. The warm Gulf Stream continues across the Atlantic Ocean. When it reaches the coast of Great Britain, it produces warm, moist air that meets the cool land surface, again forming a fog. (See Figure 17-11.)

Figure 17-11 The formation of fog.

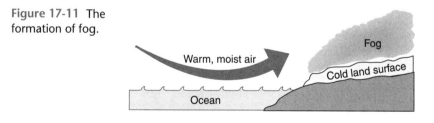

Warm, moist air

Fog

Cold land surface

Ocean

17.2 SECTION REVIEW

1. How is the relative humidity of the atmosphere determined?
2. Why does it get foggy in San Francisco and Newfoundland?
3. What is the relationship between a fog and the dew point?

17.3 STORMY WEATHER

On September 8, 1900, a storm with powerful winds from the Gulf of Mexico flooded the city of Galveston, Texas, killing about 6000 people. Although such a great loss of lives is rare these days, each year from the beginning of June to the end of November, inhabitants along the Gulf, Atlantic, and Caribbean coasts are concerned because that is their hurricane season.

Hurricanes

A **hurricane** is a coastal storm with a wind velocity that exceeds 120 km per hour. Hurricanes form in tropical seas where there is hot, moist air. Such conditions contain the heat needed to fuel a hurricane. (See Figure 17-12, which shows U.S. hurricane paths.) As the hot, moist air rises, it cools in the upper atmosphere and condenses into ring-shaped clouds about 19 km in height. During condensation, a great deal of heat energy is released, which causes more hot air to rise even quicker. This whirlwind of rising air moves in a spiral direction around a core of relatively calm air known as the "eye" of a hurricane. The eye of a hurricane may range from about 25 to 65 km in diameter, while the total diameter of a large hurricane may be more than 500 km. The worst conditions occur just outside the hurricane's eye, with heavy thunderstorms and winds of over 300 km per hour. Unfortunately, some people think that a hurricane has passed when in fact they are still in its eye, and the other half of the storm has yet to pass through their area.

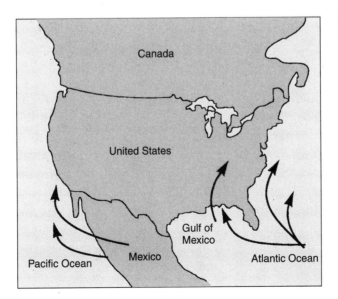

Figure 17-12 U.S. hurricane paths.

The National Hurricane Center in Florida uses radar, airplane reconnaissance, and weather satellites to track hurricanes. As soon as a hurricane is spotted far out at sea by weather satellites, hurricane "hunters" from the U.S. Air Force and the National Oceanic and Atmospheric Administration (NOAA) fly into the eye of the storm and monitor it continuously. Data on wind velocity and the speed and direction of the storm are sent to weather stations along the coast well in advance of the storm's arrival. Official hurricane warnings give people a chance to get ready for possible evacuation of their area. By instructing people to move inland to higher elevations, a good warning system can save many lives.

Damage by Hurricanes

When a hurricane does reach land, it can be very destructive. Most damage occurs when a hurricane produces a *storm surge,* which raises the sea level several meters above normal and pushes the water inland, causing flooding. A storm surge can be particularly damaging if it occurs during a high tide. As the hurricane moves inland, it slows because its winds move slower over land and there is not enough moisture to sustain it. If the storm continues to move inland, it dissipates and dies out. Hurricanes also lose power as they move over the cooler waters of the northern Atlantic because they lose the heat needed to fuel their winds. Hurricane Allen moved inland and weakened over a three-day period in August 1980. The satellite photos showed a tight, spiral mass in the Gulf of Mexico on August 10; a looser spiral on August 11 as the hurricane weakened and passed over northern Mexico and southern Texas; and finally, on August 12, all that remained of Hurricane Allen were diffuse clouds over Texas and Mexico. (See Figure 17-13.)

Hurricane intensity is rated on a scale from one to five, called the Saffir-Simpson Damage Potential Scale. A category-one hurricane causes "minimal" damage, while a category-five hurricane with winds of over 250 km per hour causes "catastrophic" damage. Hurricane Andrew in 1992 was classified as a category-four hurricane. On August 23 and 24, Hurricane Andrew killed 62 people and caused $25 billion in damage as it tore through the Bahamas, southern Florida, and Louisiana. More than 25,000 homes were destroyed

Figure 17-13 Astronaut's-eye view of a hurricane.

and 250,000 people were left homeless. Measures have been undertaken by the federal government to prepare and protect the public from hurricane damage. For example, besides having an early warning system based on satellite readings, the states of Florida and Louisiana have building codes that require homes to withstand 177 kilometer-per-hour winds.

A list of some of the most destructive hurricanes in recent U.S. history is in Table 17-1, on page 428. Note that even though Hurricane Andrew was very destructive, far fewer people were killed than in previous hurricanes, thanks to official hurricane warnings.

El Niño and La Niña

Droughts in Australia, famine in Africa, floods in California, and other climate-related disturbances in recent years have all been attributed to an unpredictable warm ocean current in the equatorial

TABLE 17-1 WORST HURRICANES IN RECENT U.S. HISTORY

Hurricane	Year	Deaths
Galveston*	1900	6,000
Florida Keys/Texas	1919	600–900
Florida	1928	1,836
New England	1938	600
Diane	1955	200
Audrey/Bertha	1957	390
Camille	1969	256
Agnes	1972	122
Andrew	1992	62

*Before 1953, hurricanes were not given names.

Pacific. This oceanic warming, which periodically causes deaths and great losses of crops, property, fisheries, and marine animals dependent on fish, is called **El Niño**. The name is Spanish for "the (Christ) child," since this ocean current flows near South and Central America during Christmastime. El Niño begins in the western Pacific Ocean, a low-pressure region where there is a large mass of warm surface water. Normally, the warm water is held in place by the trade winds, which blow westward from a high-pressure region in the eastern Pacific Ocean. However, from time to time the trade winds weaken, causing the warm water—called an equatorial countercurrent—to surge eastward toward Central and South America. The warm water invades the cold waters off the South American coast, disrupting the fisheries and bringing floods and stormy weather to North and South America. (See Figure 17-14.)

In contrast to the warm El Niño ocean currents are the unusually cool ocean currents that sometimes appear in the equatorial Pacific. These cooler surface temperatures are called *La Niña*. The impact of El Niño and La Niña can be seen in the northern hemisphere during the winter months. In the continental United States during El Niño years (when ocean currents are unusually warm in the mid-Pacific), winters are warmer than normal in the Northwest and cooler than normal in the Southeast and Southwest. In contrast, during La Niña years (when ocean currents are unusually cool

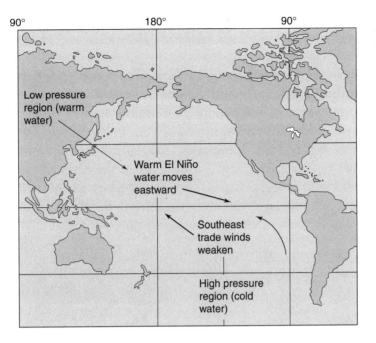

Figure 17-14 El Niño: when the trade winds weaken, warm water moves eastward.

Low pressure region (warm water)

Warm El Niño water moves eastward

Southeast trade winds weaken

High pressure region (cold water)

in the mid-Pacific), winters are cooler than normal in the Northwest and warmer than normal in the Southeast.

El Niño and La Niña are opposite phases of a cycle, or oscillation, called the *El Niño–Southern Oscillation (ENSO) event,* which usually lasts from 1 to 2 years and occurs in cycles, about every 4 to 6 years. Most oceanographers think that the ENSO is a natural event that results from a complex interplay of oceanic and atmospheric forces. Other scientists have suggested that phenomena such as volcanic activity on the seafloor or increased warming of the atmosphere may be partly responsible for generating El Niño and La Niña currents. Research continues in an attempt to better understand the dynamics of the ENSO event, so that this climatic disturbance can be predicted with greater accuracy.

17.3 SECTION REVIEW

1. Why do hurricanes occur in hot, moist (coastal) regions?

2. What is a storm surge and why is it so dangerous?

3. What are some possible causes of strong El Niños? Why do they have such a negative impact on people and marine animals?

17.4 THE GREENHOUSE EFFECT

Have you ever been to a greenhouse? Greenhouses have transparent walls and roofs, which permit sunlight to enter so plants can grow. Some of the energy in sunlight changes into heat energy inside a greenhouse, warming the air. In a somewhat similar manner, certain atmospheric gases cause Earth's atmosphere to function like a giant greenhouse. While some of the sun's rays are reflected back into space, other rays are absorbed by Earth's surface. These rays heat the land and water that absorb them. Earth's surface then radiates some of this heat back outward. (See Figure 17-15.) But the

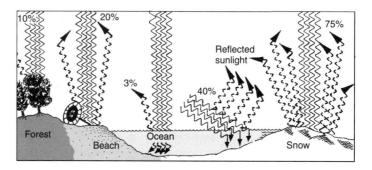

Figure 17-15 Absorption and reflection of solar radiation by different surfaces.

heat that Earth's surface radiates has a longer wavelength (infrared) than the sun's rays (mostly visible light) that pass through Earth's atmosphere. These longer-wavelength rays cannot pass through Earth's atmosphere back into space. Instead they are trapped by water vapor and carbon dioxide in the lower atmosphere, causing a warming of Earth called the **greenhouse effect**. (See Figure 17-16.)

Figure 17-16 An illustration of the greenhouse effect.

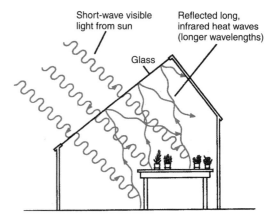

ENVIRONMENT
Rising Temperatures, Rising Sea Level

Satellite photographs taken of Antarctica in 2002 revealed the disappearance of part of its Larsen ice shelf—a loss of ice that covered an area the size of Rhode Island. An ice shelf is a mass of floating ice that is attached to a continent. Some scientists say that the melting of such a large area of ice is a result of the rapidly warming atmosphere. After all, Earth's average air temperature has risen over the past century.

The melting of an ice shelf does not increase the total volume of ocean water (since it is already part of the ocean), just as the melting of ice cubes in a glass of water does not cause the water to overflow. So, the sea level does not increase as a direct result of melting ice shelves. However, scientists worry that if the melting of Antarctica's ice shelf continues, this buffer zone between land and sea will disappear, thus exposing the ice sheets and glaciers on land. These glaciers could then move into the sea, where they would melt and cause a rise in sea level. Knowing of that possibility, researchers think that the melting ice shelves can be regarded as early warnings of climate change.

There are some climatologists (scientists who study climate changes) who reject the idea that a warming trend will cause a rise in sea level. They argue that a warming trend may actually increase Antarctic precipitation, in the form of snow. According to this hypothesis, the heavier accumulations of snow that could result would be compressed over time, thereby forming more ice. Other scientists remain neutral, not in favor of either argument until they have more information. In the meantime, orbiting satellites will continue to collect data on the Antarctic ice shelf and to monitor any decrease in glaciers or increase in sea level that may occur.

QUESTIONS

1. How might a global warming trend contribute to a rise in the sea level?

2. Why would melting glaciers, but not melting ice shelves, cause a rise in sea level?

3. Explain the argument proposed by scientists who claim that a warming trend will not cause a rise in the sea level.

Global Warming

Scientists have noticed that Earth is experiencing a warming trend; the planet's mean global temperature has increased more than 1°C over the past 100 years. It is believed that this warming trend, commonly called **global warming**, has caused some melting of the polar ice caps and a subsequent rise in sea level. (Compare the graphs in Figure 17-17.) In fact, in just the past 100 years, the sea level has risen approximately 10 cm. Scientists are taking measurements of the polar ice caps to determine if their rate of melting continues to increase.

What is causing this warming trend? Look at the graph in Figure 17-18, which shows an increase in atmospheric CO_2 (in ppm) over a recent 35-year period. Carbon dioxide is called a **greenhouse gas** because (along with water vapor) it traps heat in the atmosphere. Much of the CO_2 in our atmosphere comes from the combustion of **fossil fuels**, such as coal, oil, and gas. The Industrial Revolution, which brought on this greater use of fossil fuels, began about 100 years ago. And it was in that past century that we began to see increased CO_2 levels, a rising sea level, and atmospheric warming.

Concern about the warming trend varies. Some scientists claim that the warming trend is part of a natural cycle that will eventually reverse itself through CO_2 uptake by photosynthetic marine and terrestrial plants. Others claim that the warming of Earth's atmosphere is a real problem that can be significantly reduced by decreasing the use of fossil fuels and relying more on alternative energy sources, such as solar power and wind power. A real fear is that, if increased

Figure 17-17 Both mean global temperature and sea level have been rising.

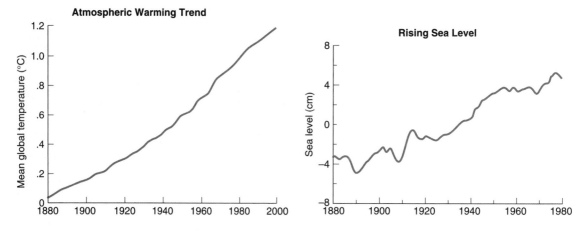

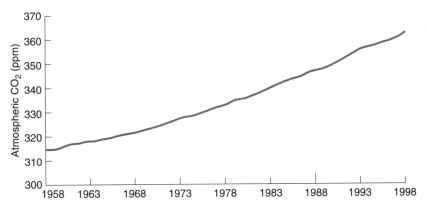

Figure 17-18 Atmospheric carbon dioxide has been increasing steadily.

warming leads to a significant rise in sea level during this century, coastal cities and island nations may be inundated. (See this chapter's feature, on page 431, about possible effects of rising temperatures on sea level.)

17.4 SECTION REVIEW

1. Explain how the greenhouse effect works in the atmosphere.

2. How does the use of fossil fuels contribute to global warming?

3. Explain how an increase in atmospheric CO_2 may lead to a rise in sea level.

Laboratory Investigation **17**

Analyzing Ocean Temperatures

PROBLEM: How can we analyze temperature differences in ocean currents?

SKILLS: Map reading; comparing and analyzing scientific data.

MATERIALS: Colored pencils, copies of the two maps of the Pacific Ocean.

PROCEDURE

1. Examine the two maps in Figure 17-19. Notice the isotherms—lines that connect locations having the same temperature. Each isotherm represents a different temperature.

2. Use the colored pencils to highlight the temperature ranges indicated by the isotherms. Prepare a temperature key that shows the appropriate colors.

3. Next, color in the temperature ranges of the isotherms on your copy of each map, using the color key as a guide.

4. Study both maps to see if any significant difference is observable.

OBSERVATIONS AND ANALYSES

1. Which year had a cooler mid-ocean temperature at the equator, 1987 or 1988?

2. Compare the temperatures of mid-ocean water at 20°S and at 20°N for the two years. Is there a noticeable difference between them?

3. Based on the ocean temperature differences between the two maps, which year probably shows an El Niño? Explain.

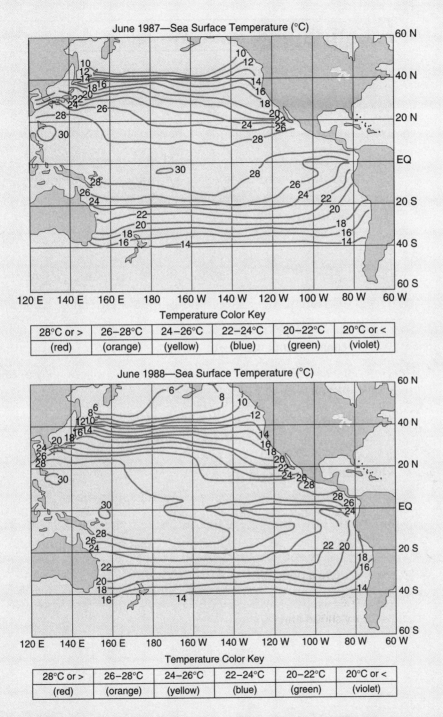

Figure 17-19 Each isotherm represents a different surface temperature.

Chapter 17 | Review

Answer the following questions on a separate sheet of paper.

Vocabulary

The following list contains all the boldface terms in this chapter.

angle of insolation, dew, dew point, El Niño, fog, fossil fuels, global warming, greenhouse effect, greenhouse gas, humidity, hurricane, hygrometers, land breeze, radiant energy, relative humidity, sea breeze, specific heat

Fill In

Use one of the vocabulary terms listed above to complete each sentence.

1. The amount of water vapor in the atmosphere is the _____.

2. Solar rays trapped in the atmosphere cause the _____.

3. Water has a greater heat storage ability, or _____, than land does.

4. Water vapor condenses into a cloud or a fog at the _____.

5. Rising hot, moist air over a tropical sea can form a _____.

Think and Write

Use the information in this chapter to respond to these items.

6. Explain how the angle of insolation, and physical properties of Earth's surface, affect the amount of solar energy absorbed.

7. Describe how and where hurricanes are formed.

8. How has industrialization caused global warming?

Inquiry

Base your answers to questions 9 through 11 on the diagram below and on your knowledge of marine science.

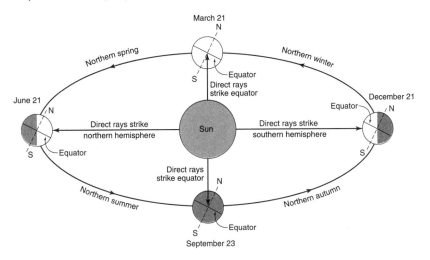

9. Ocean temperatures at the equator would be their warmest on which date?   *a.* December 21   *b.* March 21   *c.* June 21   *d.* October 21

10. Which statement is *not* accurate regarding ocean temperature and seasonal change?   *a.* The tilt of Earth's axis affects ocean temperature.   *b.* Surface ocean temperatures are affected by the angle of insolation of solar rays.   *c.* The northern oceans are warmest from December 21 to March 21.   *d.* Direct rays of light deliver more intense heat than do slanted rays.

11. Ocean temperatures in the southern hemisphere would be their coldest on   *a.* December 21   *b.* March 21   *c.* June 21   *d.* September 23.

Multiple Choice

Choose the response that best completes the sentence or answers the question.

12. Cool air that moves seaward from the land is called a
    a. sea breeze   *b.* land breeze   *c.* fog   *d.* condensation.

13. The amount of water vapor in the air as compared with the amount the air can hold is the *a.* dew point *b.* greenhouse effect *c.* relative humidity *d.* specific heat.

14. During a summer day, it is cooler at the beach than inland because of *a.* a sea breeze *b.* a land breeze *c.* cool air that moves from land to sea *d.* warm air over the ocean.

15. All of the following are true about hurricanes *except* *a.* they are fueled by warm, moist air *b.* they occur only in the Caribbean and the Gulf of Mexico *c.* they decrease in intensity as they move inland *d.* they can be very destructive.

16. The California fog is a result of *a.* a warm ocean current making contact with moist continental air *b.* warm, moist air from the ocean meeting cool continental air *c.* a cold ocean current meeting warm moist air *d.* warm continental air meeting cold continental air.

17. At night along a coast there is *a.* a sea breeze *b.* a land breeze *c.* warm air rising over the land *d.* warm air sinking over the sea.

18. Which statement is true? *a.* As latitude increases, surface ocean temperature decreases. *b.* As latitude increases, surface ocean temperature increases. *c.* Latitude affects ocean temperature only above the equator. *d.* Latitude affects ocean temperature only below the equator.

19. The photic zone is deepest at the equator because *a.* light enters the water at an angle of 45 degrees *b.* light that strikes Earth is slanted *c.* the angle of insolation is 90 degrees *d.* the light has longer wavelengths.

20. Which of the following statements is accurate regarding hurricanes? *a.* They pick up power as they move over cooler waters. *b.* They originate in tropical seas where there is hot, moist air. *c.* They originate over land and travel out to sea. *d.* They have caused more deaths in recent years than in past years.

21. Which of the following is a true statement regarding El Niño? *a.* It is a destructive hurricane. *b.* It is an ocean current from the Caribbean. *c.* It is a warm, tropical countercurrent in the Pacific. *d.* It is an active volcano in the South Pacific.

Base your answer to the following question on the information in this chapter and on the diagram, which illustrates the greenhouse effect.

22. Which statement about solar radiation is correct? *a.* Long wavelengths of heat energy can pass back out through the glass. *b.* Short wavelengths of light energy cannot pass through glass. *c.* The air temperature is colder inside the structure because heat escapes. *d.* The air temperature is warmer inside the structure because some sunlight changes into heat energy.

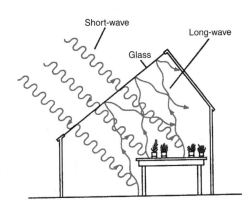

Research/Activity

■ Track the daily path of a hurricane based on satellite pictures in a newspaper or data from NOAA on the Internet. Construct a line graph and plot the wind velocity of the hurricane over the course of a week. Correlate changes in the hurricane's speed with changes in its direction (that is, as it moves over water and/or over land).

■ Use a copy of a map of the continental United States to show the effects of a rise in sea level of about 30 meters. Draw in the new coastlines with color markers. Make a list of some important cities that may end up below sea level.

■ Build a miniature greenhouse. Place one thermometer inside it and one outside it. Record the inside and outside temperatures daily for one week; calculate the average for each set of readings; explain any difference observed.

UNIT 6 ENERGY IN THE OCEAN

The ocean is a storehouse of energy. Waves and currents have kinetic energy, the energy of motion. Waves crashing on a beach produce sound energy. Hot water boiling up from hydrothermal vents on the seafloor gives off heat energy. Energy is the ability to do work, and work involves movement. For example, when a wave crashes on the beach, work is done because water is moved.

Energy is interchangeable; that is, one form can change into another form. When sunlight strikes the ocean surface, some of its energy is absorbed by the water and changed into heat. And some of the light energy is absorbed by marine plants and changed into the chemical energy stored in glucose. In this unit, you will learn how energy, temperature, and pressure affect the ocean environment and the various life-forms within it.

441

18 Temperature and Pressure

When you have completed this chapter, you should be able to:

DESCRIBE the relationship of kinetic energy to heat in the sea.

DISCUSS the effects of temperature and pressure on divers and marine organisms.

EXPLAIN how aquatic organisms regulate osmotic pressure.

If you ever have swum in the ocean, you know that the water feels cooler than the air above it. And if you have been in a lake, you probably noticed that the deeper you swim the cooler the water feels. Temperature is an important factor that affects properties of ocean water. Differences in temperature affect water density, the mixing of water layers, and the kinds of organisms that can live in different parts of the ocean. *Temperature* is a measure of the average kinetic energy possessed by the particles of a substance. In the ocean, temperatures vary from below 0°C to above 100°C.

As soon as you dip below the ocean's surface, you feel pressure on your face and body because water exerts pressure. *Pressure* is defined as the force per unit area. When you swim underwater, you can feel the water pressure all around you. Underwater pressure increases at a predictable rate with increasing depth.

Most aquatic organisms cannot survive the low temperatures and high pressures found in the great ocean depths. Humans certainly cannot withstand such conditions without the use of special equipment. However, some marine animals are equipped to live under extreme pressure and cold temperatures. In this chapter, you will study how temperature and pressure vary in the ocean and how they affect living things.

18.1 KINETIC ENERGY AND HEAT IN THE OCEAN

Due to the way that the sun heats Earth, you will recall, ocean surface temperatures at the equator are warm, whereas water temperatures at the poles are very cold. Water is warm when its molecules have more kinetic energy, and cold when its molecules have less kinetic energy. *Kinetic energy* is the energy a substance has due to the motion of its molecules. When the kinetic energy of a substance is transferred to another substance, it is called *heat*. In the ocean, the greatest amount of heat is found at the hydrothermal vents, where temperatures of 350°C and higher have been recorded. The lowest amount of heat is found at the poles, where the water temperature is at, or slightly below, the freezing point.

You may also remember learning that the ocean takes longer to heat than the land does because it has a higher specific heat, or heat capacity. The water also takes longer to cool. As a result, there is a great difference in heat capacity (also called *heat storage ability*) between the ocean and the land. *Specific heat* is the amount of heat needed to raise the temperature of one gram of a substance one degree Celsius. Heat is measured in calories. A *calorie* is the amount of heat required to raise the temperature of one gram of water one degree Celsius. Thus the specific heat of water is one calorie per gram-degree Celsius (1 cal/g-°C), which is the standard against which other specific heats are measured. (See Table 18-1.)

TABLE 18-1 **SPECIFIC HEATS OF COMMON MATERIALS (CAL/G-°C)**

Material	Specific Heat
Water	1.0
Ice	0.5
Water vapor	0.5
Dry air	0.24
Basalt	0.20
Iron	0.11
Copper	0.09
Lead	0.03

Different States of Water

The temperature in the ocean varies so much that ocean water can exist in three different phases, or states: solid (ice), liquid (water), and gas (water vapor). The temperatures at which changes of state in water occur are shown in Figure 18-1. When one gram of water changes into one gram of water vapor, 540 calories of heat are absorbed by the water. The change in state from a liquid (water) to a gas (water vapor or steam) is called vaporization; the energy absorbed when this process occurs is called the **heat of vaporization**. Normally, evaporation occurs at the ocean surface at temperatures well below the boiling point. Heat is then released into the atmosphere, along with the water vapor. In fact, most of the water vapor in the atmosphere comes from the ocean through evaporation. Where in the ocean does the water actually boil? On active volcanic islands, such as Hawaii, lava flows into the ocean. On making contact with the water, the molten lava boils it into billowing clouds of steam. Of course, the main phases of water in the ocean are liquid and solid (ice), not gaseous.

If the temperature falls below its freezing point, water changes into ice. When one gram of water freezes, 80 calories of heat are lost by the water. The change of state from a liquid (water) to a solid (ice) is called *fusion*; the energy lost in this process is called the **heat of fusion**. Likewise, when one gram of ice melts into one gram of water, 80 calories of heat are gained by the water.

Figure 18-1 Changes of state in water: from solid to liquid and from liquid to gas.

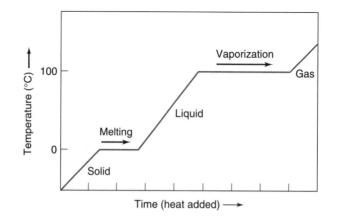

Ice in the Ocean

Much of the ocean at the North Pole and around the South Pole is covered by ice. There are two kinds of ocean ice: sea ice and icebergs. Sea ice is formed when water on the ocean surface drops below the freezing point, which is about –2°C for seawater. (The salt in ocean water lowers the water's freezing point.) Sea ice can become a hazard to navigation when coastal waterways freeze. Powerful ships called icebreakers are used to smash through sea ice up to 3 meters thick.

An iceberg is a chunk of ice that breaks off from the end of a glacier. (A glacier is a mass of moving ice formed on mountains from compacted snow.) Glaciers that reach the shore become undercut by waves. Wave action erodes the base of the glacier and pieces of ice "calve," or break off into the sea. A good-sized iceberg can be 50 to 100 meters high and several hundred meters long. (See Figure 18-2.) Icebergs can be a menace to navigation because they often float out into shipping lanes. What makes them particularly dangerous is that

Figure 18-2 An iceberg is a very large chunk of ice that has broken off from the end of a glacier at the shore. Most of an iceberg actually floats below the ocean surface.

the part you see above the water is, literally, just the "tip of the iceberg." About 80 to 85 percent of an iceberg floats below the surface, and that is the part that can split the hull of a ship. As you may recall from Chapter 1, the *Titanic* sank after colliding with an iceberg in the North Atlantic.

18.1 SECTION REVIEW

1. What is the definition of a calorie? How is it related to the specific heat of water?

2. In what different states of matter can ocean water exist?

3. For water, which process requires more calories of heat, vaporization or fusion? Why do you think this is so?

18.2 TEMPERATURE VARIATIONS IN THE OCEAN

Much of the sun's radiant energy that reaches Earth is absorbed at its surface and changed into heat. But, as you know from Chapter 17, the heating of the planet is not uniform. As a result, surface ocean temperatures vary with latitude. Surface water temperatures range from –2 to 29°C.

Variations With Depth

The temperature of the ocean also varies with depth (particularly in the middle latitudes). You may have experienced a temperature difference while diving in either an ocean or a lake. The deeper you descend, the colder it gets. The relationship between depth and temperature is shown in the graph in Figure 18-3. As depth increases, water temperature decreases. But the decrease is not uniform. There is a very steep drop in temperature between 200 and 1000 meters. This layer of ocean water is called a *thermocline*. The **thermocline** is a permanent boundary that separates the warmer water above from the colder, denser water below. Seasonal thermoclines between

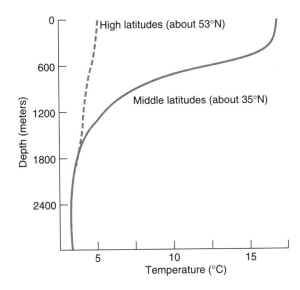

Figure 18-3 The relationship between water depth and temperature at high and middle latitudes: the steepest drop in temperature occurs (at middle latitudes) between 200 and 1000 meters in depth.

100 and 200 meters deep also occur. They are more common in the summer, when the water is heated more by the sun. Due to this heating, the surface water is less dense than the cold water below, so it floats on top in a distinct layer. As a result, there is very little mixing of water between the two layers. When the surface water cools, it sinks and displaces the bottom water, causing an "overturn" of water layers (and the minerals within them).

Oceanographers use several instruments and methods to measure seawater temperature. To get a temperature profile of the ocean, scientists use a **bathythermograph** (a narrow torpedo-shaped canister that is lowered into the ocean to make continuous temperature readings), a reversing thermometer (in a Nansen bottle), and thermistors (electrical temperature sensors towed on a cable). Sea surface temperatures are obtained from ships, floating buoys, and remote sensing by satellites (provided by the U.S. Navy, NASA, and the European Space Agency).

The Effects of Temperature on Ocean Life

Temperature affects the functioning of living things. If you have a tropical fish tank, you may have noticed that when the water temperature is high, the fish are more active than when the water temperature is cooler. This occurs because, for fish and many other ectothermic animals, when the temperature of the external environment changes, their internal body temperature changes, too. When the water temperature increases, the organism's internal energy level, or **metabolic activity**, also increases.

The metabolic activity of an animal can be determined by measuring the amount of carbon dioxide given off during respiration. As water temperature increases, the amount of CO_2 exhaled by a fish also increases, indicating an increase in metabolic activity. As a general rule, for every 10°C increase in temperature, there is a doubling of metabolic activity. However, at very high temperatures, enzymes are inactivated and metabolism decreases.

What happens to organisms in extremely cold ocean environments? Below-freezing temperatures are about as extreme as you can get. Marine biologists have discovered that several species of icefish survive in frigid Arctic and Antarctic waters because of a unique adaptation in their blood; the fish have glycoprotein, a biological "antifreeze" that lowers the freezing point of body fluids, preventing the tissues from freezing. Glycoprotein also coats ice crystals, which prevents them from enlarging in the body. Very cold temperatures can harm living tissues by destroying the enzymes that enable cellular chemical reactions. Scientists found that the icefish can breathe through its skin even while encased in ice. A rich network of blood vessels in its skin allows the fish to supplement breathing through its gills by taking in oxygen through its body wall. (See Figure 18-4.)

In shallow tropical waters, some snails have ridges on their shells. These ridges help radiate heat and keep the snails cool. Snails with light-colored shells also tend to be found in warmer waters.

Figure 18-4 The icefish can survive in frigid polar waters because it has a natural "antifreeze" in its blood.

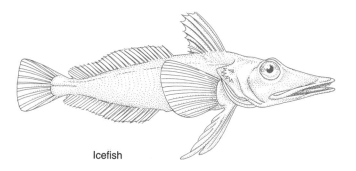

Icefish

CONSERVATION
Keeping the Chinook Chilly

A degree or two can mean the difference between life and death for the chinook salmon, an endangered species that spawns in the Sacramento River in California. The chinook can survive only in cold water; they begin to die if they are exposed to water temperatures above 14°C. In fact, when the Sacramento's water temperature rose to about 16.5°C during a 1976–1977 drought, thousands of these salmon perished. During a recent winter run, only 2000 adults were counted traveling to their spawning grounds, as compared with 117,000 that were counted making the run in 1969.

The Sacramento River has warmed due to the construction of the Shasta Dam—a 180-meter concrete barrier located about 320 km north of San Francisco. The dam, which created Lake Shasta, was built in the 1940s to provide electricity to the area. The hydroelectric system takes in and releases the warmer water from Lake Shasta; in doing so, it blocks the natural flow of colder water from the lake bottom to the salmon's spawning grounds in the river below it.

Fortunately, the hydroelectric facility and local government biologists became concerned about the decrease in the salmon population. The chinook salmon are important to both the economy and the ecology of California's river and marine communities. As a result, the Federal Bureau of Reclamation constructed an $80 million temperature-control system on the

Shasta Dam. This new water-intake system, bolted to the dam, permits colder water from the lake bottom to flow through huge louvers into the Sacramento River.

Taxpayers may complain that the $80 million project, which comes to $40,000 per fish, is too high a price to pay. But environmentalists point out that the dam has blocked the salmon from reaching their historical spawning grounds in the Cascade Mountains farther north. Thus, the Sacramento River below the dam must be maintained as a suitable habitat for the breeding population of this fish; and part of that effort means keeping the water temperature within a safe range for the salmon returning from the sea.

QUESTIONS

1. Why is the chinook salmon an endangered species of fish?
2. How is the chinook salmon affected by temperature changes?
3. Describe one measure undertaken to save the chinook salmon.

The light color reflects more sunlight than dark colors do; thus, the light color also helps to keep the snails from overheating. Temperature differences in the ocean also affect the distribution and features of certain microorganisms. For example, *Oithona* and *Calanus* are two types (genuses) of copepods. *Oithona* lives in warm water, whereas *Calanus* lives in cold water. Since warm water is less dense than cold water, objects floating in warm water tend to sink more easily. However, *Oithona* has long, frilly appendages that increase its surface area, helping to keep it afloat. In contrast, *Calanus* does not have these "extra frills," since it lives in cold water, where it is easier to stay afloat.

Marine mammals such as cetaceans and pinnipeds are adapted to survive in cold water because they have thick layers of fatty tissue (blubber) under their skin that insulate against heat loss. (In addition, pinnipeds have fur.) These defenses against the cold help whales and seals (which, as mammals, are endothermic) maintain a stable body temperature.

Humans are also endothermic. However, we do not have the special adaptations of marine mammals for retaining body heat in water. A person loses body heat 25 times faster in water than in air of the same temperature. Exposure to very cold water leads to an excessive loss of body heat, which can quickly cause a life-threatening condition called **hypothermia**. The body tries to make up for heat loss by generating heat through the involuntary contraction of its muscles, that is, by shivering. If heat loss is not stopped, a person's body temperature drops farther and the person may become unconscious. By getting out of the water, removing wet clothes, and keeping warm, people can restore their body temperature to normal.

18.2 SECTION REVIEW

1. What is a thermocline? Why are some more common during the summer?

2. How does a change in water temperature affect an animal's metabolism?

3. How is the icefish specially adapted to live in cold polar waters?

18.3 PRESSURE UNDERWATER

When you turn on the faucet, water gushes out. Water exerts pressure. Pressure is defined as a force applied over a given area, and can be calculated by using the following formula:

Pressure (P) = Force (F)/Area (A)

Pressure is measured in units called *pascals* (Pa). Force is measured in units called *newtons* (N). One pascal is equal to one newton of force per meter squared (N/m^2). One newton is the force needed to accelerate a one-kg mass one meter per second squared. The force exerted by an object equals its *weight*. Weight is defined as the product of its mass (m) times acceleration (a), or $F = ma$, where acceleration due to gravity is 9.8 m/s^2. An object with a mass of one kg has a weight or force of 9.8 newtons, as shown in the following formula:

$F = ma$

$F = (1 \text{ kg}) (9.8 \text{ m/s}^2)$

$F = 9.8$ newtons

Look at the container of water shown in Figure 18-5. Water spurts out farthest from the hole at the bottom of the can. Why?

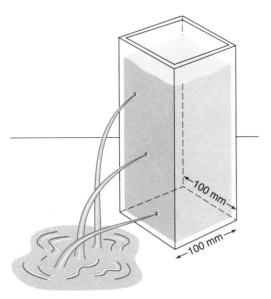

Figure 18-5 Water pressure is greatest at the bottom of the container because of the mass of water above it.

100 mm

100 mm

There is more pressure at the bottom than at the top because there is more water above the bottom hole than above the top holes. If the mass of the water in the container is 0.5 kg, what is the water pressure at the bottom of the container? First you calculate the force or weight of the water:

$F = ma$

$F = (0.5\text{ kg})(9.8\text{ m/s}^2)$

$F = 4.9$ newtons

Substituting the 4.9 newtons into the formula for pressure, you have:

$P = F/A$

$P = 4.9$ newtons$/A$

The area (A) at the base of the container is the length (100 mm or 0.1 meter) times width (100 mm or 0.1 meter), which equals 0.01 meter squared. Substituting the area into the formula you have:

$P = 4.9$ newtons$/0.01\text{ m}^2$

$P = 490$ pascals

The pressure at the bottom of the container is 490 pascals.

The water pressure in the middle of the container would be less, because there is half the mass of water pressing down at that point. Substitute 0.25 kg of water mass to calculate the force:

$F = ma$

$F = (0.25)(9.8\text{ m/s}^2)$

$F = 2.45$ newtons

The pressure in the middle of the container would be:

$P = F/A$

$P = 2.45$ newtons$/0.01\text{ m}^2$

$P = 245$ pascals

There is energy in water pressure. The water in the container has the energy of position, called potential energy. When the plugs are removed, water spurts out. When the water flows out, potential energy is changed into kinetic energy, the energy of motion.

Depth and Water Pressure

The mass of several kilometers of air exerts atmospheric pressure on Earth's surface. Under normal conditions, the atmospheric pressure at sea level, expressed as one atmosphere of pressure, is equal to approximately 101 kilopascals (kPa). Pressure exerted by the water's mass (due to its density) is called **hydrostatic pressure**. Scientists know that for every 10 meters of depth, water pressure increases by 101 kPa (one atmosphere). What is the hydrostatic pressure on a diver at a depth of 60 meters? The hydrostatic pressure would be equal to six atmospheres, or nearly 608 kPa.

The atmosphere also presses down on the diver. The total pressure, or **ambient pressure**, on the diver is the sum of the atmospheric pressure plus the hydrostatic pressure. Thus, the diver at 60 meters depth is under an ambient pressure of 709 kPa. Pressures at different depths are summarized in Table 18-2.

TABLE 18-2 OCEAN DEPTH AND WATER PRESSURE

Depth (meters)	Atmospheres*	Hydrostatic Pressure (kPa)	Ambient Pressure (kPa)
0	1	0.0	101.325
10	2	101.325	202.650
20	3	202.650	303.975
30	4	303.975	405.300
40	5	405.300	506.625
50	6	506.625	607.950
60	7	607.950	709.275

*1 atmosphere = 101.325 kPa = 14.7 lb/in.^{2}

Table 18-2 shows that as depth increases at regular intervals, pressure also increases. This relationship between water depth and pressure is shown in Figure 18-6 on page 454. As you can see from

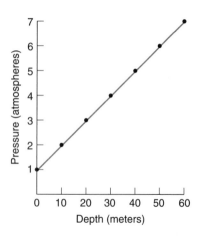

Figure 18-6 There is a direct relationship between water depth and pressure.

the graph, there is a direct relationship between pressure and depth. For every change in depth, there is a uniform change in pressure.

The Effects of Pressure on Ocean Life

Marine organisms live at many different levels in the water column. In what ways are they adapted to differences in water pressure? Diving mammals, such as dolphins, possess a very flexible rib cage that can expand and contract in response to pressure differences as the animal swims up and down. At greater depths, where the pressure is stronger, a dolphin's rib cage and lungs can collapse without damage.

Many deep-sea fish, such as the hatchetfish, cannot swim freely between the bottom and top layers of the ocean. Like many other bony fishes, the hatchetfish has an air-filled swim bladder that inflates and deflates to regulate movement through the water column. When the swim bladder takes in air, the fish rises. When the volume of air in the swim bladder decreases, the fish sinks. By regulating the size of the swim bladder, a fish can maintain a neutral buoyancy without actively swimming—an important adaptation for energy conservation. The great pressure differences, however, between the top and bottom of the ocean keep deep-sea fish like the hatchetfish "prisoners" of the depths, since their swim bladders would expand and burst if they rose to shallower waters.

Another animal whose movements are affected by differences in water pressure is the chambered nautilus. The nautilus is a mollusk whose shell contains spiral chambers filled with air. The animal rises and falls in the water column, between depths of 100 and 500 meters, by taking in and releasing water from its outermost chamber. The nautilus does not live below 500 meters, because the crushing effects of the deep-sea pressure would crack its shell.

People and Underwater Pressure

We are fascinated by the challenge and mystery of the deep ocean. The effects of pressure, however, limit the depths to which humans can descend. The Ama pearl divers of Japan attain the upper limit of human underwater endurance. They can make repeated free dives, without the aid of scuba tanks, down to 18 meters and remain underwater for as long as one minute. With the aid of scuba, however, divers have been able to descend to greater depths and stay down much longer. The current depth record for scuba diving is 132 meters. (The record free dive of 104 meters was made by Jacques Mayol in 1983.)

Scuba diving has opened up the underwater world to a variety of human activities. Scuba divers (using special gas mixtures for deep dives) carry out scientific research on the ocean floor, do salvage work on sunken ships, make repairs to ships' hulls, and install offshore oil rigs. Recreational scuba diving is also a rapidly growing industry. However, scuba diving is not without its risks. Exposure to underwater pressure can lead to various injuries.

Barotrauma

Any diving injury associated with pressure is called a **barotrauma**. There are three kinds of barotrauma: injuries occurring on descent, injuries occurring on ascent, and nitrogen narcosis.

Injuries on Descent: As soon as a diver goes underwater, ambient pressure exerts a force over the entire surface of his or her body. The body's thin membranes are the first to feel the effects of pressure. The eyes and the eardrums are pushed slightly inward. The sinuses, which are membrane-covered cavities in the bones of the face and forehead, also feel the effects of pressure. As the diver descends, pressure increases on these membranes, which may cause discomfort or pain. Pain in the ear is called ear squeeze, and pain in the forehead is called sinus squeeze.

The pain can be eliminated by relieving (equalizing) the pressure by blowing through the nose while keeping the nostrils closed. When the discomfort is eliminated, the diver can continue to descend. If ear or sinus squeeze recurs, the diver can ascend slightly and clear the sinuses again by blowing through the nose. Ear squeeze and sinus squeeze are two examples of barotrauma that can occur to both scuba divers and snorkelers.

Injuries on Ascent: Coming up from the bottom too quickly can produce a serious injury to scuba divers called the **bends**. When scuba divers breathe air under pressure, the gases dissolve in their blood at that pressure. If a diver ascends too quickly, there is a sudden decrease in pressure. This decrease, called **decompression**, can cause gases to come out of solution and form small bubbles in the blood—similar to the way bubbles appear in a bottle of soda when it is opened. The gas bubbles can travel to tissues and joints, causing the diver to bend over in pain (hence the name "the bends"). The bends is an example of a decompression illness; if severe, it can cripple or kill a diver.

Another dangerous effect of decompression illness occurs when a gas bubble, or **air embolism**, in the blood blocks a blood vessel in an important organ such as the brain. An air embolism can occur if a scuba diver ascends too quickly and mistakenly holds his or her breath during ascent. As in the bends, when the diver rises to the surface, the air inside the lungs expands as ambient pressure on the diver decreases. If the diver doesn't exhale sufficiently while ascending, the air in the lungs may rupture through the air sacs and pass into the bloodstream. Air bubbles in the blood can block circulation, cause fainting and paralysis, or even cause death.

To prevent decompression illness, scuba divers must always breathe normally while ascending, and the rate of ascent should be about 10 meters per minute to allow enough time for the dissolved gases in the bloodstream to be exhaled. Decompression illness does not happen to people who are snorkeling (that is, skin divers), because they are not breathing compressed air (air that is under pressure).

Decompression illness can be treated by placing the afflicted diver in a decompression chamber, which is made with thick steel walls. In the decompression chamber (also called recompression chamber), air pressure is first increased to redissolve the bubbles inside the person's body. Then the pressure is gradually decreased over a period of several hours, until the dissolved gases slowly come out of solution and are safely exhaled.

Nitrogen Narcosis: Scuba divers who make deep dives below 30 meters may experience what ocean explorer Jacques-Yves Cousteau called "rapture of the depths" or **nitrogen narcosis**, a kind of behavioral effect that resembles alcohol intoxication. The diver appears drunk, has difficulty concentrating, and is not able to carry out simple tasks. This confused state can pose a threat to the diver's safety. Nitrogen narcosis results from breathing nitrogen gas (N_2) under pressure.

Nitrogen gas, which makes up 78 percent of the air we breathe, is biologically inert when inhaled under normal atmospheric pressure. However, when N_2 is inhaled under pressure from a scuba tank at great depths, it can have a narcotic effect similar to that produced by nitrous oxide (laughing gas), a painkiller that some dentists give to patients. Changing the mixture of gases in the scuba tank—by removing nitrogen and adding helium, which is also biologically inert—reduces the incidence of nitrogen narcosis. This step also increases the bottom time (the amount of time a diver can stay underwater) for divers who work at depths below 40 meters.

18.3 SECTION REVIEW

1. What is the ambient pressure on a diver at a depth of 70 meters? Show your calculations.

2. Why does decompression illness occur among scuba divers but not among snorkelers?

3. What causes the bends and how can a diver prevent its occurrence?

18.4 OSMOTIC PRESSURE AND AQUATIC ADAPTATIONS

In addition to hydrostatic pressure, the water balance of an organism affects its survival. A sea star placed in a freshwater tank would die. And a goldfish placed in a marine tank would die, too. Most saltwater animals such as the sea star cannot live in freshwater. And most freshwater animals such as the goldfish cannot live in salt water. However, some aquatic organisms such as the salmon can, at different stages in their lives, live in both types of water.

Osmoregulation in Aquatic Animals

The ability of aquatic organisms to maintain a proper water balance within their bodies (in either salt water or freshwater) is called **osmoregulation**. Osmoregulation is related to the process of osmosis. As you learned in Chapter 6, osmosis is the movement of water molecules from an area of high concentration to an area of low concentration through a semipermeable membrane. If a sea star were placed in freshwater, the water molecules would move from where they are more concentrated (outside the sea star) to where they are less concentrated (inside the sea star). (See Figure 18-7.)

In freshwater, the water molecules are more concentrated outside the sea star, because the sea star contains dissolved salts within its cells that take the place of water molecules. The sea star is unable to eliminate the excess water that enters due to osmosis. The increased water pressure, or **osmotic pressure**, inside the sea star upsets cell function and causes death. The sea star is unable to adjust to waters of very different salinities and so is considered to be a poor osmoregulator. When the sea star is in its normal saltwater environment, it can regulate its osmotic pressure because the salt concentration of the sea star's body is closer to that of its external environment.

Figure 18-7 The concentration of water molecules is higher outside the sea star (than in the freshwater), so inward osmosis occurs.

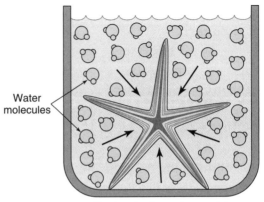

Water molecules

Freshwater (inward osmosis)

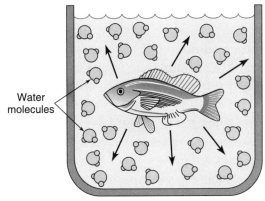

Figure 18-8 The concentration of water molecules is higher inside the goldfish (than in the salt water), so outward osmosis occurs.

Water molecules

Salt water (outward osmosis)

However, salinity changes may occur in the sea star's natural environment. In 1982, along the shores of the Gulf of California (a body of water that contains higher-than-normal salinity), the sea star population suddenly declined. Marine biologists discovered that the drop in the number of sea stars coincided with heavy rainfalls, which were unusual for this dry coastal region. Evidently, freshwater runoff from the land lowered the gulf's salinity, causing some sea stars to die and others to move to deeper, more saline waters.

The goldfish is also a poor osmoregulator when placed in a salt-water environment. (See Figure 18-8.) If a goldfish were surrounded by seawater, the concentration of water molecules would be greater inside the fish than outside, because the salt outside the fish takes the place of water molecules. Since the osmotic pressure is greater inside the fish than outside, water would leave the fish by osmosis through its gill membranes. The goldfish, unable to compensate for the water loss, would die of dehydration. In its normal freshwater environment, the goldfish can regulate its osmotic pressure—its kidneys remove excess incoming freshwater.

Osmoregulation in the Salmon

The salmon is a good osmoregulator because it can adjust to aquatic environments that vary greatly in salinity. The salmon is a migratory fish. During its life cycle, it travels from a freshwater river to the ocean and back again to spawn. Salmon are born in rivers; they swim downstream to the ocean where they spend several years maturing into adults. When the salmon are in the ocean, the salinity of their body tissues is 18 parts per thousand (ppt), while the surrounding ocean water has a salinity of 35 ppt. Since there is an imbalance in salinity, the concentration of water molecules is greater inside than outside the fish. As a consequence, outward osmosis occurs and water leaves the fish through the gill membranes. To counter this water loss, the salmon drinks seawater. To maintain a proper osmotic balance, the salmon excretes excess salt from its gills and also produces salty urine.

When mature salmon swim upstream to spawn, they encounter a salinity near zero ppt, while the salinity of their body tissues is still 18 ppt. Since the salinity is greater inside than outside the fish, the concentration of water molecules is greater outside than inside the fish. This difference in the concentration of water molecules causes water to enter the fish by inward osmosis. To counter this intake of excess freshwater, the salmon excretes water in the form of a dilute urine. The salmon is a good osmoregulator because it is capable of adjusting to large differences in salinity. However, osmoregulation in the salmon is a gradual process of adjusting to waters of different salinities. This process occurs over a period of weeks or months as the fish migrates between ocean and river.

18.4 SECTION REVIEW

1. Why would a sea star die if placed in freshwater?
2. Why can't a goldfish adapt to a marine environment?
3. How does the salmon function as a good osmoregulator?

Laboratory Investigation 18

Effects of Temperature and Salinity on Water Density

PROBLEM: How do temperature and salinity affect the density of ocean water?

SKILL: Graphing scientific data.

MATERIALS: Temperature-Salinity Diagram (Figure 18-9), ruler, pencil.

PROCEDURE

1. Two seawater samples, labeled A and B, were taken and tested for temperature and salinity. The results were plotted as two dots, A and B, on the Temperature-Salinity Diagram (Figure 18-9). Find the temperature and salinity values for A and B and record them in a copy of Table 18-3 in your notebook. (See page 462.) Record the density for each sample in your table.

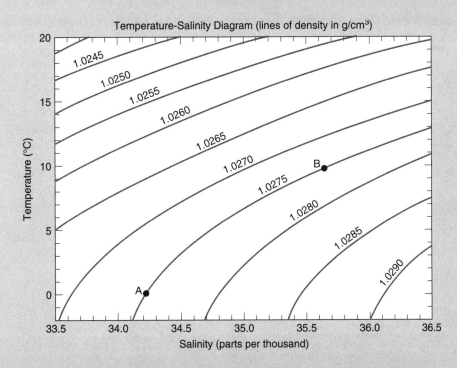

Figure 18-9 Both temperature and salinity have an effect on water density.

TABLE 18-3 TEMPERATURE/SALINITY/DENSITY OF WATER SAMPLES

Sample	Temperature (°C)	Salinity (parts per thousand)	Density (g/cm³)
A			
B			
C			

2. Record in the water sample table the temperature and salinity of water sample C that would result if equal volumes of samples A and B were mixed together. (*Hint:* Mixing one liter of 10°C water with one liter of 30°C water results in two liters of water at 20°C.)

3. Plot the new sample C by placing a dot on the Temperature-Salinity Diagram. Next, record in the water sample table the density of sample C.

4. On the Temperature-Salinity Diagram, draw a straight line between the points representing samples A and B. The point representing any possible mixture of these seawater samples, including sample C, would fall somewhere on this straight line.

OBSERVATIONS AND ANALYSES

1. How does an increase or decrease in temperature affect density?

2. How does an increase or decrease in salinity affect density?

3. Does sample C have a density that is equal to, less than, or greater than the densities of sample A and sample B prior to mixing?

4. Which water samples would sink and which would float above the others?

Chapter 18 | Review

Answer the following questions on a separate sheet of paper.

Vocabulary

The following list contains all the boldface terms in this chapter.

air embolism, ambient pressure, barotrauma, bathythermograph, bends, decompression, heat of fusion, heat of vaporization, hydrostatic pressure, hypothermia, metabolic activity, nitrogen narcosis, osmoregulation, osmotic pressure, thermocline

Fill In

Use one of the vocabulary terms listed above to complete each sentence.

1. Pressure exerted by the density of water's mass is _____.

2. Scientists use a _____ to measure ocean temperatures.

3. The total pressure on a diver is called the _____.

4. The bends is a type of _____ that can cripple a diver.

5. The boundary between warm and cold water is the _____.

Think and Write

Use the information in this chapter to respond to these items.

6. Describe some adaptations of ocean life to cold water.

7. How do dolphins adjust to changes in hydrostatic pressure?

8. Explain why the salmon needs to be a good osmoregulator.

Inquiry

Base your answers to questions 9 through 11 on the following experiment and on your knowledge of marine science.

A marine science student hypothesized that an increase in water temperature would cause an increase in cardiac activity in the shore shrimp. The heartbeat of the shrimp, as measured in beats per minute, was observed at room temperature (20°C) for the control group and at higher and lower temperatures for the experimental groups. The results are shown in the table on page 464.

Trial	Control Group (at 20°C)	Experimental Group (at 28°C)	Experimental Group (at 10°C)
1	116	235	142
2	174	396	114
3	114	377	100
4	140	300	144
5	216	276	123
6	138	285	106
Total	898	1869	729
Average	150	312	122

9. The results of this experiment show that   *a.* as water temperature decreases, cardiac activity remains the same *b.* as temperature increases, cardiac activity increases *c.* as temperature decreases, cardiac activity increases *d.* as temperature increases, cardiac activity decreases.

10. Which statement represents a valid conclusion that can be drawn from this experiment?   *a.* The student's hypothesis is supported by the data.   *b.* The hypothesis is not supported by the data.   *c.* The hypothesis could not be tested. *d.* There are insufficient data to draw any conclusion.

11. Which is an accurate statement regarding the data in the table? *a.* Temperature does not affect heartbeat rate in the shore shrimp.   *b.* The trial with the most cardiac activity occurred in the control group.   *c.* The trial with the least cardiac activity occurred in the control group.   *d.* The trial with the least cardiac activity occurred in an experimental group.

Multiple Choice

Choose the response that best completes the sentence or answers the question.

12. Aquatic organisms maintain a proper water balance by means of   *a.* hypothermia   *b.* metabolic activity *c.* osmoregulation   *d.* decompression.

13. Snorkelers do not get air embolisms on ascent because they *a.* do not breathe compressed air   *b.* hold their breath *c.* do not dive deep   *d.* come up too fast.

Base your answers to questions 14 through 16 on the graph in Figure 18-6 on page 454.

14. According to the graph, you can conclude that   *a.* as depth increases, pressure also increases   *b.* as depth increases, pressure decreases   *c.* as depth decreases, pressure increases   *d.* as depth increases, pressure remains the same.

15. According to the graph, what is the pressure in atmospheres at a depth of 50 meters?   *a.* 4   *b.* 5   *c.* 6   *d.* 7

16. There is a pressure of one atmosphere at zero meters' depth because   *a.* water exerts pressure on the atmosphere   *b.* air exerts pressure on the water surface   *c.* water pressure is pushing upward   *d.* air and water pressure cancel out.

17. In this diagram, the concentration of water molecules is higher inside the fish than outside the fish, so you could expect the occurrence of   *a.* inward diffusion   *b.* outward osmosis   *c.* inward osmosis   *d.* hypothermia.

18. Descending in the water while snorkeling or using scuba gear may cause   *a.* sinus squeeze   *b.* decompression illness   *c.* the bends   *d.* an air embolism.

19. A recreational scuba diver normally breathes   *a.* compressed air   *b.* pure oxygen   *c.* a special mixture of gases   *d.* air at atmospheric pressure.

20. Hypothermia is more likely to occur in humans than in marine mammals because humans   *a.* have less hair   *b.* are warm-blooded   *c.* are cold-blooded   *d.* have no blubber.

21. Unlike icebergs, sea ice   *a.* comes from glaciers   *b.* is formed when ocean water freezes   *c.* cannot be smashed by icebreakers   *d.* is not a menace to navigation.

22. Which statement is true?   *a.* As depth increases, temperature increases.   *b.* As depth decreases, temperature decreases.   *c.* As depth increases, temperature decreases.   *d.* As depth increases, temperature remains the same.

Research/Activity

Report on a marine organism that lives under extreme conditions of temperature and/or pressure. Use the Internet to find data on a fish or invertebrate that lives in the deep ocean or polar seas.

19 Light and Sound in the Sea

When you have completed this chapter, you should be able to:

DESCRIBE the properties of light, particularly in the ocean.

DISCUSS the importance of light and color to marine organisms.

DESCRIBE the properties of sound, particularly in the ocean.

DISCUSS the importance of sound transmission to marine animals.

The ocean pulsates with diverse forms of energy. Light and sound are two types of energy that make the ocean come alive. The deep-blue color of the water, the absorption of solar energy by phytoplankton, and light production by bioluminescent organisms are examples of how light energy has an impact on both the physical setting and living environment of the ocean. In addition, the crashing of waves, clicks and grunts of fish, and songs of whales are some of the sounds that are heard in the sea.

Light makes possible the very existence of most ocean life, since photosynthesizing plankton make up the foundation of major oceanic food chains. Within marine communities, many animals rely on sound as a means of gathering information about their environment and communicating with one another. In this chapter, you will continue your study of energy in the ocean by examining the functions and effects of light and sound in the sea.

19.1 LIGHT AND WATER

You will recall that light is a form of radiant energy that comes from the sun. (Light energy is also produced when fuels are burned or when electrons pass through the filament in a lightbulb.) All forms of solar radiation, including visible light, make up what is known as the **electromagnetic spectrum**, shown in Figure 19-1. This spectrum ranges from radiations with long wavelengths to those with short wavelengths. The **wavelength** of each type of radiant energy is the length of one complete wave cycle. The **frequency** of a particular radiation is the number of wavelengths (wave cycles) per second. The higher the frequency of a wavelength, the greater is the energy of that wavelength. Longer wavelengths have lower frequencies (and less energy), while shorter wavelengths have higher frequencies (and more energy).

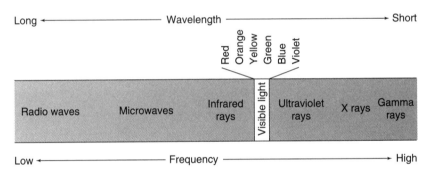

Figure 19-1 The electromagnetic spectrum (radiant energy).

Behavior of Light

Depending on atmospheric conditions, about 20 to 50 percent of the sunlight that reaches Earth either bounces off the atmosphere back into space or is absorbed by the atmosphere. When light bounces off a surface, it is called **reflected light**. When light is taken up by a substance, it is called **absorbed light**.

The remaining sunlight, which reaches Earth's surface, passes through the atmosphere. When light passes through a substance, it is called **transmitted light**. Depending on an area's physical characteristics, this transmitted light is either reflected or absorbed by

Earth's land and ocean surfaces. Thus, light rays that reach Earth undergo transmittance, reflection, and/or absorption.

How deep does light penetrate into the ocean? The area of light penetration, known as the *photic zone*, ranges from about 50 to more than 100 meters in depth, depending on how clear or turbid (murky) the waters are. The photic zone is greatest at the equator because light that strikes Earth at the equator is most direct and therefore the most penetrating. Above and below the equator, light strikes Earth at an angle that is less than 90 degrees; the light is less direct and does not travel as deep into the ocean. The greater the angle at which light strikes, the deeper the photic zone is. In general, most light energy is absorbed within the top ten meters of the ocean. In fact, more than 50 percent of the transmitted light is absorbed and changed into heat within the first meter.

Refraction of Light

When light enters water at an angle of less than 90 degrees, it does not continue in a straight path; it bends. The bending of light as it passes through substances of different densities (at an angle of less than 90 degrees) is called **refraction**. The refraction of light results from a change in the light's speed as it passes from one substance into another. Light slows when it enters a more dense substance (where the molecules are closer together) such as water from a less dense substance such as air. Note that when a beam of light passes from air into water perpendicular to the surface, no refraction occurs, since all parts of the wave front enter at the same time. However, when the beam enters water at an angle of less than 90 degrees, it bends toward an imaginary line called the *normal,* which is perpendicular to the surface. And when the light beam passes from water back into air, it bends away from the normal. The Law of Refraction states that light bends toward or away from the normal as it passes between substances of different densities at an angle of less than 90 degrees.

Refraction is responsible for a variety of optical phenomena in water. The pencil in Figure 19-2 looks broken because light reflected from the pencil underwater is bent as it passes from water into air. Should the person in Figure 19-3 aim the spear right at the fish, in front of the fish, or behind the fish? Because of refraction, the per-

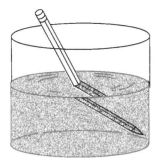

Figure 19-2 Refraction of light makes the pencil look broken.

Apparent position

True position

Figure 19-3 Refraction causes the person to see the fish in its apparent position, not its true position.

son sees the fish in its apparent position. To spear the fish, the person should aim at the true position of the fish, which is in front of its apparent position.

Color of the Ocean

Refraction also produces rainbows. A fine spray of water from a hose on a sunny day will break up sunlight into the colors of the visible spectrum, forming a rainbow. A rainbow can also be produced by passing a beam of light through a piece of glass called a *prism*. (See Figure 19-4 on page 470.) Each color represents a different wavelength of visible light, ranging from the long-wavelength red to the short-wavelength violet. When all the colors are combined, you see white light.

The color of any object is the color of the wavelength of light that is reflected from that object back to the observer's eyes. The ocean usually appears blue (or blue-green) because the blue wavelength is scattered by the water and reflected back to the observer. Blue light has a shorter wavelength and more energy (higher frequency) than

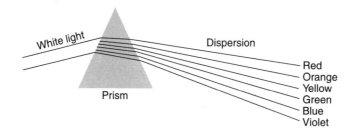

Figure 19-4 A prism breaks light into its component wavelengths.

the other colors and therefore can penetrate water more deeply—giving new meaning to the expression "deep blue sea"! The colors that have longer wavelengths and less energy (lower frequencies)—such as red, yellow, and, to a lesser extent, green—are absorbed nearer the surface. (Violet, which has the shortest wavelength, is absorbed closer to the surface than blue light.) Of course, the ocean is not always blue. It may be green, brown, and even red, depending on the kinds of substances that are suspended in the water, which also can affect light penetration. Clouds and the reflection of the sky's color also affect ocean color. See Table 19-1 for the range of depths at which light absorption occurs.

TABLE 19-1 ABSORPTION OF LIGHT BY OCEAN WATER

Color (wavelength)	Depth Absorbed (meters)
Red	5–10
Orange	10–15
Yellow	15–25
Green	30–50
Blue	60–100
Violet	10–30

19.1 SECTION REVIEW

1. What happens to sunlight that is transmitted through Earth's atmosphere to its surface?

2. What is the difference between reflection and refraction of light?

3. The anchor line on a boat appears broken when viewed from a dock. How can you explain this optical phenomenon?

19.2 LIGHT AND LIFE IN THE SEA

As a general rule, most life needs light to survive. In the ocean, algae absorb the sun's radiant energy and convert it through photosynthesis into the chemical energy of glucose. Marine animals depend on these algae, which form the base of most food chains—from microscopic zooplankton to large whales and fish.

Vertical Migration

The availability of sunlight in ocean water affects the productivity and abundance of phytoplankton. Several factors can affect the amount of light in the ocean—water depth, clarity, the time of year (season), and the time of day. You already know that light penetrates deeper in clear tropical waters than, for example, in turbid northern waters. In addition, depending on the season, waters in high or low latitudes will receive more or less sunlight.

The time of the day has an important effect on the behavior of open ocean organisms in relation to changing light levels. In a phenomenon known as **vertical migration**, many organisms move up and down in the water column each day. During the day, when the light penetrates deeper, numerous species of plankton (including jellyfish and small crustaceans) and small fish congregate in a layer from 200 to 800 meters deep. It is thought that they stay in deeper waters during the day to help avoid predation. Later in the day, these organisms follow the decreasing light up toward the surface. They feed near the surface during the night (often in moonlit waters), and return again to deeper waters at daybreak. Some animals migrate only partway to the surface, stopping at the edge of the thermocline, while others go through the thermocline to feed in the warmer waters at the surface. This vertical migration helps bring nutrients, produced in the sunlit surface waters, to the deeper layers of the ocean.

Bioluminescence

Another interesting phenomenon related to light in the ocean occurs most frequently in the region 200 to 2000 meters deep. Here, below the sunlit waters, are found numerous organisms that produce their own light. As you recall from two earlier chapters, the ability of a living organism to produce its own light is called bioluminescence. Bioluminescence occurs in various species of bacteria, phytoplankton, invertebrates, and fish.

How does an organism produce light? Light is produced in a chemical reaction that takes place inside a living cell. The reaction can be shown as follows:

$$\text{Luciferin} \; + \; \text{Luciferase (enzyme)} \longrightarrow \text{Light}$$
$$(O_2 + H_2O + ATP) \qquad\qquad\qquad (25°C)$$

The protein compound luciferin reacts with the enzyme luciferase in the presence of oxygen and water, at a temperature of about 25°C, to produce light. (Energy from ATP fuels the reaction.)

During the summer, dense concentrations of the dinoflagellate *Noctiluca* may appear on the surface of the ocean and glow in the dark. Their production of light is usually triggered by a disturbance in the water, such as the movement of a wave, a fish, or a boat. Bioluminescence also occurs in larger marine organisms. Shrimp, squid, and fish that inhabit the dimly lit waters of the mid-ocean possess patches of bioluminescent tissue called **photophores**. Photophores are sometimes located on the undersides of fish, which may serve some camouflage purpose. Many species of viperfish and anglerfish that live in deeper waters have bioluminescent lures that glow in the dark, an adaptation for attracting prey. Organisms may also use bioluminescence to see in the dark, attract mates, defend territories, or confuse predators.

The flashlight fish *(Photoblepharun palpebratus),* which lives in tropical waters, produces light from a special organ located below each of its eyes. (See Figure 19-5.) The organ is divided into compartments that contain millions of light-emitting bacteria. The bacteria produce light (in a chemical reaction also involving luciferin and luciferase) when they metabolize glucose during cellular respiration. Some of the energy in the glucose is converted into light. The flashlight fish can "turn" the light on and off by raising and lowering a dark lid located at the base of their light organ. Marine scientists think the flashlight fish use the blinking of light as a means of communication and also for seeing one another in the dark.

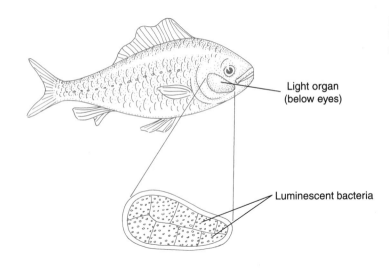

Light organ
(below eyes)

Luminescent bacteria

Color in Ocean Organisms

Some of the most colorful living things on Earth are tropical reef fish. These fish, unlike most fish that school in open waters, wear a rainbow of different bright colors. How can we explain such color in fish? Color is due to the presence of chemical compounds called pigments, found in the cells of the skin. Some common skin pigments are listed in Table 19-2.

Cells that contain pigments are called *chromatophores*; these cells make up a thin layer found either over or under the fish's scales. Each chromatophore contains one type of color pigment. One of the most common skin pigments is **melanin**. Figure 19-6 on page 474 shows two chromatophores, one with the melanin granules branched and one with them contracted. When the granules are branched, the pigment is dispersed throughout the cell, making it look darker; when

TABLE 19-2 COMMON SKIN PIGMENTS

Colors	Pigments
Black and brown	Melanin
Red and yellow	Carotenoids
Blue and green	Protein, carotenoids
White (silver)	Guanine crystals

Light and Sound in the Sea   473

Branched granules
(dark chromatophore)

Contracted granules
(light chromatophore)

Figure 19-6 Fish chromatophores, with the melanin granules dispersed (top) and contracted (bottom).

contracted, the granules are only in the center of the cell, making the cell look lighter. Depending on how these groups of cells are arranged, a fish can look lighter, darker, or blotchy.

What is the adaptive value of having skin cells that can be either light or dark? The flounder has a blotchy skin pattern that matches its background on the seafloor. (Refer to Figure 12-15 on page 298.) This ability to blend in with the color and texture of the background, called *camouflage*, gives a fish such as the flounder an advantage, since it can better conceal itself from enemies. Camouflage of this sort is also seen in some mollusks (for example, the octopus) and crustaceans.

Fixed Color Patterns

Some light-and-dark color patterns in fish are fixed and do not change. The sardine, for example, has darker skin on top and shiny lighter skin below, a pattern called **countershading**. Predatory fish swimming above the sardine have difficulty spotting it, because the sardine's dorsal surface matches the dark background of the depths. Likewise, predators looking up from below have trouble seeing the sardine against the lighter background. The shiny, silvery appearance of many fish, such as the sardine, is due to the presence of pigment cells that contain granules that reflect light, just like tiny mirrors. These cells reflect different colors, too.

Strong patterns of coloration, called color contrast, cause an organism to stand out in contrast to its surroundings. In a marine environment such as a coral reef, where there are so many nooks and crannies and territories to claim, color contrast can help identify an organism so that it can find mates and defend its territory. However, starting at about 500 meters deep—where light is severely limited—the body colors of fish and invertebrates tend to get darker, ranging from deep red and purple to brown, gray, and black. In addition, below 2500 meters deep (where fewer creatures exhibit bioluminescence), eyesight is diminished in some creatures.

19.2 SECTION REVIEW

1. Describe vertical migration of marine organisms. What controls it and how does it benefit the animals that do it?

2. Compare and contrast a photophore and a chromatophore.

3. Explain how the ability to change color helps a fish to survive.

19.3 SOUND IN THE SEA

The undersea environment was once thought to be silent. However, scientists have come to realize that there is an abundance of sound underwater—from the splashing of waves overhead to the noises that are generated by living things carrying out their life functions. In fact, sound travels faster in water than in air. However, before learning specifically about sound in the sea, you need to know what sound is and how it travels.

Understanding Sound

Do you really hear the sound of ocean waves when you hold a coiled seashell up to your ear? It sounds like the sea, but what you hear is actually the sound of air molecules vibrating inside the coil of the shell. The first requirement for sound to occur is that the object producing a sound must vibrate, like a guitar string when it is plucked. A vibration is a rapid back-and-forth motion. When an object such as a tuning fork is struck, it vibrates back and forth. When the object vibrates, air molecules that are in contact with it are squeezed closer, forming **compressions**. The sound wave that results from the movement of air molecules is a compression (or pressure) wave made up of cycles of compressed and expanded air. The parts of the wave in which the air molecules are farther apart are called **rarefactions**. (See Figure 19-7 on page 476.) As the tuning fork continues to be struck, more compression waves are sent out. Unlike light, which moves in transverse waves, perpendicular to the direction of travel, sound moves in longitudinal waves, parallel to the direction of travel.

Scientists can record sounds on an instrument called an *oscilloscope*, which is a kind of TV monitor. Sound is recorded as sound waves, which look like a series of hills and valleys. The highest point of a sound wave is called the **crest**, which corresponds to the

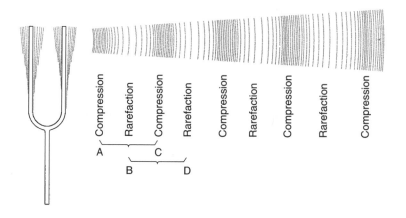

Figure 19-7 Features of a sound wave: it has cycles of compressed and expanded air.

compression part of the wave. The lowest point of a sound wave is called the **trough**, which corresponds to the rarefaction, or expanded, part of the wave. One of the characteristics of a sound is its loudness (or volume), which is represented by the height, or **amplitude**, of the wave. The amplitude is the vertical distance from the midpoint of the wave to the crest. Loud sounds have high amplitude; quiet sounds have low amplitude. (See Figure 19-8.)

Another characteristic of a sound wave is its wavelength. The wavelength is the distance from the crest of one wave to the crest of the next wave (that is, from one wave cycle to the next). Wavelength is related to frequency, which is the number of crests that pass a given point per second. Frequency is measured in cycles per second (cps) or *hertz* (Hz), named after Heinrich Hertz (1857–1894), a German physicist. One cycle per second is equal to one hertz. High-frequency waves have short wavelengths and high-pitch

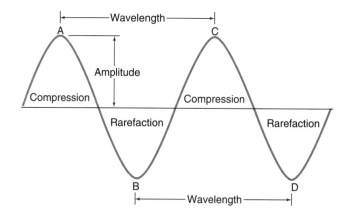

Figure 19-8 The characteristics of a sound wave include amplitude and wavelength.

TECHNOLOGY

Sound Surveillance in the Sea

The ocean was once thought to be a quiet environment. But it is now known that sounds abound in the sea. Many physical events produce sounds in the ocean: icebergs crack, undersea volcanoes rumble, waves pound on the shore, and rain patters on the sea surface. In addition, marine animals such as fish, whales, and dolphins are known to produce a variety of sounds. For instance, beluga whales (like the one shown at right) are sometimes called "canaries of the sea" because of their broad repertoire of whistles and trills. Add to these natural sounds the noises that are produced by humans in boating and shipping activities, shoreline construction, and deep-sea drilling, and you can begin to see that the ocean is a very noisy place.

Many of these sounds can now be detected hundreds and even thousands of km away from their source through the use of a formerly secret sound system called *Sosus* (*So*und *Su*rveillance *S*ystem). Sosus consists of a global network of underwater microphones that were installed by the U.S. Navy to monitor enemy ships and submarines during the Cold War. Elements of Sosus have now been made available to research groups as part of the so-called peace dividend made possible by the end of the Cold War.

Now, civilian scientists using Sosus have been able to identify the voiceprints of minke whales, finbacks, humpbacks, and blue whales. The system was even used to track the move-

ments of a single blue whale for 43 days, over a distance of 3700 km, as it swam in a huge circle in the waters near Bermuda. Environmental scientists use Sosus to track the whereabouts of right whales, the most endangered of the whale species, in order to protect them from collisions with ships.

This sophisticated acoustical recording system can detect nuclear blasts underwater and can monitor nuclear testing. NOAA scientists are using Sosus to identify ships that use drift nets, which have been banned by global agreement. Seaquakes also have been detected and monitored by the use of Sosus—an ability that may prove to be important for the safety of coastal and island populations.

QUESTIONS

1. What are some factors that cause the ocean to be a noisy environment?

2. Describe how Sosus works. What was its original purpose?

3. Give two examples of how Sosus is used to benefit marine life and/or people.

TABLE 19-3 SPEED OF SOUND THROUGH DIFFERENT SUBSTANCES (M/SEC AT 25°C)

Substance	State or Phase	Speed
Iron	Solid	5200
Glass	Solid	4540
Water	Liquid	1497
Air	Gas	346

sounds, while low-frequency waves have long wavelengths and low-pitch (deep) sounds.

The second condition needed for sound to occur is that the compression wave be transmitted through a substance such as air, metal, wood, or water. Unlike light, sound cannot travel through empty space, such as the vacuum of outer space. It has to be transmitted through the movement of molecules. Table 19-3 lists the speed of sound waves through different substances. Sound moves faster in water than in air because the molecules are closer together in water. (Recall from the section on refraction that water is denser than air.) The speed of sound in water ranges between 1400 to 1550 meters per second (m/s), whereas the speed of sound in air is about 345 m/s. Pressure, temperature, and salinity all affect how fast sound moves in water. A change in any of these factors causes a change in the speed at which sound is transmitted.

The final condition required for sound to occur is that the waves reach a sound receptor such as an ear to produce the sensation of sound, or sound reception. To summarize, the three conditions required for sound to occur are: vibration, transmission, and reception. Based on your understanding of sound, how would you answer this question: Does a crashing wave produce a sound if no one is around to hear it?

Sound, Echolocation, and Sonar

Marine biologists study the sounds produced by whales, seals, and dolphins in order to understand how they communicate. Scientists assume that these animals can hear the same range of sounds (frequencies) that they produce. Dolphins and porpoises usually produce high-frequency sounds, such as squeaks and whistles, whereas the larger whales produce a wider range of mostly low-frequency sounds. Researchers have found that there is a "sound channel" in the ocean—about 300 to 600 meters deep—through which whales communicate by means of low-frequency sounds, which can travel for hundreds of kilometers. Many fish produce sounds to communicate with members of their own species, too. Some fish make sounds with their teeth, others with their swim bladders. In fact, the species of fish called "grunts" are so named because that is the sound they produce! In the depths of the ocean, where there is little or no light, marine animals use sound to communicate, to navigate, to locate prey, and to avoid predators.

The use of sound for navigation and for locating objects underwater is called *echolocation*. When marine mammals echolocate, they send out sound waves and receive the returning sound waves, called *echoes*, which bounce off objects. Echoes are actually reflected sound. In dolphins, the outgoing sound consists of a series of clicks that are produced by moving air within the animal's nasal passages, directed outward through the fatty melon in the forehead. Marine scientists think the melon focuses the sound waves and then sends them in specific directions. After the outgoing sound wave hits its target, the wave bounces back and returns to the dolphin as an echo. The dolphin has no external ear. Dolphins "hear" sound waves through their melon and through their lower jaw. Nerve receptors in the lower jaw receive the echo and convert it into electrical signals that are sent to the brain, where the signals are then interpreted.

Scientists have discovered that marine mammals often produce low-frequency sounds when they want to scan the environment and emit high-frequency sounds when they wish to focus on a particular object. As mentioned in Chapter 14, not only can cetaceans produce sounds to communicate with one another, but they can also send out sound waves that are strong enough to stun other animals, thus helping them catch prey and deter enemies.

People have learned to use an artificial form of echolocation, called *sonar*, to locate objects in the ocean environment. By listening to sounds in the sea, people can determine their direction of origin. By actively bouncing sound waves off objects in the water and recording their echoes, people can determine with great precision the location of living and nonliving things. This technology has been used for many decades to locate submarines (especially enemy submarines), whales, fish, and other organisms; to map features of the ocean floor; and even to determine ocean depth (based on the speed of sound in the sea). Unfortunately, the use of powerful low-frequency sonar by the U.S. Navy can have a harmful effect on cetaceans and other marine life. Because whales and dolphins rely on sound for their own communication and activities, they are vulnerable to the effects of intense sonar signals. In 2000, several whales died when they beached themselves in the Bahamas after exposure to Navy sonar; the powerful signal had caused bleeding around their brains and ear bones (possibly due to burst eardrums). The Navy is now required to turn off their sonar equipment if any cetaceans or sea turtles are observed nearby.

19.3 SECTION REVIEW

1. Why does sound travel faster through water than through air?
2. Explain how cetaceans hear even though they lack external ears.
3. How is sonar used to investigate the marine environment?

Laboratory Investigation **19**

Identifying Pigments in Algae

PROBLEM: How can the pigments in marine algae be identified?

SKILL: Using filter-paper chromatography technique.

MATERIALS: Test tube, test tube rack or holder, solvent, filter paper, pin, cork stopper, chlorophyll extract (from *Ulva*), dissecting needle.

PROCEDURE

1. Pour about 10 mL of solvent into a test tube. Cut a piece of filter paper. Use a pin to attach the top end of it to a cork stopper. (See Figure 19-9.)

2. Dip the tip of a dissecting needle into the chlorophyll extract. Apply the extract to the filter paper at its bottom end. Wait one minute for it to dry; then repeat twice.

3. After the chlorophyll spot dries, insert the filter paper (with cork) into the test tube so that the bottom end just touches the solvent. Within 10 minutes you should see bands of different colors appearing on the paper, as the solvent migrates to the top of it.

4. Remove the filter paper and observe the colored bands. Carotene (the fastest-moving pigment molecule) should be at the top. What are the positions of the other bands? Write a report indicating your findings; attach the filter paper to your report.

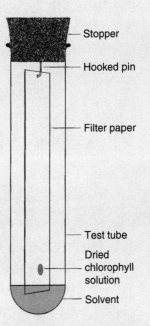

Figure 19-9 Identifying pigments in *Ulva* extract.

Labels: Stopper, Hooked pin, Filter paper, Test tube, Dried chlorophyll solution, Solvent

OBSERVATIONS AND ANALYSES

1. Identify the pigments you found in your extract from *Ulva*.

2. Why do the pigments separate into different-colored bands?

3. Which pigment had the fastest migration through the paper? Which had the slowest?

Chapter 19 Review

Answer the following questions on a separate sheet of paper.

Vocabulary

The following list contains all the boldface terms in this chapter.

absorbed light, amplitude, compressions, countershading, crest, electromagnetic spectrum, frequency, melanin, photophores, rarefactions, reflected light, refraction, transmitted light, trough, vertical migration, wavelength

Fill In

Use one of the vocabulary terms listed above to complete each sentence.

1. Some animals have bioluminescent tissue called _____.
2. All forms of solar radiation make up the _____.
3. Volume of a sound is represented by the _____ of its wave.
4. The number of wave cycles per second is the _____.
5. Bending of light as it passes through different substances is _____.

Think and Write

Use the information in this chapter to respond to these items.

6. Explain why blue light penetrates more deeply into the ocean than other colors of light do.
7. How do marine creatures produce bioluminescence? What are some benefits to the organisms?
8. How does a marine mammal use sound to locate and capture prey?

Inquiry

Base your answers to questions 9 through 11 on the graph on page 483, which represents the frequency ranges of sounds (vocalizations) made by humans and by six species of whales, and on your knowledge of marine science.

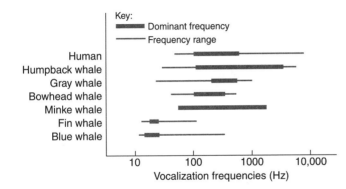

Key:
- Dominant frequency
- Frequency range

9. According to the graph, which two whales use the broadest range of frequencies in their vocalizations?   *a.* minke and gray   *b.* gray and bowhead   *c.* fin and blue   *d.* humpback and minke

10. The two species of whales that produce the deepest (most low-pitched) sounds are the   *a.* gray and bowhead   *b.* humpback and minke   *c.* fin and blue   *d.* humpback and bowhead.

11. The dominant frequency range for human vocalizations is closest to that of the   *a.* humpback whale   *b.* bowhead whale   *c.* minke whale   *d.* gray whale.

Multiple Choice

Choose the response that best completes the sentence or answers the question.

12. In which of the following organisms does the chemical reaction "luciferin + luciferase ⟶ bioluminescence" *not* take place?   *a.* bacteria   *b.* phytoplankton   *c.* dolphins   *d.* deep-sea fish

13. The daily up-and-down movements of organisms in the ocean's water column is called   *a.* compression   *b.* refraction   *c.* countershading   *d.* vertical migration.

14. The ocean usually looks blue because   *a.* blue light is absorbed near the surface   *b.* blue light is reflected from the water   *c.* blue light has a low frequency   *d.* the sky is blue.

15. The position of an object underwater looks different than its true position due to the behavior of light called   *a.* reflection   *b.* refraction   *c.* radiation   *d.* absorption.

16. Which of these statements about marine algae is true?
    a. They convert chemical energy into radiant energy.
    b. They convert radiant energy into chemical energy.
    c. They carry out chemosynthesis.
    d. They carry out photosynthesis at night.

17. Which statement is most valid?   *a.* As depth increases, levels of light increase.   *b.* As depth increases, marine algae are more productive.   *c.* As depth decreases, marine algae are more productive.   *d.* There is no relationship between depth and algae production.

18. The penetration of light into the sea depends on such factors as water   *a.* depth, pressure, and temperature   *b.* depth, clarity, and the time of day   *c.* clarity, pressure, and salinity   *d.* depth, pressure, and time of day.

19. Sound travels about 1460 meters per second in water. How deep is a wreck on the seafloor if it takes about 4 seconds for a sonar signal to return following transmission?   *a.* 2920 meters   *b.* 1460 meters   *c.* 5840 meters   *d.* 365 meters

20. The structure used by dolphins for echolocation is the   *a.* lateral line   *b.* swim bladder   *c.* external ear   *d.* melon.

21. In a dolphin, nerve receptors in which structure convert sound energy into electrical impulses?   *a.* melon   *b.* vocal cords   *c.* lateral line   *d.* lower jaw

22. All the following affect the speed of sound underwater *except*   *a.* temperature   *b.* light   *c.* pressure   *d.* salinity.

23. The porpoise squeak is higher pitched than the whale song because it has a   *a.* longer wavelength   *b.* shorter wavelength   *c.* higher amplitude   *d.* lower amplitude.

Research/Activity

Research the importance of body color or light in fish behavior. Pick a few fish from different marine habitats and make a chart that shows how their behaviors are related to these factors.

20 Tides, Waves, and Currents

When you have finished this chapter, you should be able to:

EXPLAIN the forces that cause the different kinds of tides.

DISCUSS the effects of changing tides on marine organisms.

DESCRIBE wave characteristics and different types of waves.

DISCUSS the causes, and importance, of ocean currents.

The ocean is always in motion. Massive currents of water circulate around the globe. Waves crash on the beach. Seawater is pushed and pulled by the tides.

What makes the ocean water move so much? The tides, waves, and currents result from the interaction of many forces and factors, such as conditions in Earth's atmosphere, the movement of our planet in space, and even Earth's interactions with other bodies in our solar system. In this chapter, you will learn about tides, waves, and currents; and you will see the strong effect they have on life in marine and coastal environments.

20.1 THE TIDES

Did you ever go to the beach on a hot day, drop your towel on the sand near the shore, and run into the water for a cool dip? Then, after swimming for a while, you returned to your towel and found it soaked with seawater. What happened? While you were in the water, you did not realize that the tide was coming in.

The **tide** is the daily rise and fall of the ocean as seen along the shore. When the ocean reaches its highest elevation on the shore, it is called *high tide*. The lowest elevation reached by the ocean along the shore is called *low tide*. The vertical distance between low tides and high tides is called the **tidal range**. The tidal range varies, depending on the shape and depth of the coastline. Along much of the East Coast of North America, the tidal range is about one to two meters high. However, in Canada's Bay of Fundy, a boat docked at a pier may rise as much as 20 meters from low tide to high tide, giving this bay a very high tidal range. (See Figure 20-1.) A particularly dramatic tidal range occurs around the island of Mont-Saint-Michel, which lies less than two km off the coast of France. At low tide, you can walk to the island over the tidal flats (the area between the two tide levels). At high tide, of course, that area is submerged. (See Figure 20-2.)

It is important to know when high tides and low tides occur.

Figure 20-1 A floating dock can adjust to a large tidal range.

For example, the best time to schedule a field trip to observe living creatures along the shore is at low tide. Storms hitting the coast at high tide cause much more damage than they do at low tide. But ships coming into port at low tide have to wait in deeper water for high tide before docking, to prevent their running aground.

Time and Tides

How do you know when it is high tide or low tide? If you live near the ocean, you can check the tide tables published in your local newspaper. Look at the tide table shown in Table 20-1. Pick one of the locations and determine the time difference between high tide and low tide. The time between consecutive high tides and low tides is approximately six hours and 20 minutes. (This time may vary slightly from place to place because of differences in the shape of

TABLE 20-1 TIDE TABLE

Location	High Tide	Low Tide	High Tide	Low Tide
Sandy Hook, N.J.	12:48 A.M.	6:33 A.M.	1:29 P.M.	7:49 P.M.
Port Jefferson, N.Y.	4:36 A.M.	10:56 A.M.	4:59 P.M.	11:19 P.M.
Shinnecock Inlet, N.Y.	1:14 A.M.	7:34 A.M.	1:56 P.M.	8:16 P.M.
Fire Island, N.Y.	1:15 A.M.	7:35 A.M.	1:51 P.M.	8:11 P.M.
Montauk Point, N.Y.	2:26 A.M.	8:46 A.M.	2:52 P.M.	9:12 P.M.
Stamford, Conn.	4:35 A.M.	10:55 A.M.	5:00 P.M.	11:20 P.M.

the coastline.) According to the table, there are two highs and two lows during a 24-hour period. A tide that has two highs and two lows each day is called a *semidiurnal tide*. The East Coast of the United States has a semidiurnal tidal pattern. The Gulf Coast has one high tide and one low tide each day; it is called a *diurnal tide*. The West Coast has a tidal pattern with features of both, called a *mixed tide*; it has two highs and two lows, but the first set of tides is stronger than the second set of tides each day. The U.S. government publishes tide table predictions each year.

Gravity and Tides

What causes high tides and low tides? Look at Figure 20-3, which shows the ocean bulging out on the side of Earth facing the moon and on the side facing away from it. This phenomenon is called a *tidal bulge*, and it represents a high tide. The ocean bulges out because the moon pulls on Earth with a force called *gravity*. The British scientist Sir Isaac Newton (1642–1727) discovered that gravity was a pulling force exerted between any two bodies in space. Earth pulls on the moon and the moon pulls on Earth. Since our planet's mass is much greater than that of the moon, Earth is not pulled very much toward the moon. However, the moon's pull is strong enough to cause the ocean water facing the moon to be pulled toward it, producing the high tide. The smaller high tide on the side of Earth facing away from the moon occurs because the solid part of Earth is pulling away from the ocean (and slightly toward the moon). Just as the two bulges are the high tides, the shallow areas between the bulges are the low tides. Each tidal location is approximately 6¼ hours apart.

The sun also exerts a gravitational pull on Earth. Although the sun is much larger than the moon, its gravitational pull on Earth is

Figure 20-3 High tides and low tides.

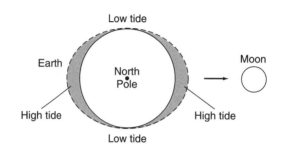

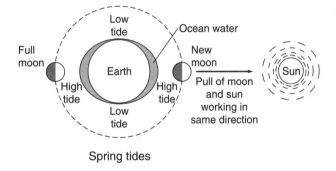

Figure 20-4 Spring tides.

Spring tides

much less than that of the moon, because it is farther away from Earth. However, twice each month, when the sun, moon, and Earth are aligned, the pulling forces are combined to produce the highest and lowest tides, called **spring tides**. (See Figure 20-4.) The spring tides get their name not from that season of the year, but from the old German word *springan*, which means "to rise or leap." Spring tides occur during the new moon and full moon, which are exactly two weeks apart.

When the moon is in first-quarter and third-quarter phases, between the new moon and the full moon, the pull of the moon and pull of the sun on Earth are at right angles, and as a result, the tides are not as high or low as at other times of the month. These weaker tides are called **neap tides**. (See Figure 20-5.)

When is the tide highest during the year? The moon's orbit around Earth is not a perfect circle, but is elliptical. The point at which the moon comes closest to Earth is called the *perigee*, and the point at which it is farthest away is called the *apogee*. The highest tides occur when the moon is at perigee. If the moon is at perigee during the new moon or full moon, the tide is at its highest. In its

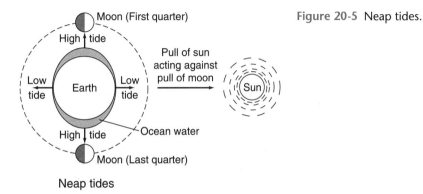

Figure 20-5 Neap tides.

Neap tides

orbit around the sun, Earth moves closest to the sun in January, called *perihelion*, and farthest away in July, called *aphelion*. During perihelion, when Earth is closest to the sun, the tides are higher. The combination of perigee and perihelion produces the very highest of all tides.

Life Cycles and Tides

The incoming tide signals the final chapter in the life cycle of many marine organisms as their remains are washed up on the shore. But the rising tide also heralds the beginning of life for other life-forms. For the grunion *(Leuresthes tenuis),* a small fish (15 cm) that inhabits coastal waters in southern California, life begins at high tide. During the spring and summer, thousands of these silvery fish swim up onto the sandy beaches, carried in by the high tide. This so-called *grunion run* occurs at night during the new moon and full moon when the tide is highest. People flock to the beaches and wait with flashlights for the grunions to appear between 10:00 P.M. and 2:00 A.M. The female grunions wiggle into the sand and lay thousands of eggs as the males deposit sperm around them. Afterward, the fish are swept back into the sea by the water. The spawning is timed so exactly that it occurs only on the second, third, and fourth days that follow a new or full moon.

After the grunion eggs are fertilized, they incubate in the sand for two weeks until the next new or full moon occurs. At that time, the waters of the high spring tides will reach the eggs and wash them out of the sand. The eggs then begin hatching into tiny grunions as they are carried seaward by the outgoing tide.

Another marine animal whose life cycle is timed to the rhythm of the tides is the horseshoe crab *(Limulus polyphemus).* (See Figure 20-6.) During late May and early June, vast numbers of horseshoe crabs congregate in shallow bays, marshes, and inlets along the Atlantic and Gulf coasts, waiting to come up on the beaches to spawn. The signal to begin is provided by the moon. During the new and full moons, when the tide is at its highest, the horseshoe crabs come ashore. They are usually in pairs, with the smaller male attached to the back of the female's abdomen. The female produces a cluster of several hundred tiny pale green eggs on her abdomen. The eggs are fertilized by the male's sperm, and then deposited in a

Figure 20-6 Horseshoe crabs come ashore during a high tide to mate and lay their eggs.

nest hollowed out in the sand by the female. During the next two weeks, the eggs incubate. As with the grunion, two weeks later during a new or full moon, the high water reaches the eggs and they hatch. The outgoing tide carries the hatchlings out to sea.

Tides are important to marine life because they transport nutrients and organisms between the shore and the deeper offshore waters, and they dictate the lives of organisms that are adapted for living in the harsh intertidal zone.

20.1 SECTION REVIEW

1. How do the moon and the sun cause Earth's tides?

2. Why are the spring tides the highest tides each month?

3. Describe how the life cycles of the grunion and the horseshoe crab are timed to the rhythm of the tides.

20.2 OCEAN WAVES

When you blow on a cup of hot tea, you see ripples in the liquid. A *ripple* is a small wave. Nature's "breath," the wind, also produces

Figure 20-7 Wave ripples
along the shore are caused
by the wind.

waves when it blows across the surface of the ocean. (See Figure
20-7.) A **wave** is an up-and-down movement of the ocean surface.
Most waves are caused by wind. How does the wind produce a wave?
When the wind blows, it pushes on the ocean surface, causing the
water to lift. A gentle wind produces a small wave; a strong wind produces a bigger wave. In general, the greater the wind's speed, length
of time it blows, and distance over which it blows, the greater are
the size and speed of the waves that it generates. (See Figure 20-8.)

Figure 20-8 The effect of
wind speed on wave size:
the faster the wind, the
larger the wave.

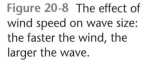

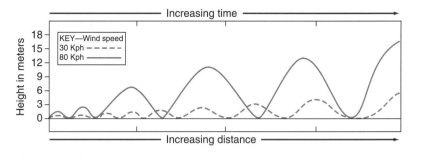

Measuring a Wave

One of the biggest waves ever recorded at sea was observed by someone aboard the U.S. Navy tanker U.S.S. *Ramapo* during a storm in the
South Pacific in February 1933. The wave was calculated to be about

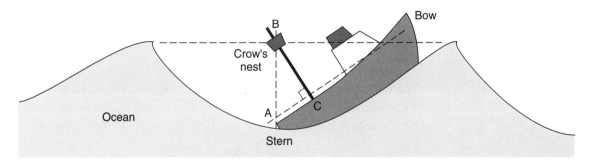

Figure 20-9 You can use the Pythagorean theorem to calculate wave height.

34 meters high. How is the height of an ocean wave calculated? The **wave height** is the vertical distance between the top of a wave, or crest, and the bottom of the preceding wave, or trough. Wave height can be measured (by use of basic geometry) either when the wave is perpendicular to the ship's direction (when the ship's stern is down in the trough) or when the wave is parallel to the ship.

When the stern of a ship is in the trough, an observer on the deck can line up the crow's nest with the crest of the wave, as shown in Figure 20-9. When the reference points are connected by straight lines, they form a right triangle, where BC is the height of the crow's nest, AC is the length from the stern to the base of the mast, and AB is the hypotenuse. The hypotenuse, which is the side of the right triangle opposite the right angle (AB), would equal the height of the wave. You can use the Pythagorean theorem ($a^2 + b^2 = c^2$), where BC is a and AC is b, to calculate the height of side AB, which would be c, as follows:

Given: $BC = 12$ meters, $AC = 5$ meters

Find: side AB

Solution: $a^2 + b^2 = c^2$

$$(12)^2 + (5)^2 = c^2$$

$$144 + 25 = c^2$$

$$169 = c^2, 13 \text{ meters} = c$$

Side $AB = 13$ meters

Wave Action

When a steady wind blows, a wave train is produced. A **wave train** is a series of waves, one followed by the other, moving in the same direction. How fast do the waves move? You can calculate the speed (velocity) of a wave if you know its wavelength and period. The wavelength is the distance between two successive crests or troughs. The period is the time it takes for one wave to pass a given point. (The wave frequency is the number of waves that pass a given point in a given amount of time, such as the number of waves per minute.) Use the following formula to calculate the velocity of a wave:

Velocity (V) = wavelength (W)/period (P)

If W = 10 meters and P = 5 seconds,

V = 10 meters/5 seconds

V = 2 meters/second

After observing waves and calculating their speed, you may be left with the impression that the water, which goes up and down as waves, is also moving horizontally. Although waves look like they are moving along, this is only an illusion. A floating object such as a boat or the cork on a fishing line does not move forward in a wave train but moves up and down with each passing wave. This is because a wave is a form of energy that moves across the water—*not* the water itself moving along.

Wave action is like the snapping of a rope. (See Figure 20-10.) When you snap a rope, the rope itself does not move forward. The movement of your hand produces mechanical energy that is transferred in waves along the length of the rope. Similarly, a wave starts with the energy of the wind pushing on the water. Mechanical energy is transferred to each successive wave. When waves extend beyond the windy area in which they are generated, they have longer periods and more rounded crests, and are called **swells**. The swells may travel for thousands of kilometers across the ocean, until they reach a distant shore where the energy is released in the form of a crashing wave.

Figure 20-10 A vibrating rope illustrates wave action.

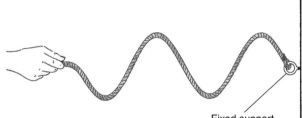

Fixed support

RECREATION
The Science of Surfing

Did you ever try "catching a wave" on a boogie board or surfboard, or by bodysurfing, and then riding it in to the shore? Surfing is such an exhilarating sport because you can feel the power of the wave as it propels you forward through the water. The ability to catch a wave requires being at the right place in front of the crest of the wave before it breaks. But you also need to paddle or swim rapidly in order to catch the wave before it starts breaking.

Waves break on beaches when the bottom of an advancing wave makes contact with the shallow seafloor. Friction results, causing the bottom of the wave to slow and the top of the wave to rear up, pitch forward, and crash. Riding a wave involves the dynamic interplay of three main forces: the upward surge of the wave, the downward pull of gravity (as the wave rises and falls), and the forward momentum of the wave (as it pitches forward). Timing is critical. If you slide down the face of a wave too quickly, you will reach the bottom, or trough, of the wave and the ride will be over. If you are too late in catching a wave, it will outrun you.

Waves are caused by the wind. The big waves that surfers love to ride are spawned kilometers away by strong winds that create waves with long periods and rounded crests, called *swells*. The distance over which a wind blows, called the *fetch*, affects the size of a swell. The longer the fetch is, the bigger the swell will be. When swells reach the shore, they rise up as steep waves and crash, thereby releasing their energy.

Now, due to advances in science and technology, surfers have extra help in trying to chase the big waves. For example, information is available from NASA's *Quik Scat* satellite, which records wind speed across the ocean surface. Offshore buoys measure wave height and wave period; the data gathered are posted on the Internet. Surfers can even taxi out in motorized craft called *Wave Runners,* to help them catch that perfect wave. In addition, thanks to protective wet suits, the "surf's up" year-round—surfers can finally enjoy the excitement of an endless summer of surfing the wave.

QUESTIONS

1. Describe how surfers should catch a wave. What happens if they miss?

2. Define the terms *swell* and *fetch*. How does one affect the other?

3. Explain how advances in science and technology can help surfers.

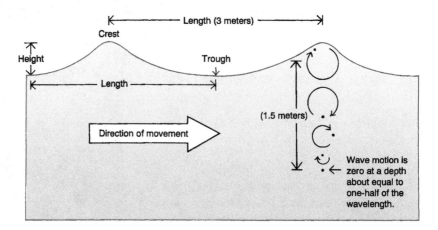

Figure 20-11 The characteristics of a breaking wave.

Length (3 meters)

Crest

Height

Trough

Length

Direction of movement

(1.5 meters)

Wave motion is zero at a depth about equal to one-half of the wavelength.

Breaking Waves and "Tidal Waves"

What causes a wave to crash, or break, on the beach? As a wave approaches the shore, it enters shallow waters. As the bottom of the wave makes contact with the seafloor, the wave slows (due to friction), which decreases its wavelength, too. This occurs when the water depth is about one-half the wave's wavelength. (See Figure 20-11.) When the water depth is less than one-half the wavelength, the top of the wave—which moves faster than its bottom—pitches forward and crashes (See Figure 20-12.) This action produces a type of wave known as a *breaker*. (See Figure 20-13.)

Waves can also break on the open seas. Strong winds produce steep waves with narrow crests. The narrow crests are easily blown off by the winds, creating a mixture of air and water known as a **whitecap**. When a ship's crew spots whitecaps ahead, they know they are in for rough weather. Every now and then, huge lone waves, with very high crests and low troughs, are encountered at sea. These tall waves, known as **rogue waves**, are formed either

Figure 20-12 An illustration of how a breaker forms.

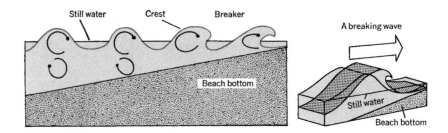

Still water   Crest   Breaker

Beach bottom

A breaking wave

Still water

Beach bottom

Figure 20-13 Breaking waves on a beach.

when two or more large waves from a storm unite, or when waves meet opposing currents. Rogue waves are dangerous and have caused the loss of many ships at sea.

In some rivers, the energy of the incoming tide can create a solitary wave called a **tidal bore**. This occurs where the seafloor at the mouth of the river slopes gently and the tidal range is greater than 5 meters. A strong tidal bore may reach several meters in height and rapidly advance many kilometers upriver. In 1976, a wave 6 meters high surged up the Penobscot River in Maine and flooded the town of Bangor, 25 km upriver from the sea. Tidal bores occur in rivers and estuaries around the world. Dwarfing that of the Penobscot, the tidal bore in the Amazon normally extends hundreds of kilometers upriver.

Tidal bores can also prove fatal. In 1843, the daughter of French novelist and poet Victor Hugo was drowned when a tidal bore in the Seine River capsized her boat; she was memorialized in one of his poems. Fortunately, tidal bores are limited to the relatively small number of rivers that have the unusual combination of a high intertidal range and a gently sloping river mouth.

Tsunamis

On April 24, 1971, a massive wave "attack" struck a chain of islands south of Japan. On one of these islands, a wave 84 meters high was observed—the highest wave ever recorded. One of the waves lifted a massive chunk of coral weighing three-quarters of a million

Figure 20-14 A famous Japanese print showing a tsunami.

kilograms and tossed it inland a distance of 0.8 km. These giant waves, which are often incorrectly called *tidal waves*, are in fact not related to tides at all. The Japanese have a more accurate name for this type of wave—they call it **tsunami** (pronounced soo-NAM-e), which means "harbor" (*tsu*) "wave" (*nami*). (See Figure 20-14.)

What causes a tsunami? Tsunamis are generated by a sudden disturbance in Earth's crust, that is, by seismic activity such as an undersea earthquake, a landslide on the ocean floor, or a volcanic eruption. An earthquake on the seafloor was responsible for another tsunami off Japan in 1993. The point of origin of an earthquake is called the *epicenter*. An underwater earthquake releases a great deal of energy, which is transmitted through the water column. When this energy reaches the ocean surface, it generates high-velocity waves. Some tsunamis have been clocked at more than 800 km per hour. The waves are also characterized by long wavelengths (some being more than 200 km long) and a long period. Contrary to what you might expect, the waves at the epicenter are only a meter or two high. Not until they reach shallow waters do they grow to great heights.

When a tsunami wave train going hundreds of kilometers per hour approaches a shore, its wave speed slows. On slowing, the wave's energy of forward motion is converted into a lifting force that creates a giant wave, as much as 20 to 30 meters high. Just before the wave hits, water along the beach is suddenly sucked away and then the giant wave approaches with a loud noise. After the tsunami breaks on the shore, there is another tremendous rush of water back to the sea. In years past, many people, thinking that the tsunami was over, would go down to the beach to take advantage of the unexpected harvest of stranded fish. To their horror, they discovered another tsunami ready to crash down on them. Many people lost their lives because they did not know that these giant waves often come in a succession of three or more high crests that arrive 15 to 60 minutes apart. A tsunami wave train may strike a coast for the better part of a day before it ends.

Since tsunamis are unexpected and can be so destructive, the U.S. Coast and Geodetic Service has placed seismic recorders at various locations in the Pacific Ocean, where most tsunamis occur (because of the frequency of undersea seismic activity). These recorders can detect disturbances on the seafloor that might cause tsunamis. The information is then relayed to coastal stations and analyzed. Because the Pacific Ocean is so large, this early warning system can give coastal populations enough time to move inland before a tsunami reaches their shores.

20.2 SECTION REVIEW

1. How are waves formed? What determines their size and speed?

2. How are ocean swells different from whitecaps?

3. What is the difference between a rogue wave and a tsunami?

20.3 OCEAN CURRENTS

The continents may be far apart, but highways of moving water called *ocean currents* connect these separate landmasses. If you were to catch one of these currents while onboard a sailboat, you would be able to travel a great distance—as did explorer Thor Heyerdahl on his journey across the Pacific Ocean. (See the feature about Thor Heyerdahl's voyages, on page 12 in Chapter 1.) But it is not necessary to set sail to find ocean currents. The evidence for far-ranging currents can be found on a beach, perhaps in the form of a coconut from distant islands, driftwood from an offshore wreck, or even a bottle with a message in it from a potential pen pal across the sea.

Global Ocean Currents

A **current** is a large mass of continuously moving ocean water. The largest currents that move across the ocean are called *global ocean currents*. These currents are like rivers that travel great distances. Locate, in Figure 20-15, the South Equatorial Current, the global ocean current that carried Thor Heyerdahl across the Pacific Ocean. As you can see, it moves from east to west, just south of the equator.

Locate the Gulf Stream, another global ocean current, in Figure 20-15. As you may recall, the Gulf Stream was first described in detail by the American statesman Benjamin Franklin. You can see that the Gulf Stream flows up from the Gulf of Mexico along the East Coast of the United States and then moves across the Atlantic Ocean. The average surface speed of the Gulf Stream is about 8 km per hour. It is approximately 160 km wide and more than 100 meters deep. The surface temperature of the Gulf Stream is about

Figure 20-15 The major world ocean currents.

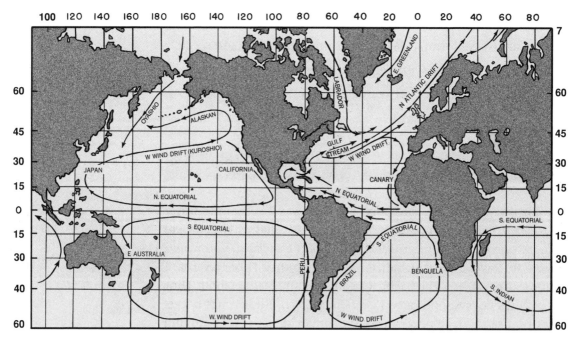

24°C, warm enough to significantly affect the climate of two countries that are along its path, England and Ireland. Where the warm Gulf Stream flows across the Atlantic toward Europe, it becomes known as the North Atlantic Current. On reaching the shores of England and Ireland, the North Atlantic Current warms the coastal water temperature of these two countries by as much as 15°C. This produces a moderate climate of warm summers and not very cold winters. The warm, moist air brought in by the current meets the colder air coming from the north and condenses to produce the rain and fog that are typical of this region.

Another global ocean current is the California Current, shown in Figure 20-15. Compare the California Current with the Gulf Stream. Notice that they move in opposite directions. The Gulf Stream flows from south to north, and the California Current flows north to south. Which one is warmer? The Gulf Stream is warmer because it originates in tropical waters; and the California Current is cold because it comes from the north. The Gulf Stream warms the beaches along the East Coast from Florida to Massachusetts, making it possible to swim comfortably during the summer. However, because of the cold California Current, bathers encounter colder ocean water from Washington to California.

The Coriolis Effect

Notice in Figure 20-15 that the ocean currents are deflected to the east in the northern hemisphere and to the west in the southern hemisphere. What causes ocean currents to move in these directions? This circular drift of the oceans was first studied by the French physicist Gaspard Coriolis (1792–1843) and has come to be known as the *Coriolis effect*. The **Coriolis effect** states that the spinning Earth causes the winds and surface waters to move in a clockwise direction in the northern hemisphere and in a counterclockwise direction in the southern hemisphere. (The winds help drive the movement of the ocean's surface waters.) The continents deflect the ocean currents, causing them to move in giant circles called **gyres**. These wind-driven ocean currents are also called *surface currents*. Such currents are important to marine life because they move the drifting plankton thousands of kilometers across the ocean.

Vertical Ocean Currents

Global ocean currents move horizontally across the ocean's surface. There are also subsurface currents, parts of which may move in a vertical direction. Figure 20-16 shows a profile of the Mediterranean Sea and Atlantic Ocean near Spain. There is a salinity difference between the two bodies of water. The Mediterranean Sea has a higher salinity (about 3.9 percent) than does the Atlantic Ocean (about 3.5 percent). Why is the Mediterranean saltier? The climate in the Mediterranean region is hot and dry; so, the sea, which is enclosed by land, is warmer than the Atlantic and its water evaporates faster. When water evaporates, salt is left behind, which increases salinity. The saltier water in the Mediterranean is denser, so it sinks below the cold waters of the Atlantic Ocean and flows out as a subsurface current. The less salty (and thus less dense) waters of the Atlantic flow into the Mediterranean at the surface. (See Figure 20-17.)

Temperature differences also can produce vertical water currents. The experiment illustrated in Figure 20-18 shows this. Dye is added to a beaker of water. The beaker is heated underneath on one side. The flow of the dye shows that warm water rises and cold water sinks (unless the salinity differences are great, as in the case of the Mediterranean subsurface current discussed above). Cold water sinks because its molecules are closer together, making the water denser.

Figure 20-16 The Gibraltar Current is a subsurface current. During World War II, submarines used this current to drift undetected into the Atlantic Ocean.

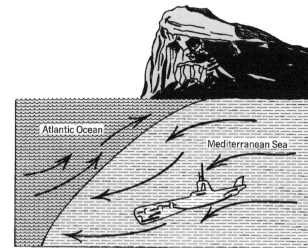

Atlantic Ocean

Mediterranean Sea

Figure 20-17 The current caused by the salinity difference between the Atlantic Ocean and the Mediterranean Sea is visible from space. The less-dense Atlantic surface waters can be seen flowing into the Mediterranean.

Warm water rises because its molecules are in motion and are spaced farther apart, making the water less dense.

Picture the ocean as a giant pool that is heated by the sun. (See Figure 20-19.) At the equator, the water is warmer, so it rises. At the poles, the water is colder, so it sinks. As the warmer water rises at the equator, the colder water from the poles flows in to take its place. This creates giant cycles of flowing water from the poles to the equator. The spreading of heat energy that results from the rising of warm water and the sinking of cold water is an example of a *convection current*. These currents also bring oxygen from the surface waters to the deeper waters. (Recall that convection currents also occur when warm air rises and cool air sinks, and when magma flows within Earth's mantle.)

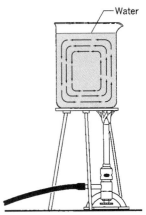

Figure 20-18 Vertical currents can be caused by temperature differences.

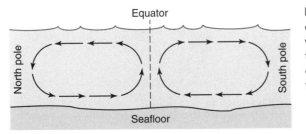

Figure 20-19 Global convection currents: warm water rises at the equator, and cold water sinks at the poles.

Deep Ocean Currents

Warm water rises at the equator, flows until it cools, and sinks at the poles. Cold-water currents (from the poles) that replace the rising currents at the equator flow below the surface. In recent years, scientists have found evidence for deep **countercurrents**, which are slow horizontal ocean currents that flow in a direction generally opposite to the wind-driven currents at the surface.

Not all subsurface currents are slow. One of the fastest types is called a **turbidity current**. Turbidity currents are found along the continental slope, where the seafloor around a continent drops off steeply. *Turbid* means "cloudy," and the cloudiness is due to the presence of silt, mud, and clay in the current as it rushes down a slope like an underwater avalanche. The great speed of a turbidity current, as high as 80 km per hour, is due to the steepness of the slope. Turbidity currents are powerful enough to carve out many V-shaped depressions (canyons) on the floor of a slope. Turbidity currents off the mid-Atlantic coast produced the Hudson and Baltimore canyons.

When vertical currents rise to the surface from the depths, they often contain nutrient-rich sediments from the bottom. The rising of such waters from deep in the ocean is called an **upwelling**. Upwellings are significant because nutrients such as phosphates and nitrates are important for the growth of plankton. And plankton, as you have learned, are an important food source for a variety of marine animals. Areas of significant coastal upwellings make excellent fishing grounds. (See Figure 20-20.) Peru has traditionally been

Figure 20-20 Nutrient-rich sediments brought up from the deep sea by coastal upwelling are important for the growth of plankton.

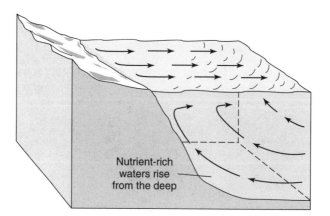

Nutrient-rich waters rise from the deep

one of the leading fishing nations in the world because of the upwelling that occurs along its coast. The tiny anchovy (*Engraulis* sp.), which feeds on the plankton that thrive in upwellings, was the backbone of the fishing industry in Peru. In good years, millions of metric tons of this fish were harvested. Unfortunately, a combination of overfishing and El Niños has caused a crash in the anchovy population since the 1970s and a reduction in catch to about 100,000 metric tons per year. Humans are not the only ones to suffer. Millions of ocean fish, invertebrates, seabirds, and marine mammals also have suffered from the loss of this food source.

Wave- and Tide-Induced Currents

After a wave breaks on a beach, the forward momentum transports water up the slope of the beach. The returning current or backwash is called the **undertow**. An undertow is an example of a current caused by wave action. As you may know from experience, an undertow has enough force to cause someone standing in the surf zone to lose his or her footing.

On beaches with heavy surf, sand eroded by wave action gets deposited a short distance from shore in a long hill called a **sandbar**. A sandbar forms parallel to the beach and acts like a dam by holding accumulated water from breaking waves. If water accumulation is too great, the pressure causes the sandbar to break, producing a rush of water seaward. This fast, narrow current of water seaward is called a **rip current**. (See Figure 20-21.) If you get caught

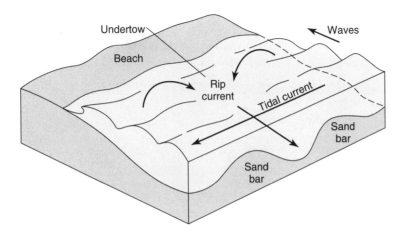

Figure 20-21 Offshore currents: an undertow is caused by wave action (backwash); a rip current forms where there is a break in a sandbar; a tidal current, which runs parallel to the shore, is produced by the tides.

in a rip current, do not fight it. Let the current carry you out a short distance, where its energy is dissipated. You can then swim back to the beach, but do so diagonally to avoid swimming into another rip current.

Tides also produce currents. When the tide enters and leaves bays and inlets, the tidal change produces swift-moving **tidal currents**, which run parallel to the shore. These currents are swiftest when the tide is changing from high to low or from low to high. Tidal currents slow during a period called *slack water*, which usually occurs at the end of each high tide and low tide. Tidal currents are important to marine life along the coasts because they carry nutrients and small organisms back and forth between the bays and the offshore waters.

An interesting, and potentially dangerous, phenomenon usually caused by tidal currents that move past each other in coastal waters is a whirlpool. A **whirlpool** (also called an *eddy*) is the rapid movement of surface waters in a circle. Whirlpools, which often form between islands, may also result from strong winds or when ocean currents flow against tides or unusual coastal features. At the center of the whirling water is a depression. Larger whirlpools can pose a danger to boats and people, because the water's movement is strong enough to draw large objects into the whirlpool's center.

Sea and Shore Interactions

The land and sea are always involved in a game of give and take. Tides, waves, and currents remove sediments from the shore, a process called *erosion*. These same movements of ocean water also deposit sand along beaches, a process called *deposition*. Along every shoreline, these two dynamic processes occur. When people choose to live along the shore, they must contend with the natural forces that may cause a beach to grow in one area, while it erodes in another.

Every time a wave breaks on the beach, it dislodges sand. Waves that break at an angle produce a current that moves parallel to the beach, called a **longshore current** (or *littoral current*). The movement of beach sand along the shore, pushed by waves and currents, is called *littoral drift*. During storms, when waves and currents have much more energy, littoral drift increases significantly.

Each year, government agencies spend billions of dollars trying to decrease erosion and preserve beachfronts by constructing barriers. One kind of barrier, called a **groin**, is made of wood or rock and extends straight out from the sand into the water at regular intervals. Although groins cause a build-up of sand on one side, erosion still occurs on the other side. Another barrier, called a **jetty**, traps sand and prevents it from accumulating in a channel. Other barriers, called *breakwaters*, are placed offshore to reduce the erosive power of wave action on the shore.

Pumping sand from a nearby seafloor and dumping it onto the beach, a process called *dredging*, also can be used to slow beach erosion. However, dredging is just a temporary solution to beach erosion because, in time, the ocean will reclaim the sand. Structures called *seawalls* are also built along shores to prevent property from being flooded during storms. But, eventually, wave action undermines seawalls, causing them to collapse into the water.

20.3 SECTION REVIEW

1. What causes global ocean (surface) currents?

2. What causes deep ocean currents (countercurrents)?

3. How do temperature and salinity differences cause vertical ocean currents?

Laboratory Investigation 20

Measuring Ocean Waves

PROBLEM: How are ocean waves measured?

SKILLS: Interpreting diagrams; making calculations.

MATERIALS: A copy of Table 20-2 (below).

PROCEDURE

1. Examine the Wave Characteristics diagram (Figure 20-22). One of the characteristics of a wave is wave height. Wave height (C) is the vertical distance from the crest (A) to the trough (B). Copy Table 20-2 and Figure 20-22 into your notebook, and label parts A, B, C, and D in the figure.

2. Wavelength (D) is the horizontal distance between two successive wave crests or wave troughs. Use the scale in Figure 20-22 to measure the wavelength in the diagram. Record your answer in the table.

3. The wave period is the time required for two successive crests to pass a fixed point. If it takes 100 seconds for 10 waves to pass a given point, what is the period? Record your answer in the table.

4. The wave speed is the distance a wave travels divided by the time it takes to travel that distance, or speed = wavelength/wave period. Calculate the wave speed and record your answer in the table. Fill in the correct "generating factor" under the *Wind* column in the table.

TABLE 20-2 WAVE CHARACTERISTICS AND WAVE TYPES

Wave Characteristics	Wind	Tide	Tsunami
Wavelength	_____ (m)	_____ (km)	_____ (km)
Wave period	_____ (sec)	_____ (hr)	_____ (hr)
Wave speed	_____ (m/sec)	_____ (km/hr)	_____ (km/hr)
Generating factor (wind, earthquake, gravitation)	_____	_____	_____

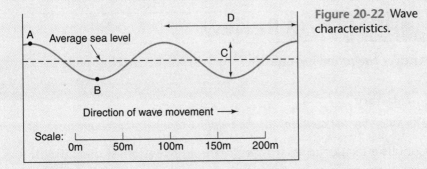

Figure 20-22 Wave characteristics.

5. The crest of a global ocean wave represents high tide and the trough is low tide. Where two high tides occur each day, the wave period is about 12.5 hours. If the wavelength is about 20,000 km, calculate the wave speed and record your answer in the table. Fill in the correct "generating factor" under the *Tide* column in the table.

6. If a tsunami has a period of 0.5 hour and a wavelength of about 200 km, calculate its speed. Record your answer in the table. Fill in the correct "generating factor" under the *Tsunami* column in the table.

OBSERVATIONS AND ANALYSES

1. Suppose an earthquake occurred off Alaska. Using the wave speed from Table 20-2, calculate how long it would take a tsunami to travel to Hawaii, a distance of about 4000 km.

2. Explain, based on your data, why the term *tidal wave* is not an accurate description of a tsunami.

3. Why do waves produced by tides have longer wave periods than those of tsunamis?

Chapter 20 Review

Answer the following questions on a separate sheet of paper.

Vocabulary

The following list contains all the boldface terms in this chapter.

Coriolis effect, countercurrents, current, gyres, longshore current, neap tides, rip current, rogue waves, sandbar, spring tides, swells, tidal bore, tidal currents, tidal range, tide, tsunami, turbidity current, undertow, upwelling, wave, wave height, wave train, whirlpool, whitecap

Fill In

Use one of the vocabulary terms listed above to complete each sentence.

1. Vertical distance between low and high tides is the _____.
2. A steady wind can produce a series of waves, or a _____.
3. Undersea earthquakes can generate a large wave, or _____.
4. The rising of nutrient-rich, deep waters is called an _____.
5. A fast, narrow current that moves seaward is called a

   _____.

Think and Write

Use the information in this chapter to respond to these items.

6. Why does the moon have a stronger influence on Earth's tides than the sun does?
7. Explain why a wave breaks on the shore. What is the importance of its wavelength to this process?
8. How do upwellings develop from vertical ocean currents?

Inquiry

Base your answers to questions 9 through 11 on the diagram on page 511, and on your knowledge of marine science and mathematics.

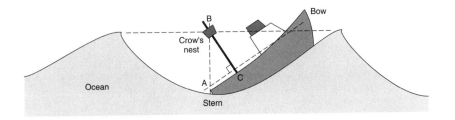

9. Use the Pythagorean theorem ($a^2 + b^2 = c^2$) to calculate the height of the wave in the diagram, given that side BC = 12 meters and side AC = 5 meters. Show all work.

10. As a result of wave action, in which direction will the boat move? *a.* up and down   *b.* forward only   *c.* backward only   *d.* side to side

11. Which statement about the diagram is correct?   *a.* The wavelength and the wave height have the same magnitude. *b.* The wavelength is equal to the amplitude.   *c.* The ship's stern is in the trough of the wave.   *d.* The crest is stable in strong winds.

Multiple Choice

Choose the response that best completes the sentence or answers the question.

12. Global ocean currents move   *a.* counterclockwise north of the equator and clockwise south of the equator   *b.* clockwise north of the equator and counterclockwise south of the equator   *c.* clockwise north and south of the equator   *d.* counterclockwise north and south of the equator.

13. Convection currents in the ocean are characterized by *a.* cold water rising at the poles   *b.* warm water sinking at the equator   *c.* warm water moving from the equator to the poles   *d.* cold water moving from the equator to the poles.

14. Which statement is correct?   *a.* A rip current is caused by the wind.   *b.* An undertow is caused by strong winds. *c.* An undertow is caused by wave action.   *d.* A longshore current is caused by the wind.

15. All of the following can cause a tsunami *except*   *a.* an earthquake   *b.* a volcanic eruption   *c.* hurricane winds   *d.* an undersea landslide.

16. A wave will break on the beach when   *a.* the water depth is less than half its wavelength   *b.* the water depth is twice its wavelength   *c.* the water depth is equal to its wavelength   *d.* the winds are very strong.

17. A coast that has two high tides and low tides each day, with the first set of tides being stronger than the second, has   *a.* diurnal tides   *b.* mixed tides   *c.* semidiurnal tides   *d.* neap tides.

18. Large waves that form when waves meet opposing currents are   *a.* rogue waves   *b.* swells   *c.* tidal waves   *d.* tidal bores.

19. A large current that moves across the ocean surface is called a   *a.* countercurrent   *b.* turbidity current   *c.* global ocean current   *d.* longshore current.

20. Currents that move swiftly down the continental slope are called   *a.* rip currents   *b.* turbidity currents   *c.* longshore currents   *d.* gyres.

21. The highest tides can occur during the   *a.* new moon only   *b.* full moon only   *c.* neap tide   *d.* new moon and full moon.

22. According to the diagram below of the moon and Earth, what would be the approximate time difference between one high tide and the next high tide?   *a.* 24 hours   *b.* 18 hours   *c.* 12 hours   *d.* 6 hours

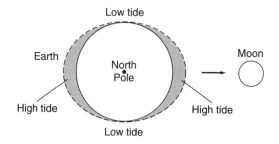

23. The whitecaps seen on waves at sea are caused by   *a.* strong winds mixing sea salt with water   *b.* strong winds creating a mixture of air and water   *c.* vibrations from undersea volcanic eruptions   *d.* ocean surface currents.

Research/Activity

Measure the speed of a current in a body of water alongside a dock or pier. Cast out a float on a fishing line and measure the speed of the float as it moves through the water. (Recall that speed = distance/time.) You can take several readings during the day in order to compare the current's speed at different times of the tidal cycle. (*Note:* Stay safely on the dock at all times.)

UNIT 7 MARINE ECOLOGY

There are few, if any, places in the ocean where life does not exist. The place in which a group of organisms lives is called a *habitat*. The inhabitants of a place such as a coral reef or mangrove swamp form a *community*—a group of plants, animals, and other organisms that interact with one another and their environment. The study of such relationships among living things and their environment is called *ecology*. The study of such relationships in the ocean environment is called *marine ecology*.

How do marine organisms interact with one another to produce a successful and thriving community? What impact do human activities have on marine life? In this unit, you will study marine ecology and the role that humans play in marine communities.

515

21 Interdependence in the Sea

When you have finished this chapter, you should be able to:

IDENTIFY important nutrient cycles in the ocean environment.

EXPLAIN characteristics of oceanic food chains and food webs.

DISCUSS the importance of symbiotic relationships in the sea.

DESCRIBE how succession occurs in marine environments.

Some day you will be on your own. But you will not be completely independent. You will depend on others to help you meet your needs, and people will depend on you, too. You will need to get a job in order to provide food, clothing, and shelter for yourself and your family. Your employer will pay you a salary. Medical and dental care also may be provided. There will be other expenses that you will have to pay. Life consists of such relationships in which individuals interact with one another.

In the marine world, there are also many relationships among organisms. A relationship in which organisms interact in a mutually dependent way is called *interdependence*. Some relationships are based on the need for food. In other relationships, organisms group together for mutual protection, such as fish swimming in a school. Yet, not all interdependent relationships are mutually beneficial. In this chapter, you will study the interactions among marine organisms to see how these relationships illustrate interdependence.

21.1 CYCLES IN THE SEA

Humans have long used natural resources, although not always wisely. Some of these resources—such as trees, minerals, topsoil, fish—have been seriously depleted because they or the products made from them have been used up or discarded. However, we can learn an important lesson from nature. There are processes in the natural world that help conserve resources. One of these processes involves the recycling of materials. Recycling is a way of conserving natural resources by using them over and over again. Many substances in the marine environment are naturally recycled, too.

Marine Ecology

Imagine that you are on a field trip exploring tide pools along a rocky coast. In such an environment, you could find snails, mussels, sea stars, and fish in the water. And there would be seaweed clinging to the rocks. The tide pool is like a miniature ocean, where you can observe the activities of living things in their habitat. The study of the interactions of living things with each other and their environment is called **ecology**. The study of such interactions within the ocean is called **marine ecology**.

The living things in the environment are called **biotic factors**. The biotic factors would include all the organisms in the tide pool. These organisms require the proper environmental conditions to carry out their life functions. The nonliving factors in the environment are called **abiotic factors**. Some abiotic factors in the tide pool are the water, temperature, sunlight, minerals, substrate, and dissolved gases. Together, the interacting biotic and abiotic factors within an environment make up an **ecosystem**.

Within an ecosystem, many resources are recycled. Algae use energy from sunlight, along with carbon dioxide and water, to perform photosynthesis and produce food. You may recall that organisms, such as algae, that make food are called *producers*. In turn, the algae are eaten by snails. Animals that eat other organisms are called *consumers*. The snail, sea star, mussel, and fish in the tide pool are consumers. All producers and consumers need certain nutrients that cycle through the ecosystem. Among the most important nutrients that are naturally recycled are carbon, oxygen, nitrogen, sulfur, and phosphorus.

The Carbon Cycle

By definition, organic compounds contain the element carbon. Photosynthesizing plants and algae take in carbon, in the form of carbon dioxide (an inorganic compound), and produce carbohydrates (organic compounds). Animals take in carbon compounds when they eat plants. Animals give off this carbon, again in the form of carbon dioxide, as a waste product of cellular respiration. This carbon then becomes available to plants for photosynthesis. The movement of carbon through living things in an ecosystem is called the *carbon cycle*. (See Figure 21-1.) In the marine environment, the carbon in dissolved carbon dioxide may be taken up by plants and by animals (such as corals and mollusks) and combined with the calcium in seawater to produce calcium carbonate shells and body parts. The carbon in these organisms is released into the marine environment when the organisms die and decay.

Figure 21-1 The carbon cycle.

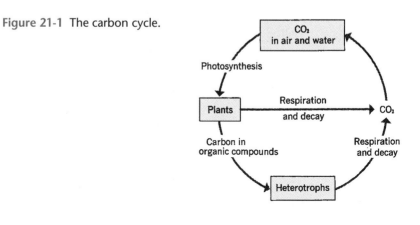

The Oxygen Cycle

Oxygen is the crucial element needed for cellular respiration in both plants and animals. Plants (and algae) release oxygen into the atmosphere when they split water molecules during photosynthesis. Animals, as well as plants, take in oxygen to use in respiration. Just

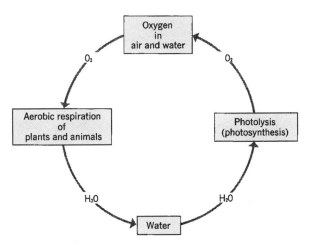

Figure 21-2 The oxygen cycle.

as plants need the carbon dioxide that animals give off, animals depend on the release of oxygen by photosynthesizing plants and algae. This movement of oxygen molecules between plants (algae) and animals is called the *oxygen cycle*. (See Figure 21-2.) Together, the carbon and oxygen cycles are often referred to as the *carbon dioxide–oxygen cycle*. Oxygen is also involved in the cycles of other nutrients in the environment, such as nitrogen, phosphorus, sulfur, and silicon.

The Nitrogen Cycle

All living things grow. To grow, animals and plants must make chemical substances called *proteins*. Important proteins include hemoglobin, myoglobin, enzymes, and hormones. The main elements that combine to form proteins are carbon (C), hydrogen (H), oxygen (O), and nitrogen (N). The formation of proteins begins with algae and plants, which produce the amino acids that build the proteins. Nitrogen, which is needed to make up the amino acids, is made available to living things in an ecosystem by a process called the *nitrogen cycle*. (See Figure 21-3 on page 520.)

Compounds that contain nitrogen, such as urea and ammonia, are excreted as metabolic wastes by animals and are used by bacteria. When plants and animals die, their bodies are decomposed by different kinds of bacteria. Decay bacteria break down the proteins in wastes and tissues and produce ammonia (NH_3), a toxic compound

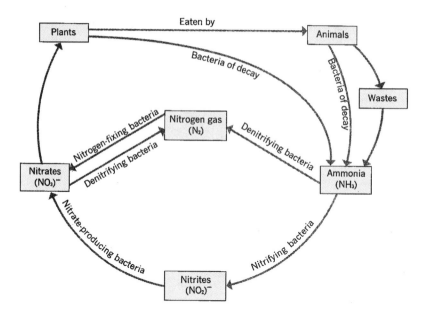

Figure 21-3 The nitrogen cycle.

Plants — Eaten by → Animals

Bacteria of decay

Bacteria of decay

Wastes

Nitrogen-fixing bacteria

Nitrogen gas (N₂)

Denitrifying bacteria

Denitrifying bacteria

Nitrates (NO₃)⁻

Ammonia (NH₃)

Nitrate-producing bacteria

Nitrifying bacteria

Nitrites (NO₂)⁻

that cannot be used by plants. Some decay bacteria convert ammonia into the nontoxic ammonium ion (NH_4^+). Denitrifying bacteria convert the ammonia into free nitrogen (N_2), while nitrifying bacteria oxidize the ammonia to form nitrites (NO_2^-) and then nitrates (NO_3^-). Nitrogen-fixing bacteria, such as blue-green bacteria, also oxidize the free nitrogen to nitrates (while denitrifying bacteria reverse that process).

The process of producing nitrates from atmospheric nitrogen is called **nitrogen fixation**. Algae and marine plants are able to take up ammonium ions, nitrites, and nitrates and use them in organic compounds. The nitrogen in marine plants and algae that are eaten by animals are used to make proteins. When these plants and animals decay, their proteins are broken down and the nitrogen is recycled. Lightning in the atmosphere also converts free nitrogen in the air into usable nitrates that fall to the ground in the rain.

The Sulfur Cycle

Another important element that is recycled in the marine environment is *sulfur*. Many large protein molecules with long amino acid chains contain sulfur. For example, the protein hemoglobin, which binds oxygen in red blood cells, has eight sulfur atoms. The sulfur is

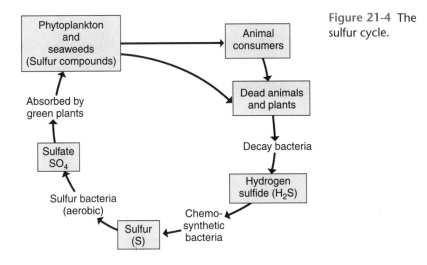

Figure 21-4 The sulfur cycle.

essential because it acts as a bridge that allows the molecule to bend and twist, giving it its shape.

Sulfur is recycled from dead matter in the sea back into marine plants (algae) and animals through a series of chemical reactions controlled by bacteria. This process is called the *sulfur cycle*. (See Figure 21-4.) In the first reaction, decay bacteria break down wastes into hydrogen sulfide (H_2S). Hydrogen sulfide (which smells like rotten eggs) is toxic, so plants and algae cannot use it. Chemosynthetic bacteria convert the hydrogen sulfide into sulfur. But this form is still not suitable for absorption by algae. Special sulfur bacteria, however, convert sulfur into sulfate (SO_4^{2-}), which is absorbed by the algae and used in the manufacture of proteins. Animals get the sulfur they need when they eat algae, plants, and other animals.

The Phosphorus Cycle

Living things need energy to do work. Energy is found in molecules such as glucose. Inside cells, when glucose is oxidized during cellular respiration, energy is liberated and converted into a more usable form called *adenosine triphosphate*, or *ATP*. The energy in ATP is stored in its phosphate bonds, which contain the elements phosphorus and oxygen. Phosphorus is also present in deoxyribonucleic acid, or DNA, the genetic blueprint found in all living cells. Thus, without phosphorus, there is no DNA or ATP, two of the most important chemicals in living things.

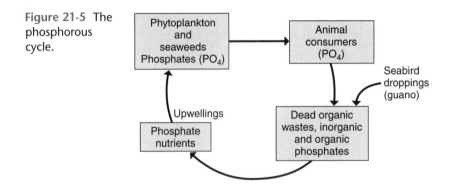

Figure 21-5 The phosphorous cycle.

Phytoplankton and seaweeds Phosphates (PO$_4$)

Animal consumers (PO$_4$)

Seabird droppings (guano)

Upwellings

Phosphate nutrients

Dead organic wastes, inorganic and organic phosphates

Phosphorus is supplied to all living things through the *phosphorus cycle.* (See Figure 21-5.) Decay bacteria decompose dead matter on the seafloor. The molecules that contain phosphorus, such as DNA and ATP, are broken down and phosphate (PO$_4^{3-}$), a product of decay, is released into the water. Phosphates from human activities and from natural sources such as weathered rocks and soil also enter the marine environment. Plants and algae in the ocean absorb the phosphates. Marine animals get their supply of phosphorus when they eat these plants and algae or other marine animals. Another source of phosphates for marine life is the droppings from seagulls and other seabirds. Seabirds nest along coasts and on islands, where thick deposits of their droppings accumulate. These droppings, called *guano,* are rich in phosphorus. Since phosphorus is also an important nutrient for the growth of land plants, guano is often harvested and sold as fertilizer.

21.1 SECTION REVIEW

1. How are the carbon cycle and the oxygen cycle interrelated?

2. Why are various bacteria so important for the nitrogen cycle?

3. How is phosphorus cycled in the marine environment?

21.2 FOOD RELATIONSHIPS IN THE SEA

Relationships between organisms that are based on nutritional needs are called *food relationships.* In the marine environment, just as on land, several types of food relationships can be observed. You will study some of these food relationships next.

Food Chains

Look at the food relationship shown in Figure 21-6. This kind of relationship is called a **food chain**, because one living organism serves as food for another organism, which serves as food for the next organism in the chain. All food chains begin with a producer. As you can see in the figure, the producers are the microscopic phytoplankton. The next organism in the food chain is a consumer, usually an animal, such as the zooplankton. The zooplankton is called a **primary consumer**, since it is the first animal in the food chain. The shrimp, a **secondary consumer**, feeds on the zooplankton. In turn, the fish or the squid, a **tertiary** (third-level) **consumer**, eats the shrimp. As each organism feeds on another, there is a transfer of chemical energy—from the producer through each level in the food chain. Each feeding level is called a **trophic level**.

Food chains vary in length. All food chains must have a producer and at least one consumer. How many consumers are shown in the food chain in Figure 21-6? As you can see, there are three consumers. Consumers vary in their food requirements. Animals such as zooplankton that feed only on algae or plantlike organisms are called *herbivores*. The other animals pictured in the food chain (the shrimp and the herring) are *carnivores*. An animal that consumes both animals and plants is called an **omnivore**. Humans are omnivores, since we eat both plant and animal foods.

When producers and consumers die, their remains are eaten by scavengers and decayed by decomposers—the organisms that break down dead matter into smaller particles. The **scavengers**, such as

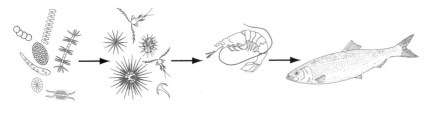

Phytoplankton   Zooplankton   Shrimp   Herring

Figure 21-6 Feeding levels in a marine food chain; the first organism is always a producer.

crabs and mud snails, eat the remains of already dead plants and animals. The most common of the decomposers are the bacteria, which are found everywhere in the marine environment. However, most are concentrated in the bottom sediments because the dead matter on which they feed settles there and accumulates. Some types of fungi, such as molds, are also decomposers. Another term used to describe decomposer bacteria and fungi is saprophytes.

The decomposers break down the dead matter into organic molecules and simple compounds, which are taken up by the producers. This transfer of organic matter—from producers, to consumers, to decomposers, and back to producers again—"cycles" energy and nutrients through the food chain. The recycling of organic matter is the means by which all living things in the ocean can satisfy their nutritional needs.

Food Webs

In the ocean, many food chains are interconnected, thereby forming a giant **food web**. Look at the food web shown in Figure 21-7. The consumers vary in their relationships with one another. The producers are always at the base of the food web. Most of the consumers shown in the figure are predators; they occupy trophic levels above the primary consumers. A **predator** is an animal that kills and eats another animal. Consumers may also be scavengers, eating the remains of animals that are already dead. In addition, most of the consumers serve as prey for one or more other animals. The **prey** is the animal that is eaten by a predator. Which animals in the food web are not both prey and predator? All the animals are both predator and prey, except for copepods and mosquito larvae, which graze on diatoms, and killer whales, which are the top predators that are not preyed on by other animals.

What would happen to this food web if the mosquito larvae in a saltwater marsh were suddenly wiped out by insecticide spraying along the coasts? It might appear that the killifish would decline in number (and thus disrupt the food web) because they feed on the mosquito larvae. However, killifish have another food source, the copepods, which may sustain them. In food webs, predators often feed on more than one type of prey. This increases their ability to survive if one food source becomes scarce. In addition, predators

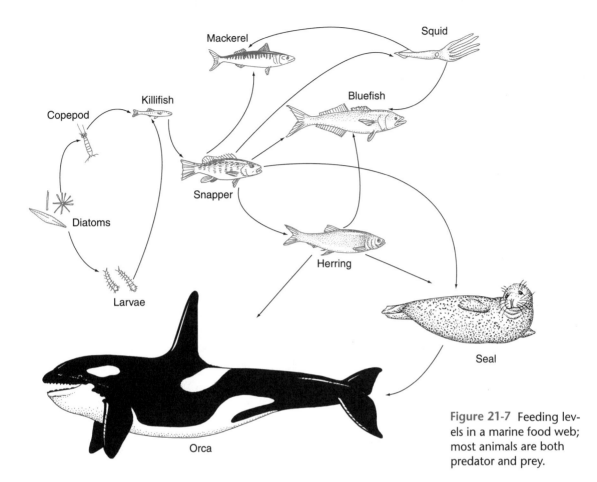

Figure 21-7 Feeding levels in a marine food web; most animals are both predator and prey.

such as the killifish can be omnivores; that is, they also eat plants. However, serious disruptions to food chains and food webs—for example, those that result from El Niño or intense overfishing—have negative effects on populations of marine organisms such as seabirds and seals. Similarly, harvesting of krill in Antarctic waters for human consumption gives cause for concern, since that could have an impact on baleen whale populations. The whales are secondary consumers that feed directly on these zooplankton.

Food Pyramids

The organisms in food chains can be arranged in a diagram called a **food pyramid**. In a food pyramid of numbers, the population at each higher trophic level is smaller than the one below it. (See

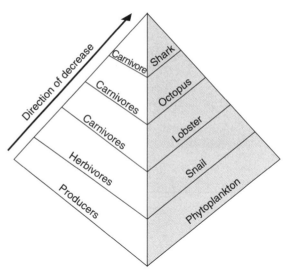

Figure 21-8 Trophic levels in a marine food pyramid; the producers (bottom level) are highest in numbers and in concentration of food energy.

Figure 21-8.) The phytoplankton (algae and plants) outnumber the snails. There are more snails than lobsters. Lobsters are more numerous than octopuses. And sharks are the least numerous of all. In other words, each prey population outnumbers the predator population above it. What would happen if predators outnumbered their prey? Some predators would die from a lack of food. Large populations of producers and prey are needed to sustain the consumers in each higher trophic level. (At times, there may seem to be more zooplankton than phytoplankton, but the high productivity of the algae supports the continual grazing by the zooplankton.)

The food pyramid also illustrates the transfer of energy from producers through each successive trophic level of consumers. In fact, this transfer of energy explains the need for large populations of producers and prey. Which organism in Figure 21-8 possesses the highest concentration of available energy? The population of algae at the base (the largest area) of the pyramid has the highest amount of available energy. The population of sharks at the top of the pyramid has the least amount of available energy.

Why do the algae have the highest concentration of energy and sharks the lowest? At each level of consumption, a great deal of energy is lost as heat. The abundant seaweed is highest in energy because it absorbs energy directly from the sun and converts it into glucose. Some of the energy is used for the seaweed's life functions and is lost as heat. When snails consume the seaweed, they use up most of the energy in the food (about 90 percent), while the rest of it is stored in the snail's body tissues.

Only about 10 percent of the food energy consumed by an organism is available as food to the next organism in the chain. Thus, when a lobster eats a snail, most of the energy is used for life processes and is lost as heat, while the remaining 10 percent or so is stored in the lobster's body. Consequently, as food energy is transferred from one trophic level to the next, a great amount is lost at each step. Within the pyramid, there is less energy available for consumption at the trophic level of the shark. To produce 10 kg of body matter, a shark has to eat about 100 kg of food. But that 100 kg of food was produced as a result of the consumption of 1000 kg of food, and so on. That is why the producers and lower-level consumers have to be more abundant than the organisms at higher trophic levels. So a food pyramid is a way of showing that there is more energy available at the bottom of a food chain than at the top.

21.2 SECTION REVIEW

1. What are the different trophic levels in a food chain?
2. Explain how an organism can be both predator and prey in a food web.
3. Why does food energy decrease with each higher trophic level?

21.3 SYMBIOSIS IN THE SEA

In the ocean, there is a constant struggle for existence as marine organisms avoid predators and compete for food, mates, and territories. Living things have evolved a variety of relationships that help them survive. One kind of relationship that may be beneficial to an organism is *symbiosis*. (See Chapter 7.) In symbiosis, organisms of different species live in close association with one another. There are three different types of symbiotic relationships, each with varying degrees of benefit to the parties involved but all related to feeding strategies.

Mutualism

Look at the symbiotic relationship shown in Figure 21-9. A clown-fish lives among the stinging tentacles of a sea anemone. This special kind of symbiosis is called **mutualism**, a relationship in which both species benefit from the association. The sea anemone benefits because it can feed on scraps of food left by the clownfish and is protected by them from the nibbling of other fish. The clownfish benefits because it is protected from predators by the stinging cells in the anemone's tentacles. The mucus coating on the clownfish's skin contains a chemical inhibitor that prevents the discharge of stinging cells in the tentacles. This kind of mutually beneficial relationship is also found between other small fish and some jellyfish, and between cleaner shrimps and sea anemones. (See the photograph on page 516.)

Other examples of mutualism can be found in the sea. Many large tropical fish, such as the grouper, cannot rid themselves of the parasites that attach to their skin and gills. However, some tiny fish, such as the cleaning wrasse, feed on fungi and invertebrates that live on the fish's skin and in its body cavities. The relationship between coral polyps and the zooxanthellae (algae) that live inside them is another example of mutually beneficial symbiosis. Both

Figure 21-9 The clownfish and sea anemone have a symbiotic relationship called mutualism.

organisms benefit because the algae get a home, carbon dioxide, nitrate, and phosphate from the polyps, while the coral polyps get glucose and oxygen from the algae. Mutually beneficial zooxanthellae are also found in the mantle tissue of the giant clam *Tridacna* of the South Pacific.

Commensalism

In another symbiotic relationship, called **commensalism**, one species benefits while the other species is, apparently, unaffected by the association. In several species of sharks, a remora (attached by a dorsal suction disk) or a small group of pilotfish may swim below or ahead of the shark as it cruises along. (See Figure 21-10.) The small fish scavenge on leftover bits of food after the sharks have fed. The shark is neither helped nor harmed by the presence of these fish; the small fish benefit by getting a free meal.

Another commensal relationship is that of a whale and the barnacles that live on its back. The barnacles benefit by being attached to a substrate from which they can filter-feed as the whale swims through nutrient-rich waters. The whale is, for the most part, unaffected by the presence of the barnacles. Organisms such as barnacles, sea anemones, and slipper shells that attach to the bodies of larger animals without causing them harm are called *commensal organisms*.

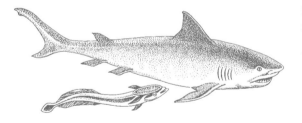

Figure 21-10 The shark and the remora have a symbiotic relationship called commensalism.

Parasitism

A third type of symbiosis, called **parasitism**, involves a relationship in which one species benefits while the other species is harmed. An organism that lives in or on another creature and feeds off its tissues is called a *parasite*. (See Chapter 8.) The animal that is infected is called a *host*. A parasite can survive only if it is attached to its host. If the host dies, so does the parasite, unless it is able to find a new host.

Because of the loss of commercially valuable fish, parasitism can be costly to the economy. In the Great Lakes, the lake trout is in decline due to the accidental introduction of the sea lamprey, which parasitizes lake trout. (Refer to Figure 12-3, on page 283.) With its suckerlike mouth and rough tongue, the sea lamprey attaches to a trout, then scrapes a hole in the trout's skin and feeds on its body fluids. The lamprey may feed on the trout for a few hours or several days, then move on to another trout to continue feeding. The deep-sea hagfish takes its parasitism a step farther—it burrows into the bodies of larger fish and feeds on their tissues from inside.

The life cycle of a parasite often involves more than one type of host. Black spot disease is a kind of parasitism that occurs in some marine fish and is caused by a trematode, a type of flatworm. (Refer to Figure 8-1, on page 194.) Sea gulls that feed on fish with black spot disease eat the parasitic worms as well. The flatworms reproduce in the intestines of the sea gull, and the eggs are eliminated with the bird's droppings. The eggs develop into small swimming larvae in the shallow waters of the intertidal zone. The larvae then enter the first intermediate host, usually a small marine snail, where they develop into juvenile worms. They leave the snail and attach themselves to the second intermediate host, a passing fish, where they form black cysts on the skin and then mature. A serious infection can weaken the fish to the point where it dies. Sea gulls feeding on these fish consume the mature flatworms, and the cycle continues.

Although parasitism is harmful for the host involved, it is an important feature of life in animal populations. Most major groups of marine organisms have internal parasites, usually some type of flatworm. Many ocean fish (such as the grouper mentioned above) also have external parasites. These are usually tiny crustaceans, such as isopods and copepods, that attach to the host's skin and gills and obtain their nutrients from the fish's blood. Environmental conditions can also play a role in determining an animal's susceptibility to parasitism. Scientists have noticed that poor water quality increases stress in marine organisms and lowers their resistance to disease, making them more vulnerable to parasitic infections.

1. How do symbiotic relationships illustrate interdependence?

2. How does the clownfish survive among the stinging tentacles of the sea anemone? Why is that an example of mutualism?

3. Explain the kind of symbiosis that sharks and pilotfish have.

21.4 SUCCESSION IN MARINE ENVIRONMENTS

What was it like growing up on the street where you live? Do you remember new neighbors arriving and old friends moving away? New buildings may have been constructed on your block or others remodeled. Over time, neighbors and neighborhoods change.

Change also occurs in the marine environment. You can place a hard object, such as a rock, in a tide pool at the beach and examine it over a period of time. At first, a thin coat of algae will begin to cover the rock's surface, making it feel slimy. If you leave the rock in a tide pool long enough, you may see its coat of algae replaced by barnacles. Later on, seaweed, mussels, and other small organisms may join the barnacles that coat the rock. The same process also occurs on the hull of a boat. In fact, boats sometimes need to have their hulls scraped clean of barnacles and mussels, because they affect the vessel's movement through the water.

Ecological Succession

The process by which one community of organisms gradually replaces another community of living things over time is called **ecological succession**, or *biological succession*. Succession occurs in a variety of marine habitats, such as sandy beach dunes, rocky shores, and coral reefs. (See Figure 21-11 on page 532.)

What causes one community of organisms to replace another community? When the chemical and physical conditions in an area are no longer suitable for the existing life-forms, a new community

Figure 21-11 Succession on a sandy dune: beach grasses are followed by shrubs.

of living things slowly takes over. The first group of organisms to appear in an area devoid of life—such as on a new wharf piling or new volcanic island—is called the **pioneer community**. The pioneer organisms represent the first stage in ecological succession. When the pioneering algae settle down on a rock, they begin to change their physical and chemical environment. Algae take up carbon dioxide, make their own food, and produce oxygen. Changes brought about by the algae actually favor colonization by other organisms, such as barnacles. The barnacles, in turn, alter the environment and are succeeded by a more stable community, dominated by mussels. The last stable community to appear in any succession is called the **climax community**. Each climax community has a dominant species, which is the main species that finally appears there. Marine succession varies with climate. The algae-to-barnacle-to-mussel succession occurs along temperate shores in the more northern latitudes.

In tropical seas, corals form the climax community. Coral polyps attach to any hard substrate and, over a long time, cover the substrate with a coral reef. Other organisms are attracted to the corals, slowly forming a more complex and diverse community. In fact, old sunken ships are difficult to find in tropical waters because this process of succession entombs ships under coral.

ENVIRONMENT
Turf Wars off Fisherman's Wharf

San Francisco is a multicultural city where people from all over the world have settled. Paralleling this influx of people from foreign lands has been the arrival of foreign, or exotic, marine species from other oceans to the waters of San Francisco Bay. But not all the inhabitants get along. Turf wars have developed between some of the native species and the exotic invaders. So far, marine biologists have identified more than 200 exotic species in the bay, and new ones continue to arrive. These organisms are usually introduced in the ballast water that is discharged daily by ships anchored in the bay. Water from the ballast tanks contains eggs and larvae of fish and invertebrates from the ports where the ships embarked. Some species arrive packed in boxes of fish bait. Crabs from China, protozoans from Japan, and a variety of species from the East Coast—including Atlantic ribbed mussels (see photograph), striped bass, and cordgrass—have found a niche in the shallow bays and inlets of the San Francisco estuary.

This great increase in species diversity has come at a price. Competition with, and predation by, the exotic fauna and flora has decreased populations of some local species. The thicktail chub, a native to the bay, has been eliminated by the influx of foreign species. Another local fish, the delta smelt, is now an endangered species. The Atlantic green crab—originally a native of Europe that "emigrated" to New England coastal waters in the 1950s and later traveled in ballast tanks to San Francisco Bay—is a predator that attacks the juvenile Dungeness crab, a popular food item in the bay area. One exotic, a burrowing isopod from Australia, presents a different kind of problem: it tunnels into the Styrofoam blocks that support floating docks, causing them to sink. On the other hand, some of the exotic species have turned out to be valuable new food sources for the bay area.

The ultimate effects of the exotics in the bay are yet to be seen. Marine scientists think that, over time, a new ecosystem will establish itself—with a climax community of many different marine species replacing the ecological roles of former inhabitants that lost the fight for their turf off Fisherman's Wharf.

QUESTIONS

1. Identify two exotic species in San Francisco Bay. Why are they called *exotic*?
2. How did the various exotic marine species arrive in San Francisco Bay?
3. Why are some exotic species harmful to the native bay species? Give an example.

Climax communities are not necessarily permanent. Natural disturbances such as storms can strip mussels from rocks. The activities of people can have an effect, too. For example, overharvesting of mussels from rocks can wipe out a climax community. When a climax community is destroyed, the area is made available for pioneering organisms to start the process of succession over again. It takes time, perhaps many years or decades, for a new climax community to become established.

Succession on Islands

Succession occurs on land as well as in the sea. In areas of seismic activity, such as the South Pacific, tiny islands are born each year as molten lava from the seafloor reaches the surface and solidifies into rock. At first, these islands are devoid of life. But after the rock cools, conditions for life become more favorable. Spores and seeds arrive on the new islands, brought by winds, waves, and various animals. For example, seeds may be carried in the digestive tracts of migrating birds. The first pioneer organism to appear on barren rocks is usually lichen. The **lichen** is actually composed of two organisms, an alga and a fungus, growing together in a symbiotic relationship. Acids in the lichens erode the surface of rocks, producing a more porous surface and loose sediments that favor the growth of the next community, the mosses. Mosses carpet the surface of rocks and cause further erosion into rock particles and gravel. The loose sediments and dead plant matter create a thin soil that is favorable for the germination of grass seeds. Grasses take over and dominate the landscape for a number of years and then are succeeded by shrubs. The shrubs cause further changes in the soil that favor the growth of trees, the climax community (of plants) that will dominate the island. (See Figure 21-12.)

Figure 21-12 Succession on an island, from pioneer to climax plant species.

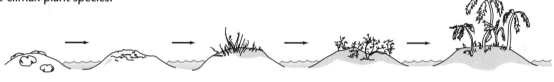

| Lichens | Mosses | Grasses | Shrubs | Trees |

21.4 SECTION REVIEW

1. What causes one community of organisms to succeed another?

2. Compare one pioneer community with one climax community.

3. How does succession occur on a rocky volcanic island?

Laboratory Investigation 21

Observing How a Barnacle Filter-Feeds

PROBLEM: How does the barnacle obtain its food?

SKILLS: Conducting an experiment; observing the behavior of organisms.

MATERIALS: Live barnacles attached to substrates, bowls or shallow containers, seawater, hand lens, watch or clock, fish food (dried plankton).

PROCEDURE

1. Put a barnacle with its attached substrate in an empty bowl or container. Notice the overlapping shells that surround and protect the barnacle. Make a sketch of the barnacle in your notebook. Observe the barnacle closely and note if it moves or responds to the environment in any way.

2. Cover the barnacle with seawater. Wait a minute for the barnacle to respond. Notice the appendages, called *cirri*, as they extend outward and then retract into the barnacle's shells. This movement is an automatic response by the barnacle to its seawater environment.

3. Count the cirri and record the number in your notebook. Notice the hairlike bristles attached to the cirri. Use your hand lens for a close-up look. Food particles get trapped in the bristles as the cirri sweep the water. The movement of the cirri in and out of the shells is the barnacle's method of filter-feeding. Make a sketch of a filter-feeding barnacle; label all parts.

4. You can measure the automatic feeding response of the barnacle by calculating the number of times the cirri beat per minute. Count the number of times the cirri move in 15 seconds and multiply by 4 to get the number of movements per minute. Record this number in the *Automatic Response* column in your copy of Table 21-1. Try to perform six trials to obtain an average number of cirri movements per minute. (This is the control group.)

TABLE 21-1 COMPARING FEEDING RESPONSES OF THE BARNACLE (MOVEMENTS/MINUTE)

Trials	Automatic Response	Response to Food
1		
2		
3		
4		
5		
6		
Total		
Average		

5. Now sprinkle some fish food in the water, near the barnacle. Again, count the movements for 15 seconds and then calculate the number per minute. Record the number in the *Response to Food* column in your copy of Table 21-1. Try to perform six trials to obtain an average number of cirri movements per minute. (This is the experimental group.) Compare the results from both sets of trials to see if there are any significant (measurable) differences.

OBSERVATIONS AND ANALYSES

1. Does the barnacle show any feeding response when it is *not* covered with seawater (that is, when it is exposed to the air)?

2. Compare the feeding responses of the barnacle in the presence and in the absence of food. Is there a measurable difference?

3. Explain how the barnacle is adapted for filter-feeding. What special structures and functions did you observe that help the barnacle survive?

Answer the following questions on a separate sheet of paper.

Vocabulary

The following list contains all the boldface terms in this chapter.

abiotic factors, biotic factors, climax community, commensalism, ecological succession, ecology, ecosystem, food chain, food pyramid, food web, lichen, marine ecology, mutualism, nitrogen fixation, omnivore, parasitism, pioneer community, predator, prey, primary consumer, scavengers, secondary consumer, tertiary consumer, trophic level

Fill In

Use one of the vocabulary terms listed above to complete each sentence.

1. A relationship in which both species benefit is called
   _____.

2. A _____ shows the trophic levels in a food chain.

3. The _____ is the first group of living things in an area.

4. Zooplankton would be the _____ in an ocean food chain.

5. The nonliving factors in an environment are called _____.

Think and Write

Use the information in this chapter to respond to these items.

6. What is the difference between mutualism and commensalism?

7. Why are producers more abundant than consumers in a habitat?

8. Explain why a climax community is not always permanent.

Inquiry

Base your answers to questions 9 through 11 on Figure 21-7 on page 525, which shows a marine food web, and on your knowledge of marine science.

9. In terms of numbers, which is the most abundant organism in the food web? Explain.

10. Which organism in the food web is the least numerous? Explain why.

11. How would a sudden decline in the herring population affect the food web?

Multiple Choice

Choose the response that best completes the sentence or answers the question.

12. Which statement is most correct about the relationship between the two fish shown here?
    a. Both fish benefit equally.
    b. The shark benefits and the remora is harmed.   *c.* The remora benefits and the shark is harmed.   *d.* The remora benefits and the shark is unaffected.

13. The ecological cycle shown here provides plants in the marine ecosystem with
    a. carbon to carry out photosynthesis
    b. phosphorus for ATP production   *c.* oxygen to carry on aerobic respiration   *d.* a source of nitrogen for protein synthesis.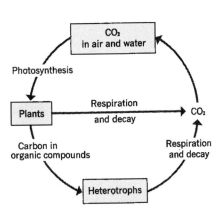

14. A producer–consumer interaction in the marine ecosystem is illustrated by   *a.* bacteria feeding on detritus   *b.* sea gulls scavenging on dead clams   *c.* snails grazing on algae   *d.* a fish eating a shrimp.

15. The nitrogen cycle makes it possible for marine algae to
    a. carry on photosynthesis   *b.* breathe   *c.* produce ATP   *d.* synthesize proteins.

16. The flow of energy in a food chain occurs as follows:
a. consumer to producer to producer   *b.* decomposer to producer to consumer to producer   *c.* producer to consumer to decomposer to consumer   *d.* producer to consumer to decomposer to producer.

17. The carbon cycle and the oxygen cycle directly involve
a. consumers and decomposers   *b.* producers and consumers   *c.* predators and prey   *d.* predators and consumers.

Base your answers to questions 18 through 21 on the following food relationship: algae ⟶ shrimp ⟶ lobster ⟶ octopus.

18. This sequence illustrates a   *a.* food web   *b.* food pyramid   *c.* food chain   *d.* food cycle.

19. The population of which organism has the highest food energy?   *a.* algae   *b.* shrimp   *c.* lobster   *d.* octopus

20. Which organism is the most numerous in its habitat?
a. algae   *b.* shrimp   *c.* lobster   *d.* octopus

21. Which organism is both predator and prey here?   *a.* algae   *b.* shrimp   *c.* lobster   *d.* octopus

22. The cleaning wrasse that eats parasites living on a grouper's skin provides an example of   *a.* commensalism   *b.* succession   *c.* mutualism   *d.* parasitism.

23. The correct sequence for plant communities on an island is
a. mosses, lichens, shrubs, trees, grasses   *b.* lichens, mosses, grasses, shrubs, trees   *c.* grasses, mosses, lichens, trees, shrubs   *d.* lichens, grasses, mosses, trees, shrubs.

Research/Activity

With a team of classmates, set up a saltwater aquarium tank containing some local marine species, such as killifish, snails, seaweed, shore shrimp, and mussels. Identify the producers, consumers, and decomposers (scavengers) in the aquarium. Prepare a poster board that shows the food relationships among the different organisms. (*Remember:* Do not overstock the tank, and be sure to use proper filtration and aeration.)

22 Pollution in the Ocean

When you have finished this chapter, you should be able to:

DISCUSS the impact of sewage pollution on aquatic ecosystems.

DESCRIBE the effects of toxic chemicals on marine organisms.

EXPLAIN the importance of clean waters to aquatic life-forms.

DISCUSS the problem of solid wastes in marine ecosystems.

For years the ocean has been used as a dumping ground for the disposal of human wastes and garbage. We now find wastes produced by industrial societies washing up on remote tropical islands. Toxic chemicals have been found in the bodies of ocean animals. Many people are now aware of the hazards of polluting the water and are becoming involved in both local and large-scale efforts to clean up the marine environment.

This change in attitude and behavior occurred when people realized that marine pollution can produce harmful effects in living things, including humans. Although the ocean is vast and seems capable of absorbing great quantities of wastes, it is not unaffected by the activities of humans. Fortunately, the ocean is important to people—as a source of food, as a place for recreation, and for its natural beauty. How polluted is the ocean? In this chapter, you will learn about different kinds of pollution in the marine environment, and about attempts to protect the ocean and improve its water quality.

22.1
**Sewage
Pollution**

22.2
**Toxic
Chemicals**

22.3
Clean, Clear Waters

22.4
Litter in the Water

22.1 SEWAGE POLLUTION

Tens of thousands of years ago, people lived in small groups that traveled from one site to another as they hunted and gathered food. Garbage and wastes rarely accumulated, since populations were small and people never settled in one place for too long (and their wastes were from natural materials that degraded easily). People were able to move away from the messes they made. As they settled in more permanent communities, and populations grew, wastes began to accumulate.

Pollution of Water

Most human populations lived along rivers and coasts. In these areas, there is access to water, food resources, and trade. Bodies of water have always been convenient places for people to dispose of wastes. In pre-industrial times, the water could dilute and wash away most of the wastes that were dumped into it. However, human populations along the world's coasts have increased dramatically. In the United States, about three-fourths of the population lives within 80 km of a coast. As a result, there is a greater concentration of wastes along the coasts, and these wastes may be dumped into the sea. The quantity and quality of these wastes can, in some instances, exceed the ocean's capacity to store or dispose of them without being damaged.

The natural by-products of living systems are wastes. Marine organisms certainly produce their share of natural wastes; and, as you learned in Chapter 21, ecosystems handle these wastes through the processes of decomposition and nutrient recycling. However, when people introduce substances into the environment that harm living things and damage water quality, it results in the condition known as **pollution**. The substances that have a harmful effect are called **pollutants**. Pollution not only threatens the well-being of other living things but also endangers human health, since we depend on Earth's natural resources for our survival.

Sewage and Sewage Treatment

Unlike the pristine tropical beach shown in Figure 22-1, many U.S. beaches are closed periodically because they are contaminated, or polluted, with sewage. Sewage is made up of the human intestinal (fecal) wastes that are discharged into our waterways. Water that is contaminated with sewage poses a serious public health problem. Such water may contain a variety of harmful microorganisms, or *pathogens*, that can cause life-threatening diseases such as typhoid fever, cholera, dysentery, and hepatitis.

How do we know if the water is contaminated with sewage? Public health scientists routinely test water samples for fecal coliform bacteria. Fecal coliform bacteria, which are present in the large intestine of humans and normally do not cause disease, are used as indicators of sewage pollution. The presence of coliform bacteria in a water sample means that the water has been contaminated by sewage. If coliform bacteria are present, there is a good chance that disease-causing bacteria are also present. Usually, water from a harbor area has the highest bacterial count because raw sewage is dumped into many harbors in industrialized or urban areas. Much less raw sewage is dumped into the waters near public beaches and fishing areas. (*Raw sewage* is waste material that has not been treated; that is, the pollutants have not been removed from it.)

Figure 22-1 An unpolluted tropical beach.

Escherichia coli (*E. coli*) is one species of coliform bacteria found in water samples contaminated with sewage. This rod-shaped unicellular organism, only 1 micron (μm) in length, is too small to be identified in a water sample. In order to be seen, the *E. coli* bacteria have to be grown under suitable conditions until they produce enough cells to form masses called *colonies*. A special procedure is used to culture the coliform bacteria for easy identification. Bacterial growth begins with the division of a single bacterium into two cells. During a 24-hour incubation period, the cells multiply, producing millions of bacteria that form raised colonies. The colonies are counted and compared with EPA contamination standards to determine if the water sample is, or is not, in compliance with the standards.

Contamination of water by sewage can be reduced or even eliminated by sewage treatment, a process that helps kill bacteria and other harmful microorganisms. Sewage is treated in facilities called *sewage-treatment plants*. (See Figure 22-2.) Solid litter such as sticks, rags, and other debris are separated from sewage and trucked to a landfill or incinerator. Notice the large tanks that treat the incoming sewage. In some tanks, bacteria are added to digest the fecal wastes. (Remember, in nature, bacteria help break down organic wastes, including those produced by people. But the great volume of human waste that may be poured into an area can overload the natural ability of the area to deal with it. That is why sewage treatment is so important.)

Figure 22-2 A sewage-treatment plant.

Figure 22-3 A sewage effluent pipe along the shore.

During treatment, the solid part of the sewage settles to the bottom of the tanks, forming what is called **sludge** (a mixture of water and solid wastes). The sludge is pumped into tankers and delivered to other sites for further treatment. For many years, sewage sludge was dumped at sea, without additional treatment. The sludge often had a harmful effect on local marine life. Besides organic substances, sludge may contain harmful chemicals from industry and agriculture. Finally, in 1991, the federal government required that sludge be recycled. For many centuries, people in rural societies have used human wastes to fertilize their fields. Today, some sludge, which is high in organic matter, is converted into fertilizer and sold to farmers for commercial use. The liquid part of the sewage in the tanks is further treated with chlorine to kill the harmful microorganisms before it is discharged into waterways. As a result, the liquid discharge, or **effluent**, contains far fewer pathogens than the raw sewage that comes into the plant. (See Figure 22-3.)

In big cities, such as New York, there has been an upgrading of sewage-treatment plants and an improvement in the local water quality. Yet, from time to time there is a sharp rise in coliform bacteria in water samples from New York City harbors. The sudden increase in fecal coliforms coincides with periods of heavy rain. When it rains heavily, the sewage-treatment plants cannot handle the excess water that comes in from storm drains in the streets. (In cities, much of the

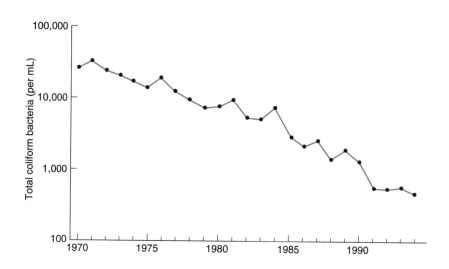

Figure 22-4 Coliform trends in New York City waters.

ground is paved over, so much less water can be absorbed by the ground.) This runoff water and the wastes it contains are diverted and dumped untreated into local waterways. (See Figure 22-4.)

In southern California, an area prone to water shortages, one facility has a productive and profitable way of treating sewage. The purified sewage water is used for irrigation and to grow aquatic plants and animals, and the leftover sludge is used for landfill. Wetlands, such as salt marshes, have the ability to purify polluted water. As a result, marshes are being explored for their potential to naturally treat sewage. Unfortunately, in many parts of the world, untreated sewage is dumped directly into the sea—so sewage is still a major source of pollution in the ocean.

22.1 SECTION REVIEW

1. Why are some waters more polluted than others?

2. How do we determine if water is contaminated with sewage?

3. Describe how a sewage-treatment plant works.

22.2 TOXIC CHEMICALS

In 1962, the American biologist Rachel Carson (1907–1964) published a book called *Silent Spring,* in which she warned about the increasing dangers of chemical pollutants in our environment. Carson, who was a respected marine scientist and writer, is generally credited as being the first person to make the public aware of the dangers of chemical contaminants to wildlife and, ultimately, to ourselves. Chemicals that are harmful to living things are known as **toxic chemicals**. Wastes or by-products from industrial, agricultural, and domestic activities often contain toxic chemicals, which are pollutants that harm the environment. Carson warned that unless we cut back on the use of toxic chemicals—especially pesticides, which can poison living things besides insects—we could wake up to a "silent spring" in which no more birds would be left to sing. Fortunately, many readers listened to her warning, and there are ongoing efforts to protect the environment. One chemical Rachel Carson warned about was DDT. The use of this chemical provides an important case study about the dangers presented by toxic chemicals to the environment.

DDT

The toxic chemical DDT (*d*ichloro*d*iphenyl*t*richloroethane) is an insecticide. For more than 20 years in the United States, it was sprayed on farms, in swamps, and in coastal areas to kill insects such as mosquitoes. DDT belongs to a group of compounds called *chlorinated hydrocarbons*. When DDT enters the marine environment, the chemical passes through the food chain where it accumulates in ever-greater concentrations, from mosquito larvae to fish to marine birds.

One marine bird, the California brown pelican, almost became extinct in the mid-1960s as a result of DDT spraying. Scientists discovered that DDT interfered with the birds' use of calcium, an important element in eggshells. DDT caused the pelicans to produce thin-shelled eggs. Many of the eggs cracked during incubation. As a result, the pelican population declined. DDT had similarly harmed other predatory birds, such as the bald eagle, peregrine falcon, and osprey. Since 1971, however, the U.S. government has banned the use of DDT, and in recent years we have seen a gradual increase in the populations of brown pelicans, ospreys, peregrine falcons, and eagles. (See Figure 22-5.) Unfortunately, the use of DDT is not banned worldwide, and its effects are seen far from the areas in which it is sprayed. DDT has been found in the fatty tissues of

Figure 22-5 Brown pelicans have made a comeback since DDT was banned.

seabirds and marine mammals in the Arctic and the Antarctic, proving that the chemical can be carried quite far by air and water.

PCBs

Factory sites are usually located near bodies of water. The water is a convenient place to dispose of chemical wastes produced by the factories. From 1950 to 1975, a company located along the Hudson River in upstate New York dumped hundreds of thousands of kilograms of another type of chlorinated hydrocarbon, called *PCBs* (*p*olychlorinated *b*iphenyls), into the river. PCBs are used in a variety of consumer products, including paints and electrical components. PCBs were found to cause cancer in laboratory animals and are suspected of causing cancer and birth defects in humans. The dangers of PCB contamination are now well documented.

After being discharged into the Hudson River, the PCBs sank to the bottom, where they remained in the sediment for a long time. Here they contaminated bottom-dwelling invertebrates. When fish ate these animals, the PCBs in them entered the food chain. As the PCBs moved up the food chain, their concentration increased. Remember that this had also occurred with DDT. The increase in concentration of a chemical substance, particularly a toxic one, as it moves up a food chain is called **biological magnification**, or *biomagnification* for short.

In 1976, the Environmental Protection Agency (EPA) tested the Hudson's striped bass for PCBs, and found the concentration to be 5 parts per million (ppm)—more than twice the permissible limit. As a result, the commercial fishing of striped bass in the Hudson River was stopped. Like many other marine organisms that hatch and develop in brackish water or freshwater, striped bass migrate between the river (an estuary) and the ocean, forming a link with the marine ecosystem. This connection is important, because PCBs have been detected in the tissues of seals, porpoises, and beluga whales.

How has PCB contamination fared in recent years? There is some good news. First, PCBs are no longer being dumped into the Hudson River. As a result, PCB levels in striped bass have dropped to less than 2 ppm. Second, now that the level of contamination has dropped, the striped bass fishing industry can start again in New York. (See the feature about striped bass and the Hudson River on page 549.)

ENVIRONMENT
Swimming with the Fishes

Who would have thought that New Yorkers would ever be able to have locally caught striped bass for dinner again? Well, the tide is finally turning against water pollution in the Hudson River. For more than two decades, commercial fishing for striped bass has been forbidden due to the high levels of cancer-causing PCBs in river sediments. The latest tests show that, as result of cleanup efforts, PCB levels are now low enough to again permit commercial fishing for "stripers" in the lower Hudson River.

There is more to this success story—you can now "swim with the fishes" in the Hudson River. The construction of new sewage-treatment plants and the upgrading of other plants have greatly reduced the levels of pathogenic bacteria in the river. Marathon swims around the island of Manhattan are now regular events each summer because the water quality has improved so dramatically. Also, when you swim, you will have lots of company. Fishermen have reported record numbers of striped bass in some parts of Lower New York Bay. During the spring, huge numbers of the adult bass swim up the Hudson River to spawn. The fish are so numerous in some areas that they bump against the fishing boats!

With stripers thriving, water quality improving, and New Yorkers discovering first-hand that Manhattan is indeed an island city, we are reminded that success stories in the battle against water pollution are possible when stewardship of the environment is taken seriously.

QUESTIONS

1. Why was commercial fishing for striped bass banned in the Hudson?

2. Explain why it is now safe for people to swim in the Hudson River again.

3. Why is a cleaner Hudson River important for the survival of the striped bass?

Mercury

Another toxic substance that has been dumped into the marine environment is the element mercury (Hg). In Minimata Bay, Japan, from the early 1950s to 1960, more than 100 people developed tremors, fell into comas, and died. Many more were stricken with a variety of nervous system ailments that included blindness, loss of hearing, insanity, and paralysis. Doctors discovered that all victims had consumed large amounts of local fish and shellfish, so they suspected that the seafood was contaminated. When the seafood was examined, high levels of mercury were found. The mercury was traced to a nearby industrial plant that manufactured plastics and chemicals. The factory's liquid wastes, which contained mercury, were being discharged into the bay.

How does mercury get into humans from a contaminated environment? Like DDT and PCB, mercury enters the food chain. Wastes that contain mercury compounds enter the water from industrial sources. The mercury settles to the ocean bottom, where it is acted on by bacteria and changed into a form that can be absorbed by other organisms. First, plankton take up the mercury; then it moves along the food chain to humans, increasing in concentration (biomagnifying) at each step. In the so-called Minamata disease that occurred in Japan, small fish contained 24 ppm of mercury, while people had 144 ppm. Doctors found that mercury binds to the proteins in nerve cells. This explains why victims suffered from various nervous system disorders.

In the 1970s, the chemical plant in Minamata finally stopped discharging mercury. Unfortunately, as in the Hudson with its PCBs, the sediments in Minamata Bay were still contaminated with mercury. The events in Minamata spurred a global investigation into mercury contamination. Findings showed that fish eaten in other countries—including the United States—were contaminated with mercury. In 1972, Congress passed the Clean Water Act, which required industrial plants to install equipment that prevents mercury from being released directly into the water. However, mercury can also enter the ocean from factories (such as coal-burning plants and paint and chemical factories) located inland. Mercury in their smoke emissions is absorbed by moisture in the atmosphere. Winds carry the tainted air over bodies of water, such as lakes and oceans. Then rain and snow deposit the mercury into the ocean.

The U.S. government's Food and Drug Administration (FDA) set a limit of no more than 0.5 ppm of mercury in fish. As a result, the occurrence of mercury poisoning from marine sources has declined significantly in this country. (By contrast, in the 1960s, some fish in the Great Lakes had mercury levels between 1 ppm and 2 ppm.) Not all countries have strict environmental laws, so the mercury levels in some food fish, such as swordfish, often exceeds the maximum level. Accordingly, the FDA has forbidden the importation of swordfish from other countries. Swordfish that is sold in this country comes from U.S. waters. The FDA has also recommended that, in the case of tuna, consumption be limited to no more than 400 grams (about four servings) per week. Tuna and swordfish are both large, predatory fish. Since these fish are high on the food chain, mercury concentrations are biomagnified in their tissues.

Other Heavy Metals

Mercury is classified among a group of elements known as **heavy metals**. Other heavy metals, such as lead, cadmium, arsenic, chromium, copper, nickel, zinc, and tin, are also discharged into the world's waterways as a result of industrial processes. Some of these metals are discharged along with sewage sludge. Although these metals may occur naturally in the ocean, they pose a problem when they are present in large amounts. Like mercury, when these metals accumulate in fish and shellfish, they can harm marine organisms and present a health risk to people who eat them. The FDA monitors the concentration of such toxic elements in seafood and offers suggestions about safe levels of consumption.

Oil Pollution

On March 24, 1989, the oil tanker *Exxon Valdez* hit a reef in Prince William Sound near Valdez, Alaska, and spilled about 50,000 tons of crude oil. The oil spill affected marine life in the area. The crude oil sank to the bottom, covering innumerable shellfish. Seabirds were coated with oil. The birds could not remove the sticky oil from their feathers, so thousands of birds froze to death, drowned, or were poisoned by the oil as they tried to clean themselves. The spill killed mammals, too. Approximately 20 killer whales, 300 harbor seals, and 2800 sea otters died. Fish, too, were covered with oil, which coated their gills and caused suffocation. Countless microscopic plankton, on which all other marine life depends, were also killed by the oil.

This oil spill was only one of many spills that have occurred over the years. The biggest spill occurred in 1978 when the tanker *Amoco Cadiz* broke up in stormy weather off the coast of France and spilled about 200,000 tons of oil into the sea. The oil continued to leak from the tanker and washed ashore for weeks following the accident. The damage the spill caused to marine life along a 300-km stretch of coastline was enormous. As in the case of the *Exxon Valdez,* thousands of fish, seabirds, plankton, and bottom-dwelling invertebrates died. The effect on the marine environment lasted for several years. Oil remained in the water and sediment, and the catch of fish in commercial fisheries declined. For millions of years, oil has been naturally seeping from sediments in parts of the marine environment. However, the effects of natural seepage are small compared to the impact of a major crude oil spill from a tanker.

In early 1996, the oil tanker *Sea Empress* ran aground off the coast of Wales and leaked more than 70,000 tons of oil. Nearly 200 km of coastline were affected, and hundreds of square kilometers of the sea were declared off-limits to the local fishing industry. In November 2002, the oil tanker *Prestige* split in half and sank off the coast of Spain, spilling at least 10,000 tons of toxic fuel oil into the stormy North Atlantic. The oil slicks that resulted from the spill damaged the region's fishing industry, killed countless seabirds, soiled more than 250 km of beaches, and harmed the fragile marine ecosystem. Unfortunately, since more than 1 billion metric tons of crude oil are transported by tankers each year, other large oil spills at sea still may occur in the future.

As devastating as these oil spills are to the environment, they account for only about 20 percent of all oceanic oil pollution. Most oil pollution originates from what are called *nonpoint sources*. Unlike a spill from a tanker, which is a specific point of origin for oil pollution, **nonpoint source pollution** includes many diffuse sources of pollution. For example, the discharge of consumer products that contain oil into sewer systems from homes, businesses, and motor vehicles is a major form of nonpoint source oil pollution. Such oil is not removed during sewage treatment, so it is discharged with the effluent into our waterways. Nonpoint source pollution is difficult to correct because it occurs everywhere along the shore, especially in harbors. Oil leaks out when ships dock and handle oil supplies, and when ships flush their tanks at sea. The practice of flushing tanks at sea is illegal, but it still occurs—often at night—because the seawater is readily available.

Efforts are underway to reduce oil contamination in the marine environment. Regulations have been issued that call for the construction of stronger, double-hulled tankers, which are less likely to spill oil in the event of a collision. The occurrence of oil spills has been decreasing. Yet slicks and globs of oil still float over the ocean's surface. In time, the oil dissipates in the environment because bacteria and wave action break it down. In addition, scientists have identified certain bacteria that break down crude oil. These oil-eating bacteria are sprayed onto oil spills to try to clean them up before they damage the marine environment. Scientists found these bacteria to be effective in cleaning up the major oil spills that were caused during the Persian Gulf War. In spite of those efforts, marine life still suffered. The large amounts of oil that suddenly enter the ocean from such spills are usually more than natural processes can handle.

Radioactive Wastes

Another type of pollutant that poses a risk to the health of the marine environment is **radioactive wastes**. Radioactive wastes, also called *nuclear wastes*, consist of radioactive isotopes, which are atoms of the same element that differ in their atomic mass. Isotopes are unstable; that is, they break down, or decay, into smaller atoms of other elements and emit high-energy rays and particles in the process. Each element has its own half-life, which is the time it takes for half the radioactive atoms present to decay into a stable nonradioactive substance. For some elements, the half-life is very short—less than a day. For others, it is thousands of years. This means that the radioactive material will be present for a very long time until it is completely decayed. Such materials can threaten the well-being of living things.

The high-energy rays and particles that are emitted during decay can pass through the tissues of organisms. As they do so, they can damage DNA and other molecules. This damage can cause cancer and genetic mutations. Energy from radioactive reactions is used for military purposes and to run power plants that produce electricity; radioactive elements are also used in medical procedures. The above-ground testing of atomic weapons, which put radioactive wastes into the atmosphere, has now been banned. However, the other processes still generate large amounts of radioactive wastes.

For a long time, it was thought that radioactive wastes could be safely stored on the ocean floor because it was thought to be calm and barren. We now know that parts of the seafloor are actually very active—geologically and biologically. Up until 1972, low-level radioactive wastes were, in fact, placed in metal drums and dumped hundreds of kilometers off the Atlantic and Pacific coasts. Some of these drums have since washed ashore on the West Coast, while others have been pulled up in trawl nets from the Atlantic. Perhaps even more frightening, other drums have rusted, causing the release of their radioactive elements into the water. Radioactive fish and invertebrates have been caught in the Pacific Ocean. Obviously, there is a risk to marine organisms and to the people who eat them.

One radioactive isotope of particular concern is strontium-90, which competes with calcium for a place in the bones. This element can enter the marine environment as fallout from weapons tests, in effluent from industrial wastes, or from leaking drums of wastes in the ocean. Although we absorb some strontium-90 from normal background radiation, exposure to it from tainted marine foods could be dangerous. At present, much less than 1 percent of the radioactive particles humans receive comes from marine resources.

Fortunately, in 1972, U.S. laws and an international agreement were passed that prohibited the dumping of high-level radioactive wastes into the ocean. In addition, clay in the ocean's sediments can absorb radioactive particles that are dispersed through the water. However, there is still a chance that the ocean floor will be used as a dumpsite for radioactive wastes, if they can be placed far below the seafloor in geologically stable regions.

Thermal Pollution

Factories and power plants are often located along estuaries so they can use seawater as a coolant for their machinery. Some plants dis-

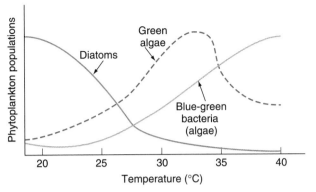

Figure 22-6 The effects of thermal pollution on algae populations.

charge the heated water back into the estuary. This creates a **thermal plume**, a flow of water with temperatures that are significantly higher than those of the water it enters. The release into natural waterways of very warm water—which can adversely affect marine life—is called **thermal pollution**. At higher temperatures, the populations of microscopic phytoplankton such as diatoms and green algae decline significantly, while the blue-green bacteria (algae) greatly increase their population. (See Figure 22-6.)

Large numbers of blue-green bacteria produce foul odors and toxic substances that are harmful to marine life. Fish avoid the high temperatures and low oxygen levels of thermal plumes, resulting in the decline of fish species in such areas. In shallow waters, benthic invertebrates are not able to escape the thermal shock produced by higher water temperatures.

Some communities have solved the problem of thermal pollution by constructing cooling towers at power plants. Others have built shallow pools on the land to cool the heated water before it is discharged into an estuary. Recently, some power-plant operators have used the cooling pools to commercially raise fish and lobsters.

22.2 SECTION REVIEW

1. Explain how mercury in the marine environment is a potential public health problem.

2. Why is dumping of radioactive wastes on the seafloor generally considered an unsafe practice?

3. Why is thermal pollution damaging to estuarine ecosystems?

As you already learned, plants and algae need light in order to carry on photosynthesis. The amount of light aquatic plants receive depends on the clarity of the water they are in and the depth of penetration of the sun's rays. Clear waters that are not clouded by suspended particles such as sand or silt, or fouled by an oil slick, enable greater penetration of light, which marine plants can then absorb. Likewise, polluted or murky waters decrease the light available for plants. The measure of the level of clarity or murkiness of water is called **turbidity**.

Turbidity

How is water turbidity measured? A simple device called a *Secchi disk*, named after the Italian scientist Father Pietro Secchi (1818–1878), is used to measure the distance light penetrates into the water. The disk, which has four alternating sections of black and white, is gradually lowered into the water from a pier or a boat until the disk can no longer be seen. The depth at which the disk is no longer visible is defined as the turbidity. A reading of less than 2 meters indicates that the water is very turbid. (See Figure 22-7.)

Cloudy or turbid water can be caused by natural conditions. In the northern latitudes, where light is less intense and high plankton populations cloud the waters, visibility is low and turbidity is high. In the tropics, where light is much more intense but there are fewer plankton, you can see more than 30 meters underwater, so turbidity is low. Other natural conditions, such as stormy weather and strong ocean currents, can stir up bottom sediments and reduce visibility.

Human activities can also cloud the waters. Sometimes, poor agricultural practices lead to soil erosion. When soil is blown or washed off the land, it can enter rivers and streams and eventually be carried out to sea. Great quantities of eroded soil cloud coastal waters and choke aquatic communities such as estuaries and coral reefs.

During several summers in the late 1980s, the coastal waters off Long Island, New York, turned so cloudy that the condition became known as *brown tide*. Marine biologists discovered that the cloudy

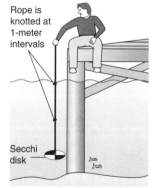

Rope is knotted at 1-meter intervals

Secchi disk

Figure 22-7 How to use a Secchi disk to measure water turbidity.

water was caused by an overpopulation of algae, called an *algal bloom*. Water turbidity was so high that bathers could hardly see their hands in front of them while swimming along the beach. The situation became so bad that, in the summer of 1987, a fish kill occurred in Hempstead Harbor, Long Island. (See Chapter 4 to review information on algal blooms, brown tides, and fish kills.)

What caused the algal bloom and the fish kill? Scientists think that pollution set into motion the chain of events that led to the fish kill. Effluent that contained sewage and chemical fertilizers from lawns and farms entered the harbor from the land. Organic matter in the sewage and fertilizers contained high levels of phosphates and nitrates, which caused an algal bloom. The large algae population used up oxygen and clouded the waters, reducing light penetration. Other microscopic algae suspended in the water and eel grass growing on the bottom died from lack of light. Dead plant matter accumulated on the bottom, where it decayed. During the process of decay, aerobic bacteria extracted still more oxygen from the water. The result was a fish kill. Fish breathe by taking oxygen out of the water through their gills. The fish suffocated because there wasn't enough oxygen dissolved in the water.

The discharge of sewage and chemical fertilizers into local waterways are major factors that contribute to the poor water quality off Long Island and elsewhere around the country. Greater efforts are needed to reduce the runoff of nutrient-rich wastes and fertilizers into rivers and estuaries.

Dissolved Oxygen

In 1976, a major fish kill occurred in the waters between New Jersey and New York. Scientific tests found that the amount of oxygen dissolved in the water was very low. The amount, or concentration, of oxygen dissolved in water is called the *dissolved oxygen (DO)*. Depending on its temperature, ocean water has a DO concentration that ranges from 1 to 12 ppm. On the day of the fish kill, the water had a DO concentration of only 2 ppm. Scientists concluded that the fish died of suffocation, because the DO level was below the critical threshold of 4 ppm, the minimum level of oxygen required for fish to breathe. This aquatic condition is known as **hypoxia**, meaning "low oxygen."

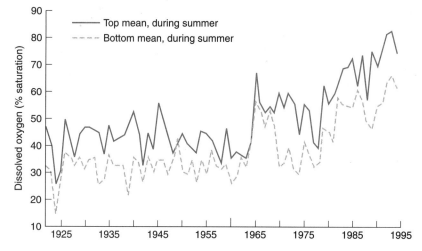

Figure 22-8 The DO levels have been increasing in the Hudson River.

What caused hypoxia in these waters? The low DO levels were found in an offshore area where the dumping of sludge had occurred for a number of years. As in the case of the Hempstead Harbor fish kill mentioned above, bacteria in the sludge were using up oxygen during the process of decay. The DO levels declined over time and then fell to 2 ppm. In light of these findings, the U.S. government and the state of New York agreed to dump sludge farther offshore and to eliminate ocean dumping entirely by 1993—which they did.

The reduction and elimination of sludge dumping in the ocean has had a beneficial effect on DO levels in coastal cities such as New York. Notice, in Figure 22-8, the overall rise in DO mean levels in the Hudson River since 1925. There is still room for improvement, and other cities around the country have not fared as well. However, the construction of new sewage-treatment plants and the upgrading of older treatment plants will greatly improve the quality of our local waterways. In addition, projects are being undertaken in many cities to recycle some of the organically rich sludge for use as fertilizer.

22.3 SECTION REVIEW

1. What is turbidity? How is it measured?

2. Explain how pollution can cause an algal bloom.

3. Describe how an algal bloom lowers DO levels and leads to a fish kill.

22.4 LITTER IN THE WATER

During the summer of 1988, a tide of solid wastes washed up on the shore, from Massachusetts to Maryland. Much of the debris consisted of medical wastes, including hypodermic syringes and other products made of plastic. There was so much garbage in the water that many beaches had to be closed. The public was frightened and outraged.

No spot is too remote for seagoing trash. On Ducie Island, an uninhabited spot in the middle of the South Pacific nearly 500 km from the nearest inhabited island, scientists have discovered litter washed up on its beaches. On one occasion, a scientist counted 953 pieces of trash, most of it plastic, along a 2.5-km stretch of beach. Much of the garbage came from ships, such as cruise ships, that had routinely dumped their trash overboard.

Litter is solid waste or garbage. Most litter consists of plastic, glass, and metal—materials that do not undergo natural decay. Litter that cannot be broken down by natural processes is called **nonbiodegradable**. A nonbiodegradable waste such as plastic may remain in the environment for hundreds of years.

Plastic wastes not only are unsightly but often pose a threat to marine life. Some animals, particularly sea turtles that eat jellyfish, mistake plastic bags for food. The turtles then die—either of starvation (with plastic bags filling their stomachs) or of suffocation (after choking on the plastic bags). Carelessly discarded plastic rings from beverage six-packs trap and choke fish, birds, and other marine life when the animals swim, or put their heads in, through the rings and are unable to get them off their bodies. And each year, thousands of fish, seabirds, turtles, and marine mammals die when they become entangled in plastic gill nets, fishing line, and huge drift nets that are discarded or lost at sea by fishing vessels.

The United States throws away more trash than does any other nation in the world. More than 150 million tons of solid wastes, or refuse, are thrown out each year—nearly 10 million tons of it into offshore waters. Among the items dumped into the sea are millions of old cars, along with old boats and military weapons. In addition, millions of glass, metal, paper, plastic, and plastic foam items are thrown into the ocean each year. Whereas many of these items may be harmful to marine life and the environment, sunken cars and

Figure 22-9 An old tire has become part of this barracuda's environment.

ships sometimes serve as artificial reefs that attract fish, which is good for ocean life and for recreational activities. (See Figure 22-9.)

Approximately 75 percent of all garbage is buried in landfill sites. But many U.S. cities and states are running out of such spaces to dump their solid wastes. Landfills are supposed to be constructed so that waste substances do not leach into the ground and contaminate groundwater. In spite of this, some landfills contain hazardous chemicals that seep through the ground into drinking water sources and into nearby waterways that are used for swimming and fishing. The proper disposal of solid wastes is a serious problem because as the U.S. population increases so too will the amount of garbage it generates.

Solutions to Pollution

What can communities do to dispose of garbage properly, without dumping it directly into the ground or the ocean? One method is incineration, the disposal of solid wastes by combustion. There are some 200 large incinerators now operating in the United States. Although the burning of wastes can be used to generate energy, it is not a perfect solution to waste disposal. Many towns cannot afford to build an incinerator. In addition, incineration of garbage can produce air pollutants, especially when plastics are burned.

There are valid environmental and economic concerns about the disposal of solid wastes in landfills and by incineration. A more ecologically sound method of handling solid wastes is *recycling*—the disposal of garbage by reusing it or by converting it into useful products. Some of the most commonly recycled items are paper, plastics, glass, and metal. The incentive of deposit payments for beverage cans and bottles has greatly diminished the number of such containers that are discarded. Billions of glass food jars and aluminum cans are recycled in the United States each year. Incentives for the recycling of paper have spawned about 2000 wastepaper dealers and brokers nationwide. Out of an estimated 27 million tons of waste paper, about 5 million tons have been recycled into packaging products and exported abroad. However, experts believe that recycling at best will involve only 25 percent of our trash. Hopefully, companies will use more biodegradable products, and the government will provide more economic incentives for companies and consumers to recycle. In Japan, 50 percent of solid wastes are recycled. Citizen participation in recycling programs is essential if we are to reduce the amount of litter we produce and discard.

There Ought to Be a Law!

Pollution affects everyone. At best, it is an eyesore; at worst, it presents a hazard to the environment and to the health of all living things. Knowing this, many governments, organizations, and individuals are working to develop solutions to pollution. These efforts often specifically concern the marine environment.

What have governments done to prevent or reduce marine litter? In 1973, the International Maritime Organization, an agency of the United Nations, formed an agreement called *MARPOL* (*Mar*ine *Pol*lution) that regulates the disposal of hazardous chemicals, sewage, and trash from ships at sea. As of 1992, fifty nations had signed the MARPOL Treaty, which prohibits the dumping of plastic trash overboard from ships at sea.

Groups and Organizations Against Pollution

Private industry, nonprofit research groups, environmental groups, and government organizations are all working to find ways to prevent and reduce debris in the marine environment. Plastics manufacturers have introduced photodegradable plastics in some consumer items. These plastics break down over time from the action of sunlight on the product. Additional work has been done to increase the feasibility of plastic recycling. The Society of Plastics Industry, a trade organization of more than 1900 members, is trying to solve the problem of plastic discarded at sea that breaks down into pellets. These small pieces of plastic disperse throughout the ocean and can be harmful to marine animals that eat them.

Federal agencies are also engaged in efforts to reduce marine debris. The Environmental Protection Agency, National Oceanic and Atmospheric Administration (NOAA), U.S. Coast Guard, Department of the Interior, and U.S. Navy have undertaken a cooperative effort to deal with the problem of garbage in the marine environment. The EPA and NOAA are cosponsors of the Center for Marine Conservation. One of their initiatives is the national Beach Cleanup Campaign, during which volunteers record the types and quantities of solid wastes that they collect along the shore. Clearly, it is better to keep a beach free of litter than to risk the health hazards, damage to wildlife, and high cleanup costs of pollution.

Individuals can make a difference in tackling ocean pollution, too. By becoming informed about the problem, people can choose to do a number of constructive things to help protect the marine environment. They can actively clean up local beaches (or at least not leave any debris on them that may damage the environment or threaten marine life); request and purchase products that have less, or recyclable, packaging on them; and lobby for the passage and ongoing enforcement of environmental laws. Consumers must be willing to assume the extra costs that may be involved in more environmentally sound manufacturing and recycling processes, because the costs involved in environmental degradation are much higher.

22.4 SECTION REVIEW

1. Explain why plastic debris is dangerous to marine life.

2. Which method of solid waste disposal is best for protecting the environment?

3. What products can be reused through recycling of materials?

Laboratory Investigation 22

Determining the pH of Water Samples

PROBLEM: How can the pH of various water samples be determined? What is the pH of ocean water?

SKILL: Using chemical indicators to measure pH levels.

MATERIALS: Tray, loose-leaf paper, red litmus paper, blue litmus paper, pH hydrion paper (wide range), medicine dropper, ocean water, rainwater, tap water, pond water.

PROCEDURE

1. Place a piece of loose-leaf paper on your tray. Open the vials containing red litmus paper and blue litmus paper. Remove four strips of red litmus paper and four strips of blue litmus paper. Place them on the loose-leaf paper, in four sets of one blue and one red each. Label each set with the type of water being tested: ocean, rain, tap, and pond water.

2. Use the medicine dropper to place one drop of each water sample on the red litmus paper and one drop on the blue litmus paper. Do one water sample at a time. Observe if there is a color change. Write the color in your copy of Table 22-1. Repeat for each of the samples.

3. To determine if the water sample is acidic, basic, or neutral, you can use the following scheme: Red litmus paper stays red in acid, but turns blue in base; blue litmus paper stays blue in base, but turns red in acid.

TABLE 22-1 TESTING THE pH OF WATER SAMPLES

Sample	Red Litmus	Blue Litmus	Acidic, Basic, or Neutral	pH level
Ocean water				
Rainwater				
Tap water				
Pond water				

4. Litmus paper is useful only for determining whether your water sample is acidic, basic, or neutral. To find the pH level, you need to use pH hydrion paper, which comes in a container with a color scale that indicates pH values.

5. Remove four strips of pH paper from the container. Put a drop of water from the first sample on one strip of pH paper. Compare the color on the strip with the color scale on the container. Note the pH and record it in Table 22-1. Repeat for the other water samples.

6. Check your results by referring to the following pH scale: 0 to 6 ranges from very to slightly acidic; 7 is neutral; 8 to 14 ranges from slightly to very basic (alkaline).

OBSERVATIONS AND ANALYSES

1. What is the pH of ocean water? Find its location on the pH scale.

2. Briefly describe how you would determine the pH of ocean water.

3. What is the advantage of using pH hydrion paper instead of, or in addition to, litmus paper?

Answer the following questions on a separate sheet of paper.

Vocabulary

The following list contains all the boldface terms in this chapter.

biological magnification, effluent, heavy metals, hypoxia, nonbiodegradable, nonpoint source pollution, pollutants, pollution, radioactive wastes, sludge, thermal plume, thermal pollution, toxic chemicals, turbidity

Fill In

Use one of the vocabulary terms listed above to complete each sentence.

1. Litter that cannot be broken down naturally is _____.
2. The solid part of the sewage that settles in tanks is _____.
3. The measure of water's level of murkiness is called _____.
4. The liquid part of sewage that is discharged is the _____.
5. The condition called _____ describes a very low DO level.

Think and Write

Use the information in this chapter to respond to these items.

6. Explain why it might be dangerous to eat shellfish that were gathered near a harbor.
7. How does oil pollution affect marine life? What are some ways that its effects can be lessened?
8. What conditions can produce a low DO? What is its effect on aquatic life?

Inquiry

Base your answers to questions 9 through 11 on the data below and on your knowledge of marine science.

A group of marine science students wanted to know if the water at their beach was clean enough to swim in. They decided to test the water for the presence of fecal coliform bacteria to see if the level was in compliance with EPA standards. (See the table below.) The students used the membrane-filter technique to culture a 10-mL sample of ocean water. A tap water sample was cultured as a control. After 24 hours, the Petri dishes were observed for the presence of fecal coliform colonies. To compare to EPA standards, the students counted the number of bacterial colonies in the dish, and then multiplied by 10 to obtain the number of coliforms that would be present in 100 mL of ocean water. (See diagram below.)

Culture of fecal coliform bacteria from ocean water

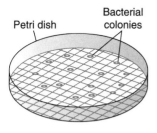

Type of Water	EPA Fecal Coliform Standards
Drinking water	Zero coliforms allowed
Shellfish water	Should not exceed 14 coliforms/100 mL
Swimming water	Should not exceed 200 coliforms/100 mL
Harbor water	Should not exceed 2000 coliforms/100 mL

9. How many colonies were found in the 10-mL sample? How many coliforms would be in a 100-mL water sample?

10. Based on the EPA standards, would this ocean water be suitable for swimming in? Explain.

11. Would this ocean water be suitable for the harvesting of shellfish? Explain why or why not.

Multiple Choice

Choose the response that best completes the statement or answers the question.

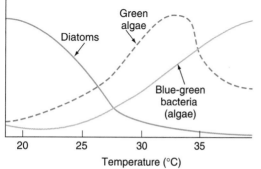

Base your answers to questions 12 and 13 on the graph and on your knowledge of marine science.

12. What would be a suitable title for this graph? *a.* The effect of overcrowding on algae populations. *b.* The effect of salinity on algae populations. *c.* The effect of temperature change on algae populations. *d.* The effect of temperature change on zooplankton populations.

13. Based on the data in the graph, which is an accurate statement? *a.* Green algae are best adapted to survive at the highest temperatures. *b.* As water temperature increases, the number of blue-green bacteria decreases. *c.* As water temperature increases, the number of diatoms decreases. *d.* Green algae are not affected by changes in temperature.

14. Which type of beach litter is biodegradeable? *a.* plastic bottles *b.* soda cans *c.* nylon fishing line *d.* pieces of wood

15. A pollutant that has biomagnified in aquatic food chains is *a.* coliform bacteria *b.* PCBs *c.* crude oil *d.* hydrogen sulfide.

16. A fish kill is most likely to occur in waters where there is *a.* a high DO level *b.* a low DO level *c.* low turbidity *d.* a high mercury content.

17. All of the following are examples of toxic substances *except* *a.* PCBs *b.* mercury *c.* DDT *d.* manganese nodules.

18. The best explanation for why some swordfish have high mercury levels is that *a.* they take in water containing mercury *b.* they absorb mercury from bottom sediments *c.* the mercury moves up the food chain to them *d.* they are exposed to atmospheric mercury.

19. Which is the greatest source of oil pollution in the ocean? *a.* oil well blow-outs *b.* nonpoint source pollution *c.* oil spills from tankers *d.* oil refinery accidents

20. The instrument used to measure water turbidity is the *a.* Secchi disk *b.* pH scale *c.* hydrometer *d.* barometer.

21. Acid rain is more of a problem in lakes than in the ocean because *a.* it rains more on land than at sea *b.* air pollution is greater inland than along the coasts *c.* the buffering action of the ocean counteracts the acidity *d.* ocean currents carry the acids away.

22. The type of pathogen that is used to indicate human sewage as the source of water pollution is *a.* blue-green bacteria *b.* coliform bacteria *c.* dinoflagellates *d.* diatoms.

Research/Activity

- Build a Secchi disk from household items. Work with a team of classmates to check water turbidity daily, weekly, or during the tidal cycle, from the safety of a local dock or pier. Record your results and prepare a graph. You can submit the completed project to your school science fair.

- Identify the types of items you are recycling that may be found in the marine environment as garbage. Report on their impact in the ocean in contrast to their use as recycled products.

23 Conservation of Resources

When you have finished this chapter, you should be able to:

LIST important living and nonliving marine resources.

IDENTIFY recent problems in worldwide ocean fisheries.

DESCRIBE different methods of farming aquatic life-forms.

DISCUSS threats to, and protection of, marine animals.

For years we have thought of the ocean as an endless source of natural resources—both living and nonliving. Fish, invertebrates, and algae are examples of living resources from the sea that are used as a source of food, medicine, fertilizer, and cosmetics for the human population. Oil, gas, and minerals are nonliving resources that are used for energy and industry.

Advances in technology have made the ocean's resources more available to people. As the size of the human population increases, the demand for those resources also grows. People will continue to look to the sea to meet many of their needs. How can we protect the ocean and still provide for the intelligent use of its resources? In this chapter, you will learn about some of the problems and progress made in the use and conservation of marine resources.

23.1
Marine Resources

23.2
Farming the Sea

23.3
Protecting Marine Animals

23.1 MARINE RESOURCES

Humans depend on materials obtained directly from the environment, called **natural resources**, to feed themselves and to make their lives safer and more comfortable. Many of these resources are obtained from the sea. Marine resources include animals, algae, fossil fuels, minerals, and even the water itself. Importantly, most resources are not endlessly available. Some items, such as fish and algae, are known as *renewable resources,* because they can reproduce and replenish their populations. Other items, such as oil and minerals, are known as *nonrenewable resources* because they are available only in limited amounts in nature. People have to wisely manage and safeguard natural resources to ensure their continued long-term availability. The careful management and protection of natural resources and the environment in which they are found is called **conservation**.

Marine Fisheries

Fish are one of the most important resources in the ocean because they are a food source for the human population. The commercial harvesting of fish and shellfish is carried out by the **fisheries** industry. People eat about two-thirds of the fish that are caught and sold; the remaining one-third is used for domestic animal feeds and other products. From the 1950s through the 1990s, the world's fishing fleets harvested increasingly larger amounts of fish. (See Figure 23-1.) Whereas the world's seafood catch was about 20 million metric tons (mmt) in 1950, by the year 2000 it reached nearly 95 mmt. Now, due to decades of intensive overfishing, that amount has decreased and the world fish catch appears to be leveling off.

Most fish, about 60 percent, are caught in the North Atlantic and North Pacific waters. Almost the entire catch is made in the neritic zone, that is, above the continental shelf. Although there are thousands of species of marine fish, only about 30 species account for most of the fish intentionally caught. The most commonly caught species of seafood (as of 1990) are listed in Table 23-1; catch numbers are in millions of metric tons.

Japan, Russia, the United States, and China are among the leading fishing nations of the world. Sixty percent of the fish eaten by

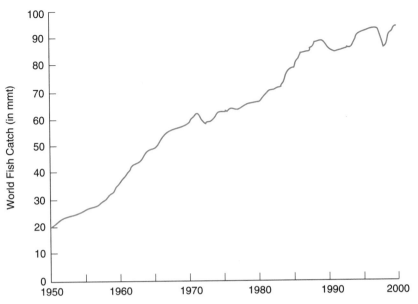

Figure 23-1 The world commercial fish catch over a period of 50 years.

people in the United States is imported from other countries, most of it from Canada and Japan. On average, a person in America consumes only about 6 kg of fish per year, whereas a person in Japan (which has the world's highest consumption of fish per person) eats several times that each year.

TABLE 23-1 **WORLD COMMERCIAL CATCH OF MARINE SPECIES**

Species Group	Catch (mmt)*
Herring, sardines, anchovies	24.6
Cods, hakes, haddocks	12.8
Miscellaneous marine fishes	11.0
Jacks, mullets, sauries	9.2
Mollusks	7.9
Redfish, basses, conger eels	5.9
Crustaceans	4.5
Tunas, bonitos, billfishes	4.0
Mackerel, snooks, cutlass fishes	3.8
Flounders, halibuts, soles	2.0

Source: FAO Yearbook of Fisheries Statistics and U.S. Department of Commerce, Fisheries, 1990.

Methods of Catching Fish

Fish are caught commercially by using either hooks and lines or a variety of nets, as shown in Figure 23-2. Long fishing lines with many hooks attached are used to catch widely dispersed surface and bottom fish in a method called **long-lining**. Some species of tuna are caught in this manner. Nets catch many more fish at a time, and are usually set on schools of fish. One type of net, the **trawling net**, is released from the rear, or stern, of the ship and is pulled through the midwater region to catch squid and schooling fish such as tuna, herring, mackerel, and anchovies. The net can also be dragged along the seafloor to catch bottom-dwelling fish such as sole, cod, halibut, and haddock.

The pelagic fish, such as tuna, mackerel, herring, and anchovies, which swim in large schools closer to the ocean surface, can also be caught in other types of nets. The **purse seine net** is used to surround and trap large schools of fish. The **gill net** intercepts a school of swimming fish, which get caught by their gills in the net's mesh. Both methods of fishing also catch unintended victims. Purse seine nets that are set on schools of tuna sometimes accidentally trap and drown dolphins (which must breathe air at the surface) that swim along with the tuna. Modifications have been made to some nets, enabling the dolphins to escape. Gill nets that are set for fish such as salmon entangle and drown thousands of marine mammals and hundreds of thousands of seabirds each year. More than 20 different seabird species are affected. Seabirds are also caught and drowned by long-line fishing hooks. Scientists have already noticed population declines in several species of seabirds and marine mammals.

There is intensive and widespread use of large fishing nets by commercial fishing fleets (by the "factory ships" that can catch and process huge amounts of fish). This has contributed to a decline in the world catch of some of the more important species of food fish. In response to overfishing by foreign vessels in our offshore waters, the U.S. Congress in 1976 passed the Magnuson Act, which forbids foreign fleets from fishing within 333 km of our coast.

Federal protection of our coastal waters from foreign fishing fleets stimulated the growth of the American fishing industry. Bigger and better-equipped fishing boats brought in record numbers of fish, many of them species that previously had been caught by other nations. However, populations of some food fish later declined. In

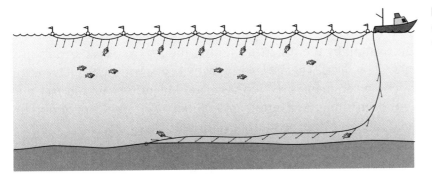

Figure 23-2 The four commercial methods of catching fish.

Hooks and long-line fishing

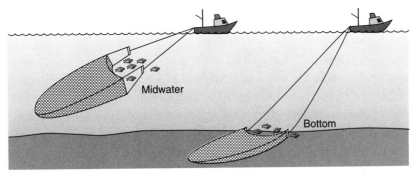

Midwater

Bottom

Trawling nets

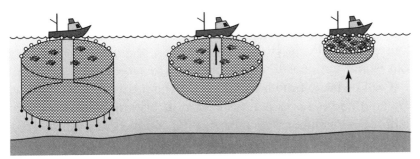

Purse seine net

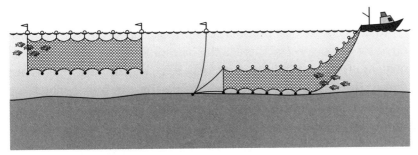

Gill nets

recent years, there has been a decline in the U.S. catch of several bottom-dwelling species, particularly cod, haddock, and flounder. Overfishing by the American fishing industry replaced overfishing by foreign fleets. There were too many boats, with high technology, catching too many fish, too fast. The Magnuson Act did not prevent overfishing. As a result, approximately 70 species of important food fish have been seriously overfished.

Limits on Commercial Fishing

Like all living things, fish need time to increase their population, or **stock**, particularly after heavy overfishing. In recent years, the federal government has initiated some important conservation measures to increase the stock of commercially valuable food fish. Some of these measures include limiting the size of the salmon catch on the West Coast; declaring the Cape Cod fishing grounds off-limits for cod, haddock, and flounder; and setting quotas and designating no-fishing zones to protect the North Atlantic swordfish. When fish are left alone, their populations can recover, although it may take years. Indeed, this has been the case with the swordfish—after 10 years of conservation measures, their numbers are increasing.

An important consideration in developing guidelines for the conservation of a species is determining a reasonable harvest size that would not deplete its population. The largest number of fish that can be taken from a population without threatening its future size (and harvest) is called its **maximum sustainable yield**. Maintaining a maximum sustainable yield for commercially valuable fish species requires regulation and cooperation between the fishing industry and the federal government. At present, three-fourths of the ocean fish that are caught are being fished at or above their sustainable yields; one-third of these species are decreasing in stock.

In addition to the problem of commercial overfishing of preferred food fish species, there is the incredible waste of fish that are incidentally caught in huge nets. Drift nets, in particular, catch tons of unwanted fish. These so-called trash or **by-catch** fish, which are not considered commercially valuable or large enough to keep, are simply discarded. In a recent year, approximately 30 mmt of marine by-catch—which could have been used to feed people—were thrown, dead and dying, back into the ocean. Populations of these

In the near future, you might be having rattail for dinner instead of flounder, or sablefish in place of haddock. These are only some of the creatures from the deep that are finding their way to the dinner table. Stocks of the well-known food fish, such as flounder, cod, and haddock, are declining rapidly worldwide as a result of overfishing. The total world catch of fish peaked by the year 2000 and has been declining ever since.

Fishing boats have been returning with empty nets from their familiar fishing grounds in the waters above the continental shelf. As a result, they are foraging farther offshore, into the deeper waters of the continental slope. Aided by the use of larger, stronger nets and faster ships, the haul of fish from the deep has increased greatly in recent years. The North Atlantic harvest of rattails, a big-eyed fish with a long tail, has reached the huge quantity of about 100,000 tons per year.

Fisheries experts claim, however, that deep-ocean fishing will not be sustainable. For example, the sablefish, which lives at a depth of 1.5 km, matures and reproduces very slowly; overfishing would limit its ability to replenish its numbers. Recently, thousands of tons of orange roughy, which feed on shrimp and squid, were being caught each year in the deep waters off New Zealand. As a result, this new fishery has already collapsed. Stocks of these deep-sea fishes have been exploited to satisfy the public's demand for fish. Scientists say their rapid decline can upset marine food chains because the fish are an important food source for other ocean animals. (See photograph above.)

In the United States, the investigation of deep-sea fisheries is encouraged by NOAA's National Marine Fisheries Service, which supports exploration of the deep ocean with millions of dollars in grants. One of the goals is to allow the depleted stocks of fish species, such as cod, haddock, and flounder, to make a comeback. The federal government is working closely with commercial fisheries to strike a balance between exploitation and conservation.

QUESTIONS

1. Why are the stocks of well-known, popular food fish in decline?

2. Explain why fishing boats are working the waters farther offshore.

3. Why are scientists concerned about deep-sea fish like the orange roughy?

unwanted fish are now in decline, just like the populations of preferred fishes. Fortunately, there are plans by several countries to phase out the use of drift nets. But what is really needed is better management of commercial fisheries.

Fossil Fuels

Your home or apartment may be heated with oil or gas. Older buildings use coal. Oil, gas, and coal are called *fossil fuels* because they come from the remains of long-dead plants and animals. Fossil fuels are formed in Earth's crust, both on land and in the ocean floor. Over time, great quantities of microscopic plankton that float near the ocean surface die and settle to the bottom, forming (along with other mineral sediments) layers hundreds of meters thick. The pressure of overlying layers of sediment fuses the bottom layers into a type of rock called *sedimentary rock,* which contains organic chemicals from the dead organisms. The chemicals in the rock, called *hydrocarbons,* consist mostly of fat compounds that come from the cells of the decayed organisms.

High temperatures and pressures from Earth's interior heat the rocks to more than 150°C, causing the solid hydrocarbons in the rock to change to oil (much like the solid fat in bacon changes to oil when it is heated on the stove). The oil that forms in sedimentary rock layers is called **petroleum**. Petroleum that is heated to higher temperatures produces gases such as methane and propane. The oil and gas under pressure seep through porous layers in the sedimentary rock until they encounter a nonporous layer called a *dome,* where they accumulate. When oil companies drill for oil, they hope to tap into petroleum-rich domes. Today, almost 70 percent of petroleum comes from land drilling and about 30 percent comes from the seafloor. However, these figures are likely to change as oil wells on land dry up and new deposits are discovered at sea.

Getting Oil from the Ocean

How are oil and gas recovered from beneath the ocean floor? Oil companies drill for oil and gas offshore, usually at the edge of the continental shelf. (Most U.S. oil drilling occurs in the Gulf of Mexico; some also occurs off the southern coasts of Alaska and California.) A large platform, about the size of a city block, is needed to support the drilling rig and to provide living quarters for workers. (See Figure 23-3.) The rig contains a hollow drill pipe with a drill bit that can bore through solid rock. The drill pipe operates inside another pipe called the riser. During drilling, mud is pumped down inside the drill pipe to produce a counter-pressure that prevents oil from gushing up through the pipe. As the drill bit cuts through the rock, pieces of rock are carried by the mud back up to the platform through the space between the drill pipe and the riser. If oil is present, it shows up in the returning chunks of sediment. When the oil amounts to more than 200 barrels a day, the drill site is considered economically worthwhile. A pipeline is then attached to the rig and the liquid petroleum is transported to storage tanks on land.

Drilling for oil is not without its problems. Sometimes the oil and gas accumulate in the rock layers under such great pressure that when a drill breaks through, the whole pipe assembly is forcefully ejected in what is called a *blowout*. One of the biggest blowouts occurred off the coast of Santa Barbara, California, in 1969, spilling millions of liters of oil into the sea. Careful monitoring is required during the drilling operation, so that adjustments can be made when a high-pressure area is encountered.

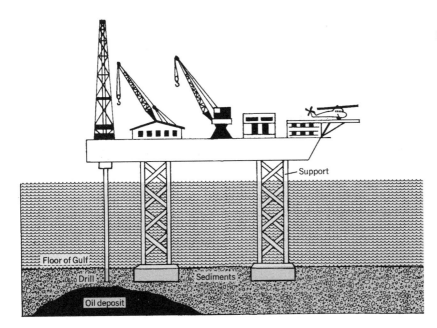

Figure 23-3 Drilling for oil beneath the ocean floor.

How is oil located beneath the seafloor? Sound or seismic technology is used to locate oil fields in the seabed. Oil companies bounce sound waves off the seafloor from ships and analyze the returning echoes to determine if there are oil deposits under the seafloor. Once they have seismic data that indicate a possible oil field, the oil company begins drilling. However, at best only one in ten drilling operations is successful. Drilling for oil is a gamble and a very costly one at that.

Recently, a new technology was developed to increase the chances of finding oil under the seafloor. An oil-hunting ship tows several rows of submerged air guns that generate sound signals aimed at the seafloor. The sound waves penetrate the seafloor and bounce off rock formations located kilometers below. The returning waves are picked up by hydrophones in fiber-optic cables that trail behind the ship. Millions of bits of seismic data are beamed up to a satellite orbiting Earth. The satellite then beams the data down to supercomputer centers, which translate the data into geological maps that reveal potential oil fields. This information is sent back to the ship, directing it to focus on particular sites on the seafloor that are most likely to have oil deposits. The use of the satellite called *ACTS* (*A*dvanced *C*ommunication *T*echnology *S*atellite) shortens the time it takes for oil-hunting ships to locate promising oil fields beneath the seafloor.

Minerals

Another natural resource is **minerals**, which are solid, inorganic substances that are formed in Earth's crust. We normally think of minerals as coming from the land. Yet the ocean also contains an abundance of minerals dissolved in the water or found in rocks. However, these elements are widely dispersed and very difficult to extract with present technology.

The most common mineral extracted from the ocean is salt. Some salt comes from shallow ponds or marshes located along tropical coasts. The ponds are flooded with seawater, either naturally or artificially. When the sun evaporates the water, salt remains behind. The salt can then be mined for commercial use. Millions of years ago, a shallow sea covered much of North America. The emergence

Figure 23-4 Manganese nodules, shown here on the seafloor, contain important minerals.

of the land and the evaporation of seawater left behind huge deposits of salt, thousands of meters deep in some places. Some salt-producing states are New York, Louisiana, Texas, Michigan, and Ohio.

One of the most valuable types of mineral deposits, called **manganese nodules**, lies on the ocean floor. (See Figure 23-4.) The potato-shaped nodules are usually more than 2000 meters deep, which makes them difficult to mine. They occur in various deep-sea locations around the world, although most are found in the Pacific Ocean around the equator. Manganese nodules have been scooped up, at great expense, and brought to the surface to be analyzed. The nodules contain not only manganese but also some iron, nickel, copper, and cobalt—all of which are important minerals for industry. When cut in two, the nodules show a ringlike pattern. Scientists are not sure how the nodules form, but they think that they grow around fossils that settled on the ocean floor. Several industrialized nations, including the United States, have already staked out exploratory claims in the Pacific Ocean.

In December 1982, at an international conference sponsored by the United Nations in Jamaica, West Indies, 119 nations gathered to sign the Law of the Sea treaty. This "constitution of the oceans" stipulated that nations with coastlines have sovereign rights over the ocean and seafloor out to 333 km. The ocean beyond that is "the common heritage of mankind." Any mining done beyond the

333-km limit is managed by the United Nations to ensure that exploitation does not occur and that mineral wealth is shared with the poorer nations. The Law of the Sea treaty also urged all nations to prevent and control marine pollution by "the best practical means at their disposal."

Freshwater and Desalination

More than 70 percent of Earth is covered by water, yet most of the world has a limited amount of available freshwater. This is because about 97 percent of all water is salt water. Most of the freshwater is frozen at the poles. In fact, of the 3 percent that is freshwater, about two-thirds is frozen. Thus, only 1 percent of all water on Earth is available as freshwater in a liquid state. Deserts, in particular, are characterized by long periods with little or no rainfall. In dry areas such as North Africa, droughts often cause massive crop failures. Without food or water, many people and animals die of thirst and starvation.

How can we increase the supply of freshwater to meet human needs? The greatest source of water on Earth is the ocean. Freshwater can be produced from ocean water by removing the salt, in a process called **desalination**. In one method of desalination, called a *solar still,* light passes through a transparent plastic or glass roof to warm the seawater inside. The seawater evaporates, leaving the salt behind. The freshwater condenses on the underside of the roof and trickles down along the sides into a gutter where it is collected and pumped through a pipe to a reservoir. (See Figure 23-5.) Only about 1 liter of freshwater is produced per day from 1 square meter of water surface. Solar stills are built along coasts that receive adequate warmth and sunshine. They are used in regions of limited freshwater, such as Israel, Peru, and West Africa. More solar stills are needed to meet increasing demands for freshwater around the world.

A more recent innovation in desalination technology is reverse osmosis. In this process, which requires energy to work, water molecules in a container of seawater are forced through a semi-permeable membrane into a freshwater tank. Florida has more than 100 desalination plants, and almost all of them use reverse osmosis.

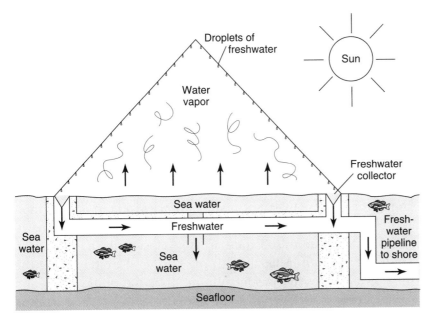

Figure 23-5 A solar still is used for the desalination of ocean water.

The nation's largest reverse osmosis desalination plant is located in Santa Barbara, California. It produces 26 million liters of freshwater per day.

23.1 SECTION REVIEW

1. Why has the catch of important food fish declined dramatically in recent years?

2. Explain why fish can be a renewable natural resource, whereas fossil fuels are not.

3. Why are nations interested in harvesting manganese nodules? Why is it a difficult business venture?

23.2 FARMING THE SEA

Humans have always been hunters and gatherers, foraging on land and fishing at sea to provide for individual and collective needs. As human society and technology progressed, people turned to raising animals and plants on land; that is, they started farming. Today, more food is needed to feed the ever-increasing human population. Since overfishing has become a serious problem, people are looking at alternative ways to obtain food from the sea. One such method is to "farm" the sea. The process of farming organisms from the sea is called **mariculture**. The farming of aquatic organisms, in general, is called **aquaculture**.

Mariculture of Invertebrates

Among the most convenient marine animals to farm are the mollusks, such as mussels and oysters. These sedentary shellfish naturally grow attached to substrates in large clusters in shallow water. Mariculture started in Japan with the raising of oysters along the shore. A female oyster produces more than 100 million eggs at a single spawning. After fertilization, the eggs develop into swimming larvae called *veligers*. The veligers attach to hard surfaces and develop into tiny oysters called spat, which are placed like seeds on a variety of substrates. After about 2 years of growth, the oysters may be transferred to rafts in oyster ponds or placed on mud flats for fattening. In another 5 years, they are ready to be harvested.

Oysters are cultivated in vast numbers in the United States, France, and Japan and are the most profitable mollusks that are farmed. How can oyster cultivation be so profitable? Oysters, which feed on phytoplankton, are sedentary and use their energy mostly for growth and reproduction. In addition, oysters have a continuous food supply, as they filter-feed on plankton that is delivered by ocean currents.

Techniques similar to those used in raising mussels are used in oyster mariculture. Mussel mariculture is particularly successful in Spain, France, Japan, and the Philippines. Mussels are also farmed in the United States. The larvae of mussels attach and grow on ropes that are placed near mussel beds in shallow areas. The ropes either dangle from wooden rafts coated with fiberglass or are wrapped around poles stuck in the mud flats. After the mussels grow for about a year, they are harvested. One acre of floating rafts can produce more than 500 kg of mussel meat a year.

It was discovered recently that mussels can be harvested from the posts of oil platforms. At least one company in California has made a profitable business out of harvesting mussels from oil platforms in the waters off Santa Barbara. Oil companies have cooperated in this venture, since the mussels damage the platforms and have to be scraped off anyway. In the United States, mollusk farming is still in its early stages. But if the demand for mollusks as a food item increases, many coastal areas can be used to raise them successfully—provided the water quality is acceptable.

The crustaceans—lobsters, crabs, and shrimp—are more difficult to farm because, when raised together, these arthropods tend to fight. You may have seen live lobsters with their claws taped shut in a restaurant tank. Normally, lobsters must be kept in separate, large compartments, which makes them more expensive to raise. Mariculturists discovered that lobsters raised in warm water grow faster than those that grow in colder water in the wild. Several lobster farms have started to use the warm water that is discharged from power plants. The mariculture of brackish and marine shrimp species has been more successful. These crustaceans grow fast and can be raised in ponds. More research is needed to develop techniques that can produce crustaceans commercially.

Fish Aquaculture

Fish aquaculture has had a long and successful history. Fish farming started in China more than 3000 years ago, with the rearing of a large edible fish called *carp*, a relative of the goldfish. Today, a 1-acre pond can produce 12,700 kg of carp meat a year. This high yield is a result of a method called **polyculture**, in which different species of carp—each consuming a different kind of food—are raised together in a pond. The grass carp feeds on rooted plants along the edge of the pond. The wastes produced by this fish cause plankton blooms. Some of the plankton remains settle to the bottom and accumulate as sediments that are fed on by invertebrates. The mud carp feeds on the invertebrates. A third species of carp feeds on plankton that lives near the water surface. (See Figure 23-6 on page 584.) Each species is successful because it eats a different type of food, thereby avoiding competition. As a result, the pond promotes maximum growth and a high rate of reproduction. Development of catfish farms in the United States is another example of fish aquaculture.

With wild stocks of saltwater fish declining each year, there is an increasing demand for fish mariculture, the farming of saltwater

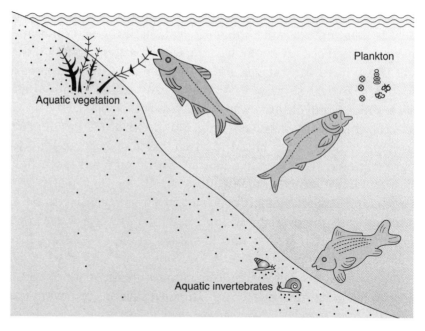

Figure 23-6 Polyculture of three different carp species in a pond.

Plankton

Aquatic vegetation

Aquatic invertebrates

fishes. The most widely cultured saltwater fish is the milkfish *(Chanos chanos)*. The milkfish is an important food item in Asia. Milkfish spawn in coastal waters. After fertilization, the eggs develop into little fish called *fry*. The fry are caught in nets and transferred to small, shallow saltwater ponds along the coast. In one month, the milkfish grow to 7 cm in length and are transferred to larger saltwater ponds for fattening. After a year, the milkfish weigh about 1 kg each. An acre of milkfish yields over 1 metric ton per year.

Freshwater fish aquaculture is generally more successful than fish mariculture, because conditions for maximum yield—from fertilization to fattening—can be controlled more easily in ponds. However, the Japanese have been able to cultivate a popular saltwater fish, the red sea bream *(Chysophrys major),* from fertilization to maturity in closed saltwater ponds. Thousands of young adult sea breams have been released into the ocean to increase the wild stock during a time of short supply.

Salmon are also farmed in the United States, Ireland, Norway, and Canada. In some cases, young salmon are reared in captivity and then released into the ocean, with the hope that they will return to spawn in the river where they were caught. However, the returns have been very low. In salmon ranching, the fish are raised in suspended cages, or pens, where they are fed well to ensure fast

Figure 23-7 Hybrid striped bass being farmed in California.

growth. This technique has been very successful. One drawback, however, is that the cages can become fouled with other organisms, and the sediments below the cages can become fouled with waste products from the salmon. In Norway, in 1990, salmon ranching production reached 100,00 metric tons per year. With such results, fish mariculture may have a very promising future. (See Figure 23-7.)

Seaweed Mariculture

While animal mariculture is a relatively new industry, the farming of seaweeds, called *seaweed mariculture*, is already big business. Who eats seaweed, you may ask? We all do, or at least we consume products that come from seaweed. Many consumer goods are made with seaweed ingredients. The chemical called *carrageenan*, extracted from Irish moss seaweed, is used to make toothpaste, ice cream, and other food items. Carrageenan prevents the ingredients in these products from separating.

Another chemical, called *algin,* is extracted from the brown alga kelp. Algin is used as a thickener and stabilizer in many foods, such as cheese and ice cream. Kelp is harvested off the California coast from large barges that have rotating blades, which cut off the tops of kelp just below the ocean surface. The same area can be continually harvested, since the kelp grows very rapidly.

Seaweeds have been farmed in China and Japan for hundreds of years. The most popular edible seaweeds are the green alga called *sea lettuce,* traditionally used by Hawaiians in salads, and the protein-rich red alga *Porphyra,* or *nori,* used to wrap boiled rice in Japanese cooking. These algae are grown and harvested in shallow coastal waters. Together with the other seaweeds, they constitute a one-billion-dollar food industry in Asia. (Refer to Chapter 5 for more information on algae and seaweed products.)

Seaweed harvesting is a growing industry in the United States. The demand for edible seaweed is increasing each year. Health food stores carry a variety of seaweed food items, and cookbooks now include seaweeds in their recipes. However, the success of seaweed mariculture may be limited in the future due to poor water quality in most coastal areas. Seaweeds cannot be raised or harvested in polluted water—another good reason to clean up coastal waters.

23.2 SECTION REVIEW

1. Describe how mollusks, such as oysters, can be successfully farmed.

2. Describe mariculture of the milkfish. How is it different from freshwater carp aquaculture?

3. Why is seaweed mariculture an important food industry?

23.3 PROTECTING MARINE ANIMALS

The great auk (*Pinguinus impennis*) was last seen alive in 1844. This marine bird stood about 60 cm tall and looked like it was part duck and part penguin. (See Figure 23-8.) Although it was once abundant along the North Atlantic coast, the great auk will not be seen again because it is extinct.

Figure 23-8 The great auk, a large seabird, was hunted to extinction.

When a species becomes **extinct**, it no longer exists anywhere in the world. How did the great auk become extinct? Great auks inhabited the shallow coastal waters near Newfoundland, Canada. Although fast-swimming while in pursuit of fish, on land these seabirds were flightless, awkward, and easy prey for hungry sailors. As a result, by the mid-1850s the great auks were exterminated—hunted into extinction.

Since the 1600s, when European settlers first came to America, more than 500 animal species have become extinct. More than three times that many are now in danger of extinction in America. Some of these species are marine animals. Table 23-2 lists a few seabird and marine mammal species that have become extinct as a result of human activities within the past 250 years.

TABLE 23-2 **RECENTLY EXTINCT SEABIRDS AND MARINE MAMMALS**

Species	Date Last Seen	Habitat
Steller's sea cow	1768	Aleutian Islands
Great auk	1844	No. Atlantic Coast
Labrador duck	1875	No. Atlantic Coast
Sea mink	1880	Coast of Maine
Caribbean monk seal	1962	Caribbean Sea
Dusky seaside sparrow	1987	Atlantic Coast

Unfortunately, the extinction of animal species continues. An example of a marine animal found in American waters that is threatened with extinction is the Florida (or West Indian) manatee. The manatee inhabits the rivers and waterways along Florida's Gulf and Atlantic coasts (and the Caribbean Sea). During the winter, these mammals move into warm coastal rivers and graze on vegetation. As summer approaches and the ocean warms, the manatees leave the protection of the inland rivers and migrate along the coast. Many manatees get struck by power boats and are injured or killed by the boat propellers. It is thought that some also die from poisoning by red tide microorganisms. The rapid development along Florida's coast also has contributed to a decline in the manatee population, mainly due to a loss of habitat. (See Chapter 14 for more information on the manatee.)

Protective Legislation and Sanctuaries

In response to a growing awareness and concern about the plight of species, such as the manatee, that are threatened with extinction, the U.S. Congress passed the Wildlife Preservation Act in 1969, the Marine Mammal Protection Act in 1972, and the Endangered Species Act in 1973. The 1972 act specifically protects such animals as whales, dolphins, seals, manatees, and sea otters. The populations of many of these animals had been decimated by human activities, such as whaling, fishing, and hunting for food and furs. These industries had threatened the survival of many species. (See Chapter 14.) The 1973 act is responsible for listing those species that are in danger of becoming extinct, so that actions can be taken to protect them. More than 100 species of marine fish have been listed. Species that are in more immediate risk of extinction are listed as **endangered species**; those that are in less immediate risk are listed as **threatened species**. Table 23-3 provides a partial list of threatened and endangered marine animals.

The Endangered Species Act forbids the import into the United States of any animal, or product from any animal, which is on the list. It further prohibits the interstate trafficking of endangered species and products obtained illegally in their country of origin. Thanks to the passage and enforcement of such laws, the popula-

TABLE 23-3 SOME THREATENED AND ENDANGERED MARINE ANIMALS

Mollusks	Fish	Reptiles	Birds	Mammals
California white abalone	Atlantic bluefin tuna	Sea turtles: loggerhead, green, hawksbill, leatherback, Kemp's ridley, olive ridley	Osprey	Whales: bowhead, sperm, gray, finback, right, humpback, sei, blue
Florida sea slug	Short-nosed sturgeon		Bald eagle	
	Round whitefish		Piping plover	Gulf of California harbor por-poise
	Chinook salmon		Roseate tern	
	Totoaba (seatrout)	Saltwater crocodile (American)	Common tern	California sea otter
	Caribbean groupers		California least tern	Seals: Mediter-ranean monk, Hawaiian monk, Guadalupe fur seal
	Great white shark		Short-tailed albatross	
	Seahorses (>30 species)		Wandering albatross	
	No. Atlantic swordfish		Eskimo curlew	Steller's sea lion
	Atlantic cod			Manatee
	Haddock			
	Coelacanth			

tions of several species that once seemed doomed have now recovered and bounced back from the edge of extinction.

One of the success stories in marine mammal conservation is that of the California gray whale. The gray whale migrates along the Pacific Coast, between its feeding grounds in the Arctic and breeding grounds in the Gulf of California. By 1910, the whaling industry had reduced the population of gray whales to approximately 5000 animals. The whale was declared an endangered species and protected from hunting. As a result, its population has returned to a much healthier number of approximately 22,000. The federal government may even remove the gray whale from the endangered species list. However, not all whale species that were formerly hunted have recovered as well. The blue whale, for example, was hunted so heavily that it may not be able to sustain its population, even though it is now protected.

In 1978, the Florida Manatee Sanctuary Act was established to stop the decline of the manatee population. One of the provisions

Figure 23-9 Sanctuaries such as this one were established to protect manatees from human activities.

of the act was the creation of sanctuaries that would eliminate or reduce human intrusion. (See Figure 23-9.) A speed limit of 8 km per hour was set for boating in sanctuary areas. In addition, efforts have been undertaken to increase the manatee population in Florida through captive breeding and release programs. Manatee tourism programs and marine reserves in Central America and the Caribbean region also have been developed to protect these animals.

Other endangered marine species would probably have become extinct if concerned citizens, local governments, and environmental groups had not taken action. Today, citizens are showing concern for the plight of endangered and threatened species. The U.S. government has responded to these concerns by establishing more sanctuaries. One of NOAA's initiatives is the National Marine Sanctuary Program, which protects unique and biologically rich coastal environments. The **marine sanctuaries** are protected areas in which no commercial activities are permitted. These sanctuaries are an important conservation measure that provides safe habitats for marine organisms. For example, a marine sanctuary for humpback whales in Hawaiian waters protects their breeding and calving grounds. See Figure 23-10, which shows 13 designated and three proposed national marine sanctuaries in U.S. waters.

One of the most significant conservation developments to occur

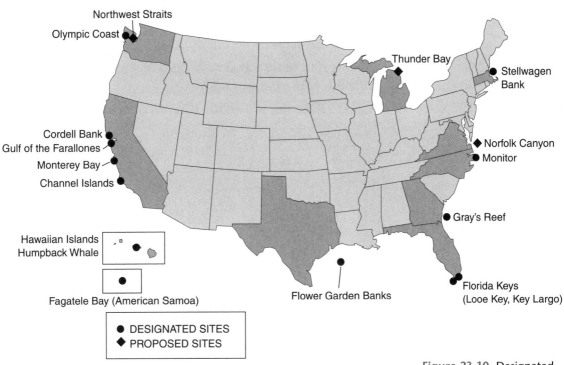

Northwest Straits

Olympic Coast

Thunder Bay

Stellwagen Bank

Cordell Bank
Gulf of the Farallones
Monterey Bay
Channel Islands

Norfolk Canyon
Monitor

Gray's Reef

Hawaiian Islands
Humpback Whale

Fagatele Bay (American Samoa)

Flower Garden Banks

Florida Keys
(Looe Key, Key Largo)

● DESIGNATED SITES
◆ PROPOSED SITES

Figure 23-10 Designated and proposed marine sanctuaries in U.S. waters.

in recent years has been the creation of *land trusts* and *conservancies,* which operate in every state. These nonprofit organizations work with landowners to protect key habitats, such as farms and wetlands, and to prevent their being developed for any commercial use other than agriculture or aquaculture. In 2002, a shellfish company donated 4455 hectares (underwater) in the Great South Bay off Long Island, New York, to the Nature Conservancy, a nonprofit organization dedicated to preserving natural habitats. This two-million-dollar gift will be used by the Conservancy to develop mariculture of mollusks such as clams and scallops.

The struggle to preserve and protect the marine environment continues. We now live in a single "global village" in which the oceans are important highways that connect us with one another. If one part of the world is exploited or polluted, other parts will be affected. More laws may be necessary to prevent the degradation of our marine environment. Respect for the environment and cooperation with other nations will help us achieve the goal of living in harmony with our marine world.

23.3 SECTION REVIEW

1. Why has the manatee become an endangered species?

2. What are some efforts that have been made to prevent the extinction of marine animals?

3. What activities threatened the survival of marine mammals?

Laboratory Investigation 23

Analyzing Fishery Data

PROBLEM: How can we determine the status of the New England fishery?

SKILLS: Analyzing data from a table; constructing a graph.

MATERIALS: Graph paper, pencil, ruler.

PROCEDURE

1. Examine the data in Table 23-4, which show the average fish catch per trawl (in kilograms) in New England waters for select years from 1963 to 1993.

2. Use the information in the table to construct a line graph on your piece of graph paper. (You may also want to construct a bar graph for further comparison.)

3. Mark an appropriate scale for the horizontal and vertical axes based on data in the table. Plot the data on the graph (you should have seven points.) Surround each point with a small circle and draw lines connecting the points.

TABLE 23-4 AVERAGE CATCH PER TRAWL (KG)

Year	Fish Catch
1963	63.6
1968	35.0
1973	27.2
1978	36.4
1983	18.1
1988	13.6
1993	14.5

OBSERVATIONS AND ANALYSES

1. How would you describe the status of the fishing industry in the New England region in 1963 versus the status in 1993?

2. Congress passed the Magnuson Act in 1976 to prevent foreign fishing boats from working our coastal waters (out to 333 km). Based on the data in the graph, how effective was this conservation measure in the short term? How effective was it in the long term?

3. After 1978, the U.S. fishing fleet grew rapidly. According to the graph, what effect did a larger and more modernized fishing fleet have on the local fish stocks?

Answer the following questions on a separate sheet of paper.

Vocabulary

The following list contains all the boldface terms in this chapter.

aquaculture, conservation, desalination, endangered species, extinct, fisheries, gill net, long-lining, manganese nodules, mariculture, marine sanctuaries, maximum sustainable yield, minerals, natural resources, polyculture, purse seine net, stock, threatened species, trawling net

Fill In

Use one of the vocabulary terms listed above to complete each sentence.

1. The process of farming organisms from the sea is _____.

2. Freshwater can be produced from ocean water by _____.

3. The careful management of natural resources is called _____.

4. Natural, inorganic substances formed in Earth's crust are _____.

5. The population of a food resource is referred to as its _____.

Think and Write

Use the information in this chapter to respond to these items.

6. What are some conservation measures that help reduce over-fishing?

7. Explain how the polyculture technique of fish farming works.

8. Why is the establishment of marine sanctuaries so important?

Inquiry

Base your answers to questions 9 through 11 on the following graph and on your knowledge of marine science.

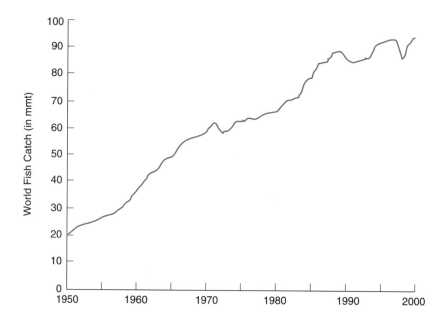

9. Based on the data in the graph, which statement is correct?
   a. Fewer tons of fish were caught in 1980 than in 1950.
   b. There was a general decrease in the amount of fish caught between 1950 and 1990.   *c.* The greatest amount of fish was caught in 1990.   *d.* The fish catch in 2000 was more than four times greater than in 1950.

10. The most appropriate title for this graph would be
    a. The increase in fish populations from 1950 to 2000.
    b. The effect of drift nets on fish populations.   *c.* Trends in worldwide fish catch from 1950 to 2000.   *d.* A comparison of fish species caught from 1950 to 2000.

11. Based on the data in the graph, which statement offers a valid conclusion?   *a.* There are unlimited numbers of fish in the sea.   *b.* Ocean fish populations have been increasing since 1950.   *c.* The worldwide fish catch peaked around the year 2000.   *d.* Ocean fish populations have been decreasing steadily since 1950.

Multiple Choice

Choose the response that best completes the sentence or answers the question.

Refer to Figure 23-5, on page 581, to answer the following question.

12. Based on the diagram of the solar still, which statement is correct?   *a.* The salinity of the seawater in the still decreases.
    b. Radiant energy is not required for desalination.
    c. Freshwater is obtained from the condensation of water vapor.   *d.* A transparent roof is usually not needed on a solar still.

13. Marine sanctuaries have been established to protect all the following *except*   *a.* safe habitats for marine organisms
    b. breeding and calving grounds of whales   *c.* U.S. fishing rights within 333 km of our coast   *d.* special areas in which no commercial activities are permitted.

14. Overall, from about 1970 to 1990, the U.S. fishery catch
    a. steadily increased   *b.* steadily decreased   *c.* increased, then decreased   *d.* remained constant.

15. Bottom-dwelling fish are usually caught by   *a.* a trawling net
    b. surface long-lining   *c.* a purse seine net   *d.* a gill net.

16. Which is an accurate statement regarding the U.S. fishery industry?   *a.* Overfishing led to a decrease in the catch of some bottom-dwelling species.   *b.* The larger boats with high technology were unable to catch more fish.   *c.* The U.S. is the leading fishing nation in the world.   *d.* The Magnuson Act increased the foreign catch in U.S. waters.

17. All of these natural resources are extracted from the ocean *except*   *a.* gas   *b.* manganese nodules   *c.* oil   *d.* coal.

18. An example of a renewable natural resource is   *a.* oil   *b.* gas
    c. flounder   *d.* manganese nodules.

19. When the maximum sustainable yield of a fish species is exceeded, the   *a.* fish stock begins to decline   *b.* size of its annual harvest increases   *c.* fishing boats catch more fish per trawl   *d.* fishing boats use nets instead of long lines.

20. An example of an extinct marine species is the   *a.* manatee
    b. blue whale   *c.* coelacanth   *d.* great auk.

21. An animal that is listed as endangered is *a.* sure to become extinct *b.* at high risk of becoming extinct *c.* at no immediate risk of becoming extinct *d.* already extinct.

22. All of the following organisms are grown by mariculture *except* *a.* seaweeds *b.* oysters *c.* seahorses *d.* salmon.

23. A method of desalination that is currently in use is *a.* melting icebergs into freshwater *b.* using filter paper to separate freshwater from salt water *c.* using solar stills to evaporate seawater *d.* boiling ocean water into freshwater at hydrothermal vents.

Research/Activity

- Prepare a chart that lists several endangered, threatened, and/or extinct marine animals. Include a picture and brief description of each animal and information on its habitat. Prepare a poster to exhibit at a school science fair.

- Do research on the technology of a fish farm or other aquaculture operation. Then, using common materials, construct a model of the farm. Write a brief description of how a fish farm works.

Glossary

abiotic factors nonliving things in the environment, e.g., water and minerals

absorbed light light that is taken up by a substance

acid a solution that contains a larger number of hydrogen ions than hydroxyl ions

acid precipitation formed when moisture in air absorbs chemicals; acid rain

acorn worm wormlike protochordate; adult has dorsal nerve cord and gill slits

adaptation any characteristic of an organism that enables it to live successfully in its environment

adductor muscles short, tough muscles that hinge together a bivalve's shells

agar chemical in red algae that is used in foods, medicines, and bacterial cultures

air embolism blockage of a blood vessel by a gas bubble in an important organ

algae plantlike protists, many single-celled, that carry out photosynthesis

algal bloom a sudden increase in the algae population in shallow waters

algin chemical in kelp that is used in foods and many different products

ambient pressure total pressure on a diver; the sum of atmospheric pressure and hydrostatic pressure

amniotic egg contains large yolk; enclosed in a case to prevent water loss; first seen in reptiles

amphipods tiny, beach-dwelling shrimplike crustaceans with flattened sides

amplitude volume of a sound, represented by height of the sound wave

ampulla bulblike structure at the top of each tube foot, used in movement

ampullae of Lorenzini in shark, nerve receptors in tiny pores in snout, which detect electric fields of other animals

angle of insolation angle at which the sun's rays strike Earth's surface

annelids *see* segmented worms

aphotic zone vast area below photic zone; no light penetrates to these depths

aquaculture process of farming aquatic organisms (marine and freshwater)

aqua-lung a tank of compressed air that is strapped to the diver's back

aquanauts scientists who explore the undersea world

Archimedes' principle states that the buoyant force on any object is equal to the weight of the liquid that the object displaces

Aristotle's lantern in sea urchins, five-toothed mouth part for eating algae

arrow worm tiny, transparent worm that lives near the ocean surface; active hunter with mouth bristles for hooking prey

arthropods animals with jointed appendages, exoskeleton, bilateral symmetry

asexual reproduction the production of offspring by one parent

atoll a string of coral islands that form a circle (around a sunken island)

autotrophs organisms that make their own food; the producers: plants and algae

bacteria single-celled organisms that lack a nuclear membrane (monerans)

baleen in whales, overlapping plates of fibrous protein used for filter feeding

baleen whales filter feeders with baleen plates; eat zooplankton and small fish

barnacle sessile crustacean with overlapping calcium carbonate plates and cirri

barotrauma any diving injury associated with pressure

barrier beaches long ridges of sand formed by deposit of sediments offshore

barrier reef a coral reef that grows about 25 kilometers offshore, separated by a channel from the mainland

base a solution (alkaline) that contains a larger number of hydroxyl (than hydrogen) ions

bathyscaphe a deep-diving vessel developed in the 1950s

bathysphere a round, deep-diving vessel with a porthole, attached to a ship by a cable

bathythermograph narrow canister used to get temperature profile of ocean

bends a decompression illness (gas bubbles in joints and tissues) that happens to divers during a rapid ascent

benthic term used to describe bottom-dwelling organisms

benthic zone life zone that includes entire ocean floor, from intertidal to basin

benthos organisms that inhabit the benthic zone

bilateral symmetry body plan in which structures on left side of body are same as those on right side of body; first in worms

binary fission asexual reproduction in which a cell divides into two cells of equal size

biodiversity the great variety of life-forms within a habitat, ecosystem, or the entire Earth

biological magnification the increase in concentration of a chemical (usually toxic) in a food chain; also called biomagnification

bioluminescence the ability of an organism to produce light in its body

biotic factors livings things in the environment, e.g., plants and animals

bivalves mollusks with two shells, e.g., clams; also called pelecypods

black skimmer seabird that catches fish with its lower jaw as it flies over ocean

bloodworm segmented worm that lives in marine sediments; has bristles

blowhole in whales and dolphins, opening on top of head used for breathing

blubber in cetaceans and pinnipeds, a thick layer of fat under the skin

bony fishes have bone skeleton, loose scales on skin, swim bladder; e.g., tuna

book gills in horseshoe crab, overlapping membranes used for breathing and locomotion

brackish a mixture of freshwater and salt water, found in estuaries

bradycardia in diving marine mammals, ability to slow the heart rate during deep dives; important part of diving response

breaching behavior in which whales leap out of water and crash back down

brittle stars solitary, nocturnal echinoderm, with long, skinny arms for moving

brown pelican large coastal bird that dives to catch fish in its big throat pouch

brown tide an algal bloom of diatoms that clouds shallow, coastal waters

bryozoan microscopic, multicellular, benthic animal that lives within a calcium carbonate or chitin compartment in colonies

budding asexual reproduction in which a smaller individual develops on, and breaks off from, the larger parent body

buffers chemicals in ocean water that help maintain a stable (neutral) pH

bulkheads walls that form watertight compartments inside a ship's hull

buoyancy upward force that supports floating objects

byssal threads tough, fibrous protein threads that attach mussels to a substrate

camouflage ability to change appearance to blend in with natural surroundings

carapace in crustaceans, part of exoskeleton that covers head and chest regions

carnivores term that describes flesh-eating animals; usually have sharp teeth

carrageenan chemical in seaweed used as a binding agent in foods and toothpaste

cartilage flexible connective tissue, composed of cells and protein

cartilaginous fishes have cartilage skeleton, gill slits, placoid scales; sharks and rays

cell theory states that all living things are made of one or more cells, perform the same basic life functions, and come from preexisting cells

cephalopods swimming mollusks, with prominent head and foot (tentacles); usually lack an external shell, have streamlined body, swim by jet propulsion

cephalothorax in crustaceans, segment that comprises head and chest regions

cetaceans order of aquatic mammals that includes all whales and dolphins

chambered nautilus deep-water cephalopod that has an external shell, with gas-filled inner compartments to regulate buoyancy

chemosynthesis process by which organisms derive energy from chemicals

chitin type of carbohydrate that makes up the exoskeleton of arthropods

chiton mollusk with overlapping shells on muscular foot; no tentacles, scrapes algae off rocks with radula, in intertidal zone

chlorophyll the green pigment; important in process of photosynthesis

chordates animals having dorsal nerve cord, notochord, and pharyngeal gill slits at some stage of development; includes all protochordates and vertebrates

chromatophores in cephalopods, special pigmented cells that expand and contract to enable camouflage

chromosomes threadlike structures made up of molecules of DNA that have instructions for reproduction

cilia microscopic hairs in and on animals, used in feeding and movement

circumnavigate to sail completely around Earth

cirri in barnacles, pairs of feathery appendages that catch food particles

claspers in male shark, pair of organs between pelvic fins that transfer sperm

climax community the last stable community to appear in any succession

cnidarians multicellular animals characterized by two cell layers, saclike digestive tract, tentacles, radial symmetry, and nerve net; e.g., jellyfish, sea anemones, and corals

cnidoblasts in cnidarians, the stinging cells contained in their tentacles

coelom fluid-filled space that separates digestive tract from skin; in annelids

collar cells sponge cells with flagella that beat back and forth to pump water

colonial animals organisms that live attached to one another by a thin membrane and so have shared nutrition; e.g., the coral polyp

color contrast in fish, a pattern of different colors that identifies the species

commensalism symbiotic relationship in which one species benefits while the other species is not affected by the association

common tern small shorebird that nests on sand and dives to catch small fish

compound substance that contains two or more kinds of atoms that are chemically joined, or bonded, together

compound light microscope research tool with two lenses for magnification

compressions in a sound wave, air molecules that are squeezed together

condensation process of cloud formation, occurs when molecules of water vapor come close enough together; part of the water cycle

conjugation among protozoans, two parents exchange parts of their micronuclei (genetic material) before undergoing binary fission

conservation careful management and protection of natural resources and the environment

consumers organisms that do not make their own food; they (animals) ingest food

continental drift theory that the continents once formed one large landmass, then broke apart and drifted to their present positions

continental rise slightly elevated region formed at base of continental slope by accumulation of mudslide sediments

continental shelf relatively shallow part of seafloor that adjoins continents

continental slope area where seafloor drops steeply at outer edge of continental shelf

contractile vacuole structure in protozoan cell that pumps out excess water

control group in an experiment, the group not exposed to the variable

controlled experiment uses an experimental group and a control group to test a hypothesis

convection current continual movement of gas or liquid in a cycle as the heated part rises and the cooler part sinks; inside Earth, causes molten magma to rise from mantle into crust; in atmosphere, causes breezes

copepod tiny, abundant crustacean; important as base of oceanic food chain

coral polyp the coral animal, usually colonial; resembles small sea anemone

coral reef stony formation built up from the seafloor by colonial coral polyps

cordgrass types of marsh grasses found in the intertidal zone (*Spartina*)

Coriolis effect states that the spin of Earth causes winds and surface waters to move clockwise in northern hemisphere and counterclockwise in southern hemisphere

cormorant common shorebird that dives to catch fish

countercurrents slow, deep horizontal ocean currents that flow in opposite direction of surface currents

countershading in fish, a fixed pattern of darker skin on top, lighter skin below

crest the highest point of a sound wave or water wave

crinoids echinoderm having many feathery arms, usually atop a jointed stalk

crocodilians large reptiles that have a four-chambered heart; alligators and crocodiles

crust Earth's surface layer (above mantle), about 40 kilometers thick

crustaceans arthropods with bilateral symmetry, exoskeleton, and two main body segments (cephalothorax and abdomen)

current a large mass of continuously moving ocean water

cuttlefish bottom-dwelling squidlike cephalopod that feeds on invertebrates

cyanobacteria the blue-green bacteria; the only photosynthetic monerans

decomposers organisms that break down and recycle dead organic matter

decompression a decrease in pressure upon ascent; if too rapid, can cause gases to come out of solution and form bubbles

delta fan-shaped feature formed by sediments that pile up at a river's mouth

density defined as mass per unit volume

desalination removal of salt from ocean water to produce freshwater

dew water that condenses (from vapor to liquid) on a solid surface

dew point temperature at which water vapor condenses as a cloud or fog

diatoms single-celled protists (algae); part of phytoplankton community

diffusion movement of molecules from an area of higher concentration to an area of lower concentration; passive transport

dinoflagellates single-celled protists (algae) that have two flagella; phytoplankton

disruptive coloration in fish, patterns that obscure body outline, for protection

dissecting microscope research tool for viewing larger specimens under magnification

dissolved oxygen (DO) the oxygen that is dissolved in bodies of water

diving chamber early chamber for diving that contained a supply of air

diving response in marine mammals, a group of structural and behavioral responses that enables deep diving; also called diving reflex

diving suit watertight canvas suit with a metal helmet, weighted boots, and air pumped through a tube

dugong herbivorous, aquatic mammal (sirenian) found mainly in warm Pacific coastal waters; also called sea cow

dunes in upper beach area, small hills of sand formed by the wind

echinoderms spiny-skinned animals with radial symmetry, internal skeleton, but no body segments

echolocation in cetaceans, a natural form of sonar used in communication and hunting

ecological succession process by which one community of organisms gradually replaces another community over time; also called biological succession

ecology study of interaction of living things with each other and environment

ecosystem the interacting biotic and abiotic factors within an environment

ectothermic term for animal in which body temperature is determined by temperature of external environment; also cold-blooded

eel grass a type of sea grass that grows along the Atlantic and Pacific coasts

effluent the liquid discharge of sewage that has been chemically treated

El Niño periodic warm ocean current that develops in middle of the Pacific

electromagnetic spectrum all forms of solar radiation, including visible light

electron microscope research tool for viewing tiny objects with highest magnification

embryo the early stage of development during which there is a rapid division of cells

encrusting organisms living things that grow over the surfaces of substrates

endangered species term for species that are in immediate risk of extinction

endoskeleton an internal skeleton; first seen in the echinoderms

endothermic term for animal that can generate its own body heat, e.g., birds and mammals; also warm-blooded

enzyme a protein that regulates the speed of a chemical reaction without itself being changed; also called organic catalysts

estuary environment formed at mouth of river, where freshwater and salt water mix

eukaryotes organisms that have nuclear material enclosed within a membrane

evaporation process by which liquid water changes to a gas

excurrent siphon in bivalves, part of siphon through which waste products exit

exoskeleton in arthropods, tough body covering, or outer skeleton, of chitin

experimental group in an experiment, the group exposed to the variable

extinct condition of a species when it no longer exists anywhere in the world

eyespots in sea stars, tiny light receptors located at the end of each arm

fault a crack in Earth's crust that occurs at the margin of two plates

feather stars crinoids that move by means of flapping their feathery arms

filter feeders term for animals that strain their food from the water

fisheries industry that commercially harvests fish and shellfish

fjord a steep, deep, narrow inlet from the sea, formed by the action of glaciers

flatworms species of worms that have a flat body form; platyhelminthes

fog air that is saturated with moisture near the ground (ground-level cloud)

food chain food relationship in which each organism serves as food for the next one

food pyramid food chain arranged as a diagram with the lowest trophic level at the base and highest trophic level at the top

food web food relationships, composed of many interconnected food chains

foraminiferan unicellular protist with a calcium carbonate shell; zooplankton

fossil fuels fuels such as coal, gas, and oil that release carbon dioxide (a greenhouse gas) into atmosphere when burned

frequency the number of wavelengths per second of a particular radiation

fringing reef a coral reef that grows a few kilometers offshore, parallel to the mainland

fungi unicellular and multicellular eukaryotic organisms that absorb nutrients

gametes reproductive cells that contain the haploid number of chromosomes; function in sexual reproduction

ganglia nerve cell clusters that act like a simple brain; found in flatworms

gastropods mollusks that have one shell, usually coiled (as in snails); univalves

genome total genetic make-up of an organism, i.e., all its genes

giant tube worms deep-sea worms that live in tubes near hot-water vents

gill net intercepts school of fish, which get caught by gills in net's mesh

gill slits in cartilaginous fish, visible openings for breathing; often ventral

global warming worldwide warming trend, possibly due to increase in greenhouse gases

graph pictorial representation of data that shows relationships

greenhouse effect a warming of Earth caused when heat is trapped in the atmosphere by water vapor and carbon dioxide

greenhouse gas substances, such as carbon dioxide, that are released when fossil fuels are burned; trap heat in the atmosphere

guyots flattened undersea structures, formed when tops of seamounts are eroded by waves and currents

gyres circular ocean currents caused by deflection of water by the continents

halocline layer of water that shows rapid increase in salinity; 100 to 200 meters deep

hatchlings baby turtles or birds, just after they break through their shells

heat of fusion energy lost when there is a change in state from liquid to solid

heat of vaporization energy absorbed when there is a change in state from liquid to gas

heavy metals elements that are discharged by industries into waterways; harmful to living things when they biomagnify in their tissues

herbivores term that describes plant-eating animals; usually have blunt teeth

heterotrophs organisms that live on food made by others; the consumers

hirudin chemical anticoagulant secreted by leech into prey to aid blood flow

holdfast tough, fibrous pad of tissue that anchors seaweed, e.g., *Fucus,* to a rock

homeostasis ability of an organism to maintain a stable internal environment

horseshoe crab arthropod with six pairs of appendages, carapace, book gills, and spiked tail (telson); lives in coastal waters

host organism that is fed on by a parasite living in or on it

hot spot an area of intense geologic activity in the crust where a seamount forms

humidity the amount of moisture in the atmosphere

hurricane coastal storm with wind velocity that exceeds 120 kilometers per hour

hydroid cnidarian; colonial animal composed of different polyps that function together as a single organism; e.g., *Obelia* and *Physalia*

hydrometer weighted glass tube used to determine the density of a liquid

hydrostatic pressure pressure exerted by the water's mass (due to its density)

hydrothermal vent an area in the rift zone at which hot springs emerge

hygrometers instruments that are used to determine relative humidity

hypothermia excessive loss of body heat caused by exposure to very cold water

hypothesis a possible solution to a problem; tested in an experiment

hypoxia meaning "low oxygen," aquatic condition in which the dissolved oxygen level is below minimum level needed by fish

incurrent siphon in bivalves, part of siphon through which water that contains food and oxygen enters

inner space term used to describe the undersea world

intertidal zone on a beach, the area between high tide and low tide

island arcs groups of volcanic islands that form an arc near an ocean trench

isopods tiny, shallow-water crustaceans that have flattened bodies, seven pairs of legs

jawless fishes parasitic, jawless, retain larval notochord (no true backbone), and lack true scales (lamprey)

kelp the largest seaweeds in the ocean, they are types of brown algae

keys (cays) small islands that form when chunks of coral stone break off from reefs and accumulate on seafloor

krill cold-water, shrimplike, planktonic crustacean; eaten by fish, whales, and seals

lancelet fishlike protochordate; adult retains all three primitive chordate traits

land breeze cool breeze that flows from the land to sea (during summer night)

larva free-living stage in the early development of an organism

lateral line organ line of sensitive sound receptors along each side of a fish's body

latitude geographic measurement lines (measured in degrees north or south) that run parallel to the equator

leech segmented worm without bristles; some free-living, some parasitic

lichen composed of an alga and a fungus growing in a symbiotic relationship; usually first pioneer organism to appear on rocks

life zone region that contains characteristic organisms that interact with one another and their environment

lobtailing behavior in which whales wave and smash tail on ocean surface

longitude geographic measurement lines (measured in degrees east or west) that run from north pole to south pole

long-lining fishing method that uses long lines with many hooks attached

longshore current a current that moves parallel to shore, produced by waves that break at an angle to the shore

magma molten material within Earth's mantle (called lava at Earth's surface)

manatee herbivorous aquatic mammal (sirenian) found in warm Atlantic coastal waters; also called sea cow

manganese nodules valuable mineral deposits found on the ocean floor

mangrove community thick growth of mangrove trees along tropical coasts

mangrove trees trees that grow in salt water along tropical shores worldwide

mantle region of geologic activity between the Earth's core and crust

mantle in bivalves, thin membrane that secretes the shell and lines insides of shell to protect internal organs; in cephalopods, mantle cavity is used in movement

mariculture process of farming marine (plant and animal) organisms

marine biology study of life in the sea; also called biological oceanography

marine ecology study of ecological interactions within the ocean

marine geologists scientists who study features of, and changes in, the seafloor

marine iguana Galápagos lizard that swims and feeds (on algae) in the ocean

marine sanctuaries protected areas in which no commercial activities are permitted

marine science study comprising marine biology and oceanography

marsh grasses a variety of plants that grow along sandy beaches of calm bays

mass the amount of matter in an object, not dependent on gravity

maximum sustainable yield largest amount of fish that can be taken from a population without threatening its future size

medusa in jellyfish, umbrella-shaped body part composed of two membranes

melanin one of the most common skin pigments (found in chromatophores)

melon in dolphins, a fatty bump on forehead that focuses sounds in the head

meniscus the curved surface of a liquid in a cylinder

mesoglea in jellyfish, jellylike mass between the two membranes of the medusa

metabolic activity internal energy (activity) level of an organism

metric system the system of measurement used in science

mid-ocean ridge undersea volcanic mountain range that encircles the globe

minerals natural, inorganic substances (resources) formed in Earth's crust

mixture contains two or more substances that can be separated by ordinary physical means

mollusks soft, bilaterally symmetrical animals with head (and brain), foot region, coelom, coiled visceral mass, and usually an external or internal shell

molting in crustaceans, process of shedding outer covering (to grow) each year

monerans bacteria and blue-green bacteria; the single-celled prokaryotes

mud flat community part of estuary, dark mud with high decay and no grasses

multicellular describes organisms made up of more than one cell

mutualism symbiotic relationship in which both species benefit from the association

myoglobin in diving marine mammals, an oxygen-binding protein in muscles that increases their oxygen-carrying capacity during deep dives

Nansen bottle used to collect water samples from different depths in water column

natural resources materials from the environment that people use and eat

neap tides weaker tides (not too high or low) produced twice each month

nekton term for marine animals that have the ability to swim

nematocyst in cnidoblasts, coiled thread with a barb, usually toxic

nematodes *see* roundworms

nephridia pair of coiled tubes, for excretion, in each body segment of annelids

neritic zone region of relatively shallow water above continental shelf; life zone beyond subtidal zone

nerve net simple nervous system, a network of nerve cells and receptor cells

neutral a solution that contains equal numbers of hydrogen ions and hydroxyl ions; having a pH of 7.0

nitrogen fixation the process of producing nitrates from atmospheric nitrogen

nitrogen narcosis a confused state of mind that may occur during deep dives

nonbiodegradable litter that cannot be broken down by natural processes

nonpoint source pollution oil pollution from many diffuse points of origin

nudibranch gastropod that lacks, or has a reduced, shell; also called a sea slug

nutrition utilization of food by living things for growth and energy

O$_2$ minimum zone area of lowest dissolved oxygen, at 1000-meters depth

ocean basin deepest part of the ocean floor (beyond continental slopes)

oceanic zone life zone beyond neritic zone that includes most of open ocean

oceanographers scientists who study the sea

oceanography study of the physical characteristics of the sea

omnivore an animal that consumes both animals and plants

operculum in snails, thick pad of tissue that closes shell opening over foot; in bony fish, flap of tissue that covers gills

organelles tiny structures in cells that carry out important functions

osculum large hole at top of sponge through which water and wastes exit

osmoregulation ability of aquatic organisms to maintain proper water balance

osmosis movement of water molecules from an area of higher concentration to an area of lower concentration across a cell membrane; passive transport

osmotic pressure increased water pressure in an aquatic organism, due to inability to osmoregulate; can upset cell function

osprey catches fish with its talons as it flies over coastal water; also called fish hawk

ostia many small holes (pores) in sponge through which food particles enter

oystercatcher large coastal bird that has long knifelike beak to catch mollusks

parapodia paddlelike appendages of sandworms, for moving through wet sand

parasite organism that obtains its food by living in or on the body of another organism

parasitism symbiotic relationship in which one species (the parasite) benefits while the other species (the host) is harmed by it

pelagic term for open-water (wider-ranging) species of fish, such as tuna

pelagic zone largest life zone in ocean, entire ocean of water above sea bottom; includes the neritic and oceanic zones

pH the degree of acidity or alkalinity of a solution (power of **hydrogen** ions)

photic zone part of the ocean that light penetrates; average depth of 100 meters

photophores patches of bioluminescent tissue in organisms in dim waters

photosynthesis the manufacture of simple carbohydrates by plants and algae

phytoplankton plantlike members of the plankton community, e.g., diatoms

pigment a coloring matter found in cells and tissues of plants and animals

pinnipeds group of marine mammals that includes seals, sea lions, walruses

pioneer community first group of organisms to appear in an area devoid of life

placoid scales in cartilaginous fish; scales are tiny teeth embedded deep in skin

planarian type of flatworm found in freshwater and marine habitats; has eyespots, ganglia, and bilateral symmetry

plankton community of organisms that float and drift near ocean surface

planula ciliated, swimming larva of some invertebrates

plate tectonics theory that Earth's crustal plates float on the mantle

plates segments of Earth's crust that float (with the continents on top) on the mantle

pod in cetaceans, the extended family group in which they live and travel

pollutants substances that have a harmful effect on living things

pollution substances introduced into the environment that harm living things

polyculture farming of different fish species (that eat different foods) together

polyp in cnidarians, body structure with mouth, tentacles, and digestive cavity; lives attached to a substrate

powder feathers in aquatic birds, they repel water to protect underlying down feathers

precipitate a solid substance that may be produced when two liquids are mixed

precipitation forms of moisture that fall to earth, such as rain and snow

predator an animal that kills and eats other animals (the prey)

preen in birds, method of grooming; the beak is used to spread oil through feathers

prey an animal that is killed and eaten by another animal (the predator)

primary consumer the first-level consumer (of producers) in a food chain, e.g., the zooplankton that eat phytoplankton

proboscis sharp, sticky extension in ribbon worm's head, used to spear food

producers organisms, such as algae and plants, that make their own food

prokaryotes organisms that lack a nuclear membrane (single-celled monerans)

prop roots arching roots that anchor mangrove trees into the muddy sand

protists kingdom of mostly single-celled organisms (protozoa and algae) that have nuclear material enclosed in a membrane

protochordates primitive invertebrate chordates: tunicate, lancelet, acorn worm

protozoa tiny animal-like protists; ingest food

pseudopods cytoplasmic extensions used for movement by some protozoans

purse seine net net that surrounds and traps large schools of fish near surface

radial symmetry body plan in which appendages are arranged around a central point; in cnidarians and echinoderms

radiant energy all forms of energy emitted from the sun

radioactive wastes unstable radioactive isotopes that emit harmful high-energy rays and particles as they break down

radiolarian a unicellular protist zooplankton with a transparent silica cell wall

radula ribbonlike toothed structure of gastropods used for scraping up food

rarefactions in a sound wave, where air molecules are spaced farther apart

red tide an algal bloom of dinoflagellates that may poison other organisms

reed grass a tall marsh grass (*Phragmites*) that has fluffy brown tassels

reflected light light that bounces off a surface

refraction the bending of light as it passes through substances of different densities

regeneration asexual reproduction in which a whole new body can be grown from parts of the parent body

relative humidity amount of water vapor in the atmosphere compared to maximum amount that air can hold at given temperature

remote sensors instruments that gather data about Earth without being in physical contact with it, e.g., satellites

rhizomes the underground stems of plants such as turtle grass

ribbon worm free-living marine worm; flat, large, unsegmented, has proboscis and more advanced body systems than in the flatworms

rift valley a depression that runs along the crest of the mid-ocean ridge

rip current fast, narrow current of water moving seaward, formed when a sandbar breaks

robots explorational vehicles that do not carry humans on board

rocky coasts shores made up of solid rock; often more steep than sandy beaches

rogue waves tall, lone waves formed when waves meet other waves or currents

rotifer microscopic, multicellular, ciliated animal that lives in moist sands

roundworms most numerous sea worms; usually small, free-living, cylindrical

salinity the amount of salt dissolved in water

salt glands in marine reptiles and birds, excrete a salty solution to rid excess body salts

salt marsh community part of estuary where grasses grow in shallow water; wetlands

sand dollar round echinoderm, with no arms and short spines covering skin

sandbar sand eroded by wave action that gets deposited in a long hill offshore

sandpiper small shorebird that has narrow bill for catching small invertebrates in sand

sandworm segmented worm that lives in intertidal and subtidal muddy sands

sandy beach environment along a shore composed of sand (loose sediments)

scaphopods mollusks with a tapering shell, burrow in sand undersea; tusk shells

scavengers animals that eat the remains of already dead plants and animals

scientific method an organized step-by-step problem-solving approach

scuba tank *see* aqua-lung; acronym for self-contained underwater breathing apparatus

sea a smaller part of an ocean, where two continents lie close together

sea anemones cnidarians; have tentacles and a nerve net; live as sessile polyps

sea breeze cool breeze that flows from the sea to land (during summer day)

sea cucumber echinoderm with soft, oblong body and tube feet arranged in five rows; has no arms, spines, or endoskeleton

sea ducks ducks that dive into ocean to eat mollusks, crustaceans, and fish

sea grass types of grass that grow in shallow subtidal zones along coasts

sea gull ocean bird that feeds on crabs, dead marine animals, and garbage

sea level the point at which the ocean surface touches the shoreline

sea lilies sessile crinoids with feathery arms; live attached by a stalk to seafloor

sea otter smallest of marine mammals, Pacific Coast member of weasel family

sea snake venomous, tropical ocean-dwelling reptile with rudderlike tail

sea stars echinoderms that usually have five arms radiating from a central body

sea turtle turtles that live and feed in sea, but come ashore to lay their eggs

sea urchin round echinoderm, with no arms and long spines covering skin

seabirds term for all birds that depend on the ocean for their survival

seafloor spreading the moving apart of Earth's plates, caused by the upward movement of magma under the mid-ocean ridge

seamounts small undersea mountains formed by lava piling up on seafloor; they form over hot spots in the mantle

seaweeds multicellular algae that live in the ocean

secondary consumer the second-level consumer (of primary consumers) in a food chain, e.g., shrimp that eat zooplankton

segmented worms characterized by body divided into segments; have coelom

sessile term for organisms that live attached to a substrate

setae hairlike bristles that stick out from the parapodia of sandworms

sexual reproduction the production of offspring by two parents

sieve plate in sea star, small filter on dorsal surface through which water enters

siphon in mollusks, water-flow passageway, or tube; used for feeding and breathing (in bivalves) and locomotion (in cephalopods)

sirenians large, docile, plant-eating marine mammals; manatees and dugongs

skin gills in sea stars, small, ciliated projections on skin, used for breathing

sludge solid part of sewage that settles to bottom of tank during treatment

snowy egret tall salt-marsh bird that has pointed bill for grabbing small fish

solute any substance that a solvent, such as water, holds in a dissolved state

solution a solvent with substances (solutes) dissolved in it, e.g., salt water

solvent substance, such as water, that can dissolve other substances

sonar a device that emits and receives sounds, used to detect underwater objects; acronym for **sound** **n**avigation **r**anging

spawning in bony fish, the release of gametes during external fertilization

specific gravity ratio of the density of a substance to the density of distilled water

specific heat the heat storage ability of a substance such as land or water

spicules in sponges, chalk or silicon spines that form a rigid skeleton

spiracles in cartilaginous fish, breathing holes on dorsal side behind each eye

spongin in sponges, an elastic framework of protein fibers

spore reproductive cell that contains the haploid number of chromosomes; functions in asexual reproduction

spring tides the highest and lowest tides, produced twice each month by strong gravitational pull when sun, moon, and Earth are aligned

spyhopping behavior in which whales raise head above water to look around

stereoscopic microscope research tool with two eyepieces for viewing

stock term that refers to population of a living resource that is harvested

stranding a behavior in which whales become beached (and die) along shores

strandline line of seaweed and debris that marks boundary between the intertidal and supratidal zones

subduction occurs when one crustal plate plunges down under another plate

submarine canyons steep, V-shaped depressions that cut through the continental shelf; extensions of sunken river valleys

submersibles manned research vehicles used for undersea scientific exploration

subtidal zone the coastal life zone that remains underwater (below low tide)

supratidal zone on a beach, the area above the intertidal zone (above high tide)

surf zone the region of crashing waves along a sandy beach

swells waves that have longer periods and more rounded crests

swim bladder in bony fish, internal gas-filled organ that lets a fish adjust its level in the water (i.e., enables neutral buoyancy)

swimmerets in crustaceans, small paddlelike appendages under the abdomen

symbionts members of two different species involved in mutually beneficial relationship

symbiosis close relationship between different species; usually beneficial

tail flukes in cetaceans, hind flippers used to propel animal through the water

talons in hawks and eagles, e.g., osprey, strong curved claws for grabbing prey

tapeworm parasitic flatworm found in intestines of fish and other animals

telson in horseshoe crab, long spiked tail that is used in locomotion

territoriality behavior in which an organism, e.g., a fish, defends its home area

tertiary consumer the third-level consumer (of secondary consumers) in a food chain, e.g., fish that eat shrimp

thallus leafy part of a seaweed; it produces the reproductive cells

thermal plume a flow of water with higher temperatures than those of the surrounding waters; usually from industrial discharge

thermal pollution the release into natural waterways of heated water

thermocline in ocean, the permanent boundary that separates warmer water above from colder water below

threatened species term for species that are in less immediate risk of extinction

tidal bore a lone wave formed by the incoming tide at sloping mouth of a river

tidal currents swift-moving currents parallel to shore, produced by tidal change

tidal range the vertical distance between low tide and high tide

tide the daily rise and fall of the ocean seen along the shore; due to gravity

tide pools small habitats formed when spaces between rocks retain water at low tide

toothed whales hunters with peglike teeth; eat fish, squid, penguins, and seals

topography study of Earth's surface features, on land and on the ocean floor

toxic chemicals pollutants that are harmful to living things and the environment; usually come from industrial and agricultural wastes

transmitted light light that passes through a substance

trawling net type of net released from stern of ship and pulled through water

trematode parasitic flatworm found in fish and other animals; also called flukes

trenches deepest, steepest depressions on ocean floor, formed by subduction

trial refers to each time that an experiment is carried out

trophic level each feeding level in a food chain (producers to consumers)

trough the lowest point of a sound wave or water wave

tsunami giant waves generated by sudden seismic activity in Earth's crust

tube feet in echinoderms, tiny feet located in groove under arms; used in feeding, breathing, and movement

tunicates sessile protochordates; larval stage has primitive chordate traits

turbidity the measure of the level of clarity or murkiness of water

turbidity current fast, subsurface current found along steep continental slope

turtle grass a type of sea grass that grows along the Gulf Coast and Florida

tusk shells *see* scaphopods

undertow the returning current of water from a wave that breaks on the beach

unicellular describes single-celled organisms

upwelling the rising of a cold, nutrient-rich current from deep in the ocean

variable any factor that can affect the outcome of an experiment

veliger swimming, ciliated larva of a snail; part of plankton community

vertebrates higher chordates; all have a skeleton, backbone, skull, and brain

vertical migration behavior in which organisms move up and down in the water column each day in response to changing levels of light (and food resources)

wandering albatross largest, most oceanic of diving seabirds; glides on air currents over the ocean for years at a time

water budget total amount of water contained in and on planet Earth

water column vertical zone of water that extends from ocean surface to bottom

water cycle continuous movement of water between ocean, atmosphere, and land

water vascular system in sea stars, a network of water-filled canals that enables movement

watershed the land area through which water passes on its way to the ocean

wave an up-and-down movement of the ocean surface, i.e, a form of energy that moves across the water

wave height the vertical distance between the top (crest) of a wave and the bottom (trough) of the preceding wave

wave train a series of waves, one followed by the other, moving in the same direction; produced when a steady wind blows

wavelength the length of one complete wave cycle of each type of radiation

wetlands area of grasses growing in shallow water; *see* salt marsh community

whirlpool rapid movement of surface waters in a circle; also called an "eddy"

whitecap a mixture of air and water from narrow wave crests blown by wind

zonation pattern of distinct bands, or zones, of habitats along a coast, each with its own distinct community

zooplankton animal and animal-like plankton that drift on ocean surface, e.g., copepods

zooxanthellae dinoflagellate algae that live symbiotically within coral polyps

zygote fertilized egg cell that contains the diploid number of chromosomes

Index

Photo Credits